CARPS

Contents

Contributors

The authors have over 120 years' total experience in fish diseases, their diagnosis and treatment, and are internationally recognised as experts, particularly in carp health, where they have undertaken extensive research and advisory work.

Professor Dave Hoole is Professor Emeritus in Fish Diseases at Keele University, UK. Since the early 1980s he has undertaken extensive scientific research in carp disease and immunology. He has published widely within the scientific literature and books, and given numerous presentations to scientific and non-scientific audiences. He has also served as a director of a fish health service that scrutinises fish health prior to consent being given for movement of fish under UK legislative requirements. He also acts as a consultant to various UK Fish Health Accreditation bodies.
School of Life Sciences, Keele University, Keele, Staffordshire, ST5 5BG
Email: d.hoole@keele.ac.uk

Professor Dieter Steinhagen is Professor Emeritus at the Fish Disease Research Unit at the University of Veterinary Medicine Hannover, Germany. Since the early 1980s, he has worked scientifically on various aspects of pathology and immunology of pathogen-related diseases in carp, on parasites as indicators of anthropogenic

stress in ecosystems and on aspects of animal welfare in fish. This work has been funded by national and international organisations and the results have been published in articles and numerous presentations to scientific and non-scientific audiences.
Fish Disease Research Unit, University of Veterinary Medicine Hannover, Bunteweg 17, 30559 Hannover, Germany
Email: steinhagenwennigsen@gmail.com

PD Dr Mikolaj Adamek is a Research Associate and Principal Investigator at the Fish Disease Research Unit of the University of Veterinary Medicine Hannover, Germany. Since the mid-2000s he has been working on various aspects of immunology, pathophysiology and diagnostics of pathogen-related diseases in carp, salmonids and tilapia. He works mainly with viral pathogens, looking for prevention and treatment measures that could lead to a reduction in the impact of the disease on fish welfare. His research is funded by national and international sources. The results have been reported in numerous scientific and non-scientific articles and presentations.
Fish Disease Research Unit, University of Veterinary Medicine Hannover, Bunteweg 17, 30559 Hannover, Germany
Email: Mikolaj.Adamek@tiho-hannover.de

PD Dr Verena Jung-Schroers has been working at the Fish Disease Research Unit of the University of Veterinary Medicine Hannover, Germany, since 2005 and has headed the institute since 2021. She is a specialist fish veterinarian and diplomate of the European College of Aquatic Animal Health and is responsible for the further training of veterinarians in the field of fish diseases. In research projects she works on bacterial infections in fish and other aquatic animals, on the importance of the microbiome in aquaculture systems and on animal welfare. She has published the results of these and other studies in numerous articles and presentations.
Fish Disease Research Unit, University of Veterinary Medicine Hannover, Bunteweg 17, 30559 Hannover, Germany
Email: Verena.Jung-Schroers@tiho-hannover.de

Preface

Carp species are perhaps the most important group of freshwater fish in the world and their diverse nature over the planet means that they are found on most continents. They have a global significance as a source of food, for sport, for their ornamental appeal and as experimental models for research. This book focuses on the prominent carp species, which include species that occur globally, such as *Cyprinus carpio* (common carp) and *Carassius* species, as well as carp species that predominate in Europe and Asia, such as *Ctenopharyngodon idella* (grass carp), *Hypophthalmichthys molitrix* (silver carp), *Hypophthalmichthys nobilis* (bighead carp) and more specialised production of, for example, *Gibelion catla* (catla) and *Labeo rohita* (rohu) in the Indian subcontinent. Carp make a significant contribution to the economic status and food security of several countries and their regions, and at a more local level to the income of a family, particularly when other forms of employment are scarce.

The major threats to carp production in either aquaculture, the ornamental trade, angling or subsistence farming are the effects of disease, which can arise from infection, nutritional disorders, environmental disturbances or inadequate management strategies. How these individual threats are recognised and diagnosed, and the techniques applied to resolve the cause of fish mortality, are fundamental to reducing the personal and economic loss that disease can have on a carp population. This user-friendly book, which specialises in a practical approach to disease diagnosis and treatment, covers both infectious and non-infectious diseases. Details are given on knowledge of disease, clinical signs, the involvement of pathogens and the application of control strategies, such as management practices, therapeutics or vaccination, which will improve general fish health and welfare. A range of disease-causing agents are considered, including viruses, bacteria, fungi and parasites, as well as non-infectious diseases. Specialised subjects include zoonosis and the impact that co-infections can have on fish health. Importantly the book describes how to analyse a fish for disease, the common pathological signs and the range of treatments available. It is hoped that this practical book will be of value, not only to the farmer producing carp, professionals involved in fish disease diagnosis, students and the academic community, but also to the angler fishing a pool, the ornamental hobbyist looking after a pond or aquarium, and the scientist carrying out research studies.

Acknowledgements

The extensive worldwide production and maintenance of the important carp group of fish has meant that the authors have made extensive use of information provided by various sources in the production of this book. In particular, we wish to acknowledge our many friends and colleagues who gave expert advice and images on a range of topics covered in the book, and who are named in the appropriate chapters. Thanks are due to all the staff at the National Fish Diseases Laboratories at the Environment Agency, UK, and in particular Chris Williams and Timor Kaya, for searching for and producing an extensive array of images. We are also grateful to John Bignell of the Centre for Environment, Fisheries and Aquaculture Sciences at Weymouth, UK, for his time in sourcing some images and his expertise in assisting in the interpretation of pathology. Within the UK we also acknowledge the contribution of images from Ian Wellby, Director of Blueroof Environmental Consultants. We are also indebted to our colleagues and friends in the veterinary section of Stiftung Tierärztliche Hochschule Hannover, Germany for providing a range of images relating to fish examination and treatment, in particular Gabi Bröcker, Henner Neuhaus and Felix Teitge.

Thanks are also due to our employers and host Institutes, i.e. Keele University, UK and Stiftung Tierärztliche Hochschule Hannover, in generously providing facilities and support during our many book meetings.

The sources of images and copyright are indicated in the appropriate figure captions. All other copyright materials (i.e. images and diagrams not credited in the captions) were provided by the authors and their appropriate Institutes.

While every effort has been made to ensure that the information in this book is accurate, no liability whatsoever can be accepted by the authors and publisher for any error or omission in the content or for any loss or damage or other accident arising from the use of any treatment method cited in this book. The reader should always consult a fish health inspector, veterinarian or appropriate professional and follow recommended manufactures' treatment protocols.

Scientific Names and Generally Accepted Common Names

The fish species names presented in this book are mainly based on recommended names on the 'Fishbase' website (FishBase : A Global Information System on Fishes).

Accepted species name	Synonym	Common name
Abramis brama		Bream
Bangana gedrosicus	*Labeo gedrosicus*	Mashkal River labeo
Barbodes gonionutus		Silver barb
Barbus barbus		Barbel
Brevoortia tyrannus		Atlantic menhaden
Carassius auratus	*Carassius auratus auratus*	Goldfish, ornamental goldfish
Carassius carassius		Crucian carp
Carassius cuvieri		Japanese white crucian carp
Carassius gibelio	*Carassius auratus gibelio*	Gibel carp, Prussian carp
Channa striata		Snakehead
Cirrhinus molitorella		Mud carp

Accepted species name	**Synonym**	**Common name**
Cirrhinus mrigala	*Cirrhina mrigala* (misspelled synonym)	Mrigal (Indian major carp)
Coreius guichenoti		Largemouth bronze gudgeon
Ctenopharyngodon idella	*Ctenopharygodon idellus* (misspelled synonym)	Grass carp
Cyprinus carpio		Common carp / koi
Cyprinus rubofuscus	*Cyprinus carpio haematopterus*	Amur carp (common carp from East Asian waters, Amur to Red River basins)
Danio rerio		Zebrafish
Esox lucius		Pike
Gibelion catla	*Catla catla, Labeo catla*	Catla
Hypophthalmichthys molitrix		Silver carp
Hypophthalmichthys nobilis	*Aristichthys nobilis*	Bighead carp
Labeo calbasu		Organgefin labeo
Labeo fimbriatus		Fringe-lipped peninsula carp
Labeo gedrosicus	*Bangana gedrosicus*	none given
Labeo gonius	*Rohita gonius*	Kuria labeo
Labeo rohita		Rohu
Leucaspius delinatus		Sunbleak
Leuciscus aspius		Asp
Monopterus albus		Swamp eel
Mugil cephalus		Estuarine mullet
Mylopharyngodon piceus		Black carp

Accepted species name	Synonym	Common name
Neotropis cornutus		Golden shiner
Osteobrama belangeri		Pengba / Belengee
Plecoglossus altevelis		Ayu
Pseudorasbora parva		Top mouth gudgeon
Rutilus rutilus		Roach
Silurus glanis		Catfish
Squaliobarbus curriculus		Barbel chub
Systomus sarana		Olive barb
Tinca tinca		Tench

Glossary

Abrasion: A scraping of the outer layers of the skin or mucous membrane.
Abscess: A localised cavity containing pus, usually produced by pyogenic microorganisms.
Acclimation: The habituation of an organism to a different climate or environment.
Acidosis: Condition caused by exposure to a pH level below the tolerance range of the species affected.
Acid-fast: Microorganisms (e.g. *Mycobacterium* spp.) not readily decolourised by acids after staining.
Actinospore: Developmental stage in the life cycle of myxozoan parasites, which develops in the invertebrate hosts and infects the fish host.
Acute: Having a short and relatively severe course (4 h to several days), e.g. acute toxicity, acute inflammation (inflammation in which progress is rapid and manifestations are immediate and pronounced).
Adenocarcinoma: A malignant neoplasm derived from glandular tissue.
Adenoma: A benign neoplasm composed of glandular tissue.
Adhesion: Uniting of two surfaces of structures that would normally be apart, for example visceral organs attached to the abdominal wall by fibrous tissue, the result of acute or chronic inflammation.

Adoral side: Part of the cell facing away from the mouth structure.
Aeration: In fish biology, the introduction of oxygen into water by agitation of the water surface or by blowing air into the water.
Aerobic: Microorganisms that grow only in the presence of oxygen.
Aerosol: The suspension of fine solid particles or small liquid droplets in air.
Aetiology: The study of the causation of diseases, or the sum of the factors of a disease; or the cause of a disease.
Aflatoxin: Toxic substance produced by moulds from the genus *Aspergillus.* Aflatoxins are considered as mutagenic (inducing alterations in the genetic material) and carcinogenic (cancer-causing).
Agglutination: Clumping of bacteria, red blood cells or other particles.
Algal bloom: A rapid growth of floating algae in a freshwater or marine water system.
Alkalosis: Condition caused by exposure to a pH above the tolerance range of the species affected.
Amoeboid: A type of locomotion of attached living eukaryotic cells by crawling movements. The cells form temporary cytoplasmic projections, the pseudopodia, towards the direction of the movement and by this the cell is pulled forward.
Anaemia: A condition characterised by a deficiency of haemoglobin and/or erythrocytes. The more important anaemias in fish are normocytic anaemia caused by acute haemorrhaging, bacterial and viral infections, and metabolic diseases resulting in red cell destruction; microcytic anaemia due to chronic haemorrhaging, e.g. caused by external parasites, iron deficiency and deficiency of certain haematopoeitic factors; and macrocytic anaemia (e.g. the absence of juvenile cells; too many mature cells) resulting from an increase in haematopoeitic activity in the spleen and kidney.
Anaerobic: Living in the absence of oxygen, e.g. anaerobic bacteria.
Angioma: A tumour comprising blood vessels or lymph vessels.
Anorexic: In zoology, a thin, emaciated and undernourished animal specimen.
Anoxia: Devoid of oxygen.
Anthelmintics: Drugs whose action is against helminth parasites, e.g. flukes, tapeworms.

Antheridium: Structure in the life cycle of fungi, oomycetes or mosses, in which male gametes develop.
Antibiotic: A chemical substance originally produced from moulds or bacteria, but now from synthetic substances. Antibiotics are capable of inhibiting the growth of, or killing, other microorganisms.
Antibiogram: Table showing susceptibility of a series of bacteria to different antimicrobials.
Antibody: A specific immunoglobulin molecule produced in vertebrates by an organism in response to an antigen. An important component of the acquired immune response.
Antigen: A substance that induces the formation of specific immunoglobulins, i.e. antibodies.
Antimicrobial: Substance that is effective against a microorganism, usually bacteria.
Apicomplexa: A group of eukaryotic parasitic protists with a complex life cycle, in which some developmental stages possess a characteristic array of subcellular structures located at the apical pole of the cell. Apicomplexa comprise several parasites with high importance in medicine, veterinary medicine and fish health as well as parasites of various groups of invertebrates.
Artificial propagation: In the context of this book, the induction of spawning of fish by the injection of gonadotropin hormones into spawner fish and the artificial fertilisation of obtained sexual products.
Ascites: Abnormal, excessive accumulation of fluid in the abdomen.
Aseptate: Without any septum (internal wall).
Asphyxia: A condition of suffocation.
Asymptomatic carrier: An animal carrying an infectious agent, but showing no overt signs of the disease.
Atresia: Congenital absence or closure of a normal body opening or tubular structure.
Atrophy: A wasting away or diminution in size of a cell, organ or part, due to disease, nutritional insufficiency, pressure or lack of innervation.
Atypical: Not corresponding to a normal specimen or condition.
Autopsy: Post-mortem examination and its description.

Autotrophic: An organism that can make its own food from simple chemical substances.

Bacteriology: The study of bacteria.
Bacteraemia: Presence of bacteria in the blood.
Bacteriophage: A virus that infects and replicates in bacteria.
Bacteriostatic: Having the ability to inhibit the growth or reproduction of bacteria.
Basophil: A type of white blood cell containing granules that stain strongly with basic dyes.
Basophilia: A condition in which the basophils and polymorphonuclear leucocytes are present in undue numbers, as in lead poisoning, leukaemia etc.
Basophilic: Substance that stains with basic dyes.
Benign: Not endangering life. However, a benign tumour is a neoplasm that is slow growing and non-invasive, but it may eventually grow to a size that may compress or occupy space, thus endangering the life of that organism.
Benthic organism: Animal that settles on the surface of sediment or submerged surfaces in an aquatic habitat.
Biofilm: An association of various microorganisms, including bacteria, protists and small multicellular organisms, which live together in a slimy extracellular matrix attached to submerged surfaces in aquatic habitats. The matrix of the biofilm protects the community of microorganisms from unfavourable environmental conditions.
Branchial: Pertaining to the gills.
Buccal cavity: The mouth structure of a protist, in particular in ciliates.

Capillaries: Small blood vessel in tissues of the vertebrates. Capillaries facilitate the exchanges of various substances, including nutrients, gases and ions, between the bloodstream and tissues.
Carcinoma: A malignant tumour derived from epithelial cells.
Cartilage: Firm elastic skeletal tissue in which cells (chondrocytes) are embedded in a matrix containing collagen fibres.
Cartilaginous: Gristly, consisting of cartilage.

Cataract: Partial or complete opacity of the crystalline lens or its capsule.
Cestoda: The taxonomic group of tapeworms.
Chemotactic: Movement of a cell or an organism in response to a chemical stimulus, swimming along a chemical gradient towards a higher concentration of a chemical stimulus.
Chemotherapy: Treatment of diseases by chemicals.
Cholangioma: Benign tumour of the biliary ducts in the liver.
Chondroma: Tumour of the cartilage cells.
Chromatin: The basophilic DNA portion of the cell nucleus.
Chromatoblastoma: A pigmented tumour.
Chromosome: A structure in the nucleus of eukaryotic cells, which is visible by light microscopy in the metaphase during cell divisions (mitosis and meiosis). The chromosomes contain DNA and proteins, in particular histones.
Chronic: Prolonged. Persisting for a long time, for example weeks, months or years.
CHSE cells: Chinook Salmon Embryo cell line, a permanently growing cell line developed from Chinook salmon (*Oncorhynchus tshawytscha*) embryo cells.
Clinical: The outward appearance of a disease in a living organism.
Clinical infection: An infection or disease generating obvious signs of pathology.
Clubbing (gills): Swelling of the tips (distal ends) of the gill filaments.
Ciliature: The arrangement of cilia on the cell surface of ciliates.
Cilium: An organelle in eukaryotic cells. It is a membrane-bound slender projection that extends from the surface of the cell with a characteristic pattern of supporting microtubules: two central microtubules are surrounded by nine pairs of double microtubules. Cilia are absent from prokaryotes and archea.
Cingulum: Equatorial cell groove in dinoflagellates.
Commensal: Living on the food from another organism without causing harm.
Complement: Non-specific immunological proteins normally present in the blood.

Condition factor: 'K Factor'. The relationship between weight (W) and length (L) of a fish, expressed usually as: $K = W^3/l$ or W^3/fish.
Congestion: Excessive or abnormal accumulation of blood in part of a tissue or organ so that the normal functions are hindered.
Conidium (pl. conidia): In fungi, an asexual non-motile spore, formed by mitosis at the tip of specialised hyphae.
Conjugation: A sexual process in ciliates that allows a recombination of genetic material.
Contractile vacuole: A membrane-bound organelle in the cells of several eukaryotic organisms from the freshwater environment. It is involved in osmoregulation and expands by collecting water from the cytosol of the cell and periodically expels its contents to the exterior by contraction.
Counterstain: A stain applied to a histological section in order to enhance the effect of an existing primary stain.
Culture: The growing of bacteria or other microorganisms in artificial media. The propagation of animals (including fish), as in farming.
Cutaneous: Pertaining to the skin.
Cyanobacteria: A group of photosynthetic Gram-negative prokaryotic microbes. Cyanobacteria are also referred to as blue-green algae, may colonise several freshwater habitats and may substantially impact water quality and the functioning of aquatic ecosystems.
Cyst: Developmental stage of a protist or a bacterium with a protective, more resistant wall, which allows the persistence of the organism in the environment for a prolonged period of time or its survival in adverse environmental conditions.
Cytopathic: Destruction of cells infected with a virus.
Cytopathology: The study of morbid changes in cells.
Cytostome: An area in the internal part of the mouth structure of ciliates. In this area the cytoplasm of the cell is covered by a simple cell membrane, and here food vacuoles are formed.

Decalcify: To deprive or rid of lime salts (calcium) in tissues, especially in reference to histological technique.
Deficiency disease: Disease caused by lack of some essential constituents in the diet.

Definitive host: Host organism in which the adult stage of a parasite develops.
Deformity: Distortion of any part of the body or general malformation of the body.
Degeneration: Any structural change in an organ or tissue or cell by which its functional power is impaired. The degenerative change may, or may not, progress to necrosis.
Dermis, dermal: The dermis is the inner layer in the skin of vertebrates and is covered by the epidermis. It is derived from the mesoderm, and contains mainly fibres and relatively few cells. The dermis may contain blood vessels, nerves, scales (in fish), pigment cells and adipose cells. In the dermis an upper stratum spongiosum with loose connective tissue and a basal stratum compactum with dense collagenous tissue can be recognised.
Deoxygenated: Containing a low concentration of oxygen.
Desquamation: The shedding of epithelial elements, chiefly of the skin, in sheets.
Detritus: In hydrobiology, organic remains in water after the decomposition of plants and animals are referred to as detritus.
Diagnosis: The determination of the nature of a given disease.
Digestive vacuole: An organelle found in cells of simple eukaryotic organisms, such as protists, that allows the uptake, processing and excretion of food particles in the cell.
Diploid: Cells possessing a paired set of chromosomes, with one set of chromosomes from each parent.
Disease: A deviation from the state of complete physical or social well-being of an organism, involving a well-defined set of signs and aetiology and leading to an impairment of its normal function.
Distal: Furthest from the centre.
Dormant cyst: The developmental stage of a microorganism, which allows the organisms to survive unfavourable environmental conditions in a resting stage.
Dorsal: Pertaining to the back of the body. In anatomy, the back or upper side of the body. In vertebrates, the side of the body where the vertebral column is located.
Dropsy: An accumulation of serous fluid in any cavity of the body.
Dysfunction: Abnormal or imperfect functioning of an organ.

Ectoparasite: A parasite that lives on the outside (e.g. body surface) of its host.
Ectozoic: Living on the surface of an animal.
Emaciated: Showing extreme weight loss.
Emboli: Clots of blood that obstruct circulation.
Encapsulate: To form a covering or surrounding capsule.
Encysted: Enclosed in a cyst.
Endemic: Presence of an organism in a particular region or country. Pertaining to a given locality or area, e.g. an endemic disease.
Endocytosis: A process that uses eukaryote cells to bring particle or liquid droplets into the cell. The process starts by surrounding the substance with an area of the cell membrane, which then invaginates into the cell, buds off and forms a vesicle inside the cell, which contains the internalised material. Particles are internalised by phagocytosis and lipid droplets by pinocytosis.
Endogenous: Produced within the organism.
Endoparasite: A parasite that lives inside its host.
Endophthalmia: Inflammation of the interior cavity of the eye.
Enteritis: Inflammation of the intestine.
Enterocytes: Cells that line the luminal (inner) surface of the intestine, form the mucosal epithelium and absorb nutrients.
Eosinophil: A type of white blood cell that contains granules and which stains with the red acidic dye, eosin.
Eosinophilic: An adjective describing a substance in a cell that has an affinity for the eosin dye.
Epicarditis: Inflammation of the visceral layer of the pericardium.
Epicellular: Living on the apical surface of a cell, in close association with the plasma membrane of the host cell.
Epidemiology: The study of diseases or health factors and their relationships with the population and the environment.
Epidermis: Outermost layer of the skin in vertebrates. It is composed of several stratified layers of cells: a basal layer of non-differentiated stem cells, a mid-layer of cells with cuboidal to columnar shape, while cells in the superficial layer are often flattened.
Epithelium: The thin outer layer of the skin, mucous membranes and the inner layer of internal cavities.

Epizootic: A disease attacking many fish in a population at the same time; widely diffused and rapidly spreading.
Erosion: Superficial loss of the epithelial tissue.
Erythrocyte: A red blood cell.
Erythrophoroma: A tumour made up of red-pigmented fibrous cells.
Erythrophores: Erythrophiles, cells that stain readily with red dyes.
Eukaryote: Organisms with cells with a true nucleus and organelles including mitochondria, endoplasmic reticulum, Golgi apparatus or lysosomes. Many eukaryote cells possess flagella with a characteristic internal microtubule structure.
Eurythermal: In zoology, refers to animals that can survive and develop in a wide range of ambient temperatures.
Exophthalmos: Abnormal protrusion of the eye.
Exoskeleton: Hard supporting structure secreted by and external to the epidermis. An external skeleton, e.g. as in insects.
Exotoxin: Peptides secreted from an organism.
Extracellular: Occurring outside the cell.
Extrasporogenic, pre-sporogenic: Referring to the life cycle of some myxozoan parasites and describing the replication of the parasite before reaching its target tissue and before spore formation.

Facultative: When related to an organism: organism that can survive in oxygenated as well as deoxygenated conditions.
FHM cells: Fathead Minnow cell line, a permanent cell line developed from the fish species fathead minnow (*Pimephalus promelas*).
Fibroblast: A type of connective tissue cell that is concerned with collagen synthesis and associated with developing or repairing tissues.
Fibroma: A benign tumour composed chiefly of fibrous tissue (pl.: fibromata).
Flagellum: An organelle in eukaryotic cells with the same structure as cilia. Flagella are long structures and move the entire cell, mainly by wavelike movements.
Fibrosarcoma: A malignant tumour of the fibrous connective tissue.
Filiform: A term used to describe threadlike organisms, e.g. *Flexibacter* sp.

Fin-rot (Fin erosion): A progressive erosion and disintegration of the fin.
Formalin: Solution of approximately 37% by weight of formaldehyde gas in water. In a diluted form it is effective in control of external parasites and fungal infections on fish and eggs. Also commonly used as a fixative for preserving tissues for subsequent histological examination.

Gametangia: Cell or organ in algae, fungi, oomycetes or plants, in which gametes develop for sexual reproduction.
Ganglioneuroma: Benign neoplasm composed of nerve fibres and mature ganglion cells.
Gas supersaturation: The solubility of gases in water is limited and increases with increasing pressure or decreasing temperature. When the temperature is increased rapidly or the external pressure is reduced, the concentration of gases in water may exceed the concentration, which can be dissolved in water under the given temperature or pressure and thus lies above saturation level. Under these conditions, gas bubbles may be formed in water as well as in the bloodstream or in tissues of fish living in this waterbody.
Gastroenteritis: Inflammation of the mucosa of the stomach and intestines.
Glomerulonephritis: Inflammation of the capillary loops in the kidney glomeruli.
Glycolysis: A series of biochemical reactions in the breakdown of glucose into pyruvate in order to generate energy without the involvement of oxygen. This process occurs in cells of most organisms.
Gonad: The reproductive organ that produces female and male gametes, the eggs and sperm, as well as some reproductive hormones of an organism.
Gram stain: A staining procedure used in identifying bacteria and classifying them as Gram-positive or Gram-negative.
Granulocyte: Class of white blood cell that characteristically contains numerous vesicles or granules, e.g. eosinophils, basophils, neutrophils.
Granuloma: A tumour of chronic inflammatory nature (pl.: Granulomata). Note: it is not a neoplasm.

Gymnospore: A spore that develops outside a sporangium or another protective envelope.
Gynogenesis: A process in which the development of an egg cell to an embryo is induced by a sperm penetrating the egg cell without subsequent fertilisation of the egg cell by the fusion of the nuclei. During gynogenesis, there is no paternal genetic contribution to the offspring, which then contains only genes from the mother. During **allogynogenesis**, activation of egg development is induced by the penetration of a heterologous sperm, from a different fish species, into the egg cell.

Habitat: An area of living for a particular species, defined by an array of physical and biological parameters that enable the survival of the given species.
Haemangiopericytoma: A tumour around blood vessels.
Haematin: A pigment formed when unbuffered formalin is used in the fixation of tissues. Also abnormal Fe^{3+} constituent of haemoglobin. A decomposition product of haemoglobin.
Haematozoic: Living in blood.
Haemocoel: Blood-filled cavity consisting of spaces between organs.
Haemoflagellate: Flagellate protozoan found in the blood of its host.
Haemoglobin: Oxygen-carrying haem protein that occurs in the red blood cells in vertebrates.
Haemolysis: Disruption (lysis) of erythrocytes and the release of haemoglobin into the blood serum.
Haemoparasite: A parasite in the bloodstream or in blood cells of vertebrates.
Haemorrhage: An escape of blood from the vessels, either through intact blood vessel walls, or by flow through ruptured walls.
Haemorrhagic petechia: Small 'pinprick' haemorrhages occurring where capillaries have ruptured, e.g. throughout musculature or in the cutaneous layers.
Haemosiderin: A yellow iron-containing substance resulting in the breakdown of red blood cells.
Haptor: Attachment organ of monogenean flukes.

Heavy metals: In chemistry, metal elements with relatively high atomic weight, such as arsenic, cadmium, chromium, copper, mercury, silver or zinc.
Hepatic: Pertaining to the liver.
Herbicide: A chemical substance used to kill unwanted plants.
Herbivorous: An animal that feeds on plant matter.
Hermaphroditic: A state where the animal or plant possesses the organs of both sexes.
Hermaphroditism: The coexistence in an individual of ovarian and testicular tissue.
Heterotrophic: Obtaining food from plants or animals.
Histology: The science of the micro-structure of tissues.
Histopathology: Pathological histology.
Histozoic: Parasitic organisms that live between the cells of the tissue of a host.
Holotrichous: The arrangement of ciliates on the body of ciliate protists and describing an even and dense ciliation that covers the whole body.
Horizontal transmission: Transmission of a pathogen from one animal to another, usually in the same environment.
Hyaline: Clear, structureless, homologous, 'glassy' material that occurs pathologically in degenerations of several tissues, and also in normal tissue.
Hybridisation: Formation of a hybrid usually by the cross-breeding of two species.
Hyperparasitism: An organism that is a parasite on, or in, another parasite.
Hyperplasia: Abnormal increase in number of cells in a tissue or organ.
Hypersecretion: Excessive production of secretions from glands or a secretory tissue.
Hyperthermia: Physiological responses induced by exposure to extreme hot conditions.
Hypha: filamentous cell of a fungus or an oomycete (pl. hyphae).
Hypodermis: The layers of tissue immediately under the skin, i.e. subcutaneous.

Hyponatraemia: Condition where the sodium levels in the blood are below normal.
Hypothermia: Physiological responses induced by exposure to extreme cold conditions.
Hypotrophy: Diminution in size.
Hypoxia: Reduction in oxygen in body tissues below physiological levels.

Immunocompromised: Having an immune response that is deficient in its operation.
Immunodiagnosis: The diagnosis of disease using immunological techniques.
Immunosuppressed: Having a suppressed immune response.
***In vitro*:** In reference to tests or experiments conducted in an artificial environment, including cell or tissue cultures.
***In vivo*:** In reference to tests or experiments conducted in or on living organisms.
Incubation; Incubation time: The period between the time of entry of a pathogen and the onset of symptoms of the disease.
Infection: The introduction or entry of a pathogen or parasite into a host, resulting in the presence of the pathogen or parasite within the body tissues or cells of the host, whether or not this results in overt disease.
Infection, latent: An infection that remains dormant for a long period of time.
Infection, opportunistic: An infection caused by a pathogen that is normally present and innocuous, but arises when conditions are favourable to that pathogen.
Infection, secondary: A pathogen or parasite that gains entry into tissues already damaged by a different infectious agent.
Infection, subclinical: An infection that does not expose itself by symptoms, but which leads to biological reaction of the organism.
Infestation: Colonisation of an organism by a parasite.
Inflammation: In fish, the reaction of the tissues to injury characterised clinically by swelling and redness. Pathologically, by vasodilation, hyperaemia, accumulation of leucocytes, exudation of fluid and deposition of fibrin.

Inoculation: The introduction of a pathogenic organism into the tissues of a living organism or culture.
Intensity: Number of individuals per given area or unit, e.g. number of parasites per host.
Interleukins: Proteins produced by various immune cells that act on other cells in the immune system.
Intermediate host: A host in which a parasite lives for part of its life cycle but in which it does not become sexually mature.
Intermyotomal: The space between muscle fibres (myotomes).
Interspecific: Between distinct species.
Intramuscular: Within the muscle tissue.
Intraperitoneal: Within the peritoneal cavity.
Iodophore: Disinfectant containing iodine.
Ion: An atom or molecule with a positive or negative electrical charge when dissolved in water.
Ionised (ammonia): A molecule that has had an electron removed or added. For example, the ionised (NH_4^+) form of ammonia that is less toxic to fish than the un-ionised form.
Iridophores: Sometimes called iridocytes. Guanine-containing granules, bodies or plates of which the reflecting, silvery or iridescent tissue of skin in fish is composed.

Karyorrhexis: Fragmentation of a nucleus.
Kinetoplast: A structure found in cells of kinetoplastid organisms. It is a specialised region of the mitochondrion that harbours an unusual form of mitochondrial DNA. The kinetoplastid DNA is organised in circular DNA molecules, forming two types of circles, minicircles and maxicircles. In the cell the kinetoplast is located in close association with the basal bodies of the flagella.
Kinetosome: A structure in eukaryotic cells with flagella or cilia. Kinetosomes are located within the cell at the base of flagella or cilia and serve as anchoring devices and organisation centres for flagellar microtubules. Kinetosomes are also referred to as basal bodies.

Leiomyoma: A benign tumour made up of smooth or involuntary muscle.
Lepidorthosis: Erection of scales.

Lesion: Any visible pathological or traumatic alteration in the normal structure of organs or tissues.
Lethargy: Morbid drowsiness, sluggishness or moving very slowly.
Leucocyte: A term referring collectively to any non-pigmented blood cell.
Leucophores: White pigmented cells.
Lipoma: A benign fatty tumour.
Liquefaction, liquefactive necrosis: A pattern of cell death (necrosis) that results in the transformation of the tissue into watery debris.
Lymphocyte: An agranular leucocyte of the peripheral blood and haemopoeitic tissue.
Lymphohaemopoietic: Generation of both lymphoid and blood cells.
Lymphoma; Lymphosarcoma: A tumour of lymphoid cells.

Macrogamete: A developmental stage in the life cycle of apicomplexan parasites that refers to the larger, immotile gamete, which is fertilised by smaller motile gametes (microgametes).
Macronucleus: The somatic nucleus, which controls vegetative functions of the cells, present in ciliates and some other protists, in addition to a nucleus that controls generative processes. In ciliates, this somatic nucleus is polyploid and of larger size.
Macrophage: A phagocytic cell belonging to the reticuloendothelial system. It is the major phagocytic cell in fish inflammation.
Malignant: Pertaining to or denoting progressive growth of certain tumours which may spread to distant sites or invade surrounding tissue and terminate in death.
Melanin: A tyrosine-derived polymeric pigment responsible for the yellow to black colouration of fish and quite often associated with parasite infestation.
Melanocyte; Melanophore: Cell in which the pigment melanin is formed. Usually found in the skin and responsible for light/dark colour changes.
Melanoma: A tumour whose parenchyma is composed of anaplastic melanocytes or any benign tumour of melanocytes.
Melanomacrophage: Macrophage that contains the pigment melanin. Thought to be involved in the immune response.

Melanosarcoma: Malignant tumour composed of melanin-pigmented cells.
Merogonic: A developmental stage in the life cycle of apicomplexan parasites describing a series of asexual mitotic divisions of the cell, which give rise to several motile daughter cells, referred to as merozoites.
Meront: A multinucleated cell that developed by multiple asexual nuclear divisions (schizogeny, merogeny), and which gives rise to several daughter cells (schizonts). Motile schizonts are referred to as merozoites. This is a stage in the life cycle of sporozoan parasites.
Merozoite: A motile cell in the life cycle of apicomplexan parasites, which developed in a series of asexual mitotic divisions.
Metaplasia: Transformation of one differentiated cell type to another differentiated cell type, often in response to an external stimulus. It may lead to a malignant transformation.
Metastasise: To change state, position, form or function, e.g. with reference to a tumour.
Methaemoglobin: Formed by the reversible oxidation of Fe^{2+} in haemoglobin to Fe^{3+}, which is no longer able to reversely bind O_2. In fish the formation of methaemoglobin can be recognised by the development of mud-coloured brown gills.
Microbiome: Collection of microbes that reside within the body or organ.
Microgamete: A developmental stage in the life cycle of apicomplexan parasites referring to small motile gametes, which as a part of the sexual development of these parasites fertilise the macrogamete.
Micronucleus: The smaller of the two types of nuclei in ciliate protozoans. Involved in sexual reproduction but does not produce RNA.
Microtomy: The cutting of thin sections of tissues for examination with a microscope.
Mitochondrion (pl. mitochondria): An organelle found in most eukaryotic cells that generates most of the chemical energy the cell needs for cellular processes. Mitochondria are membrane-bound organelles with internal membranes (cristae) that contain mitochondrial DNA and multiply by binary fission. Mitochondria are absent from prokaryotic cells.

Mitosis: In cell biology a process of cell replication, where a cell divides into two genetically identical daughter cells.
Mixotrophic: Referring to organisms that are able to use different sources of energy of their metabolism and, for instance, can generate organic compounds by using light or chemical energy from the environment (photo- or chemosynthetic) as well as break down organic material from other organisms (heterotrophic metabolism).
Monoclonal antibody: Antibody that is produced from a single clone of B lymphocytes and thus consists of a population of identical antibody molecules all specific to a single antigenic determinant on an antigen.
Monoculture: The culture or rearing of one species.
Monocyte: A large mononuclear leucocyte with a more-or-less deeply indented nucleus, slate grey cytoplasm and eosinophilic granulation. This cell is formed into a macrophage once it migrates into the tissue.
Monozoic: An unsegmented body; comprising a single body.
Morphology: The branch of biology dealing with form and structure.
Mortality: The death of an individual or a population.
Mortality rate: The number of deaths per unit of population during a specific period.
Mucin: A protein found in mucus.
Mucopolysaccarides: Types of carbohydrate (1:2 glycol groups such as starch, glycogen, mucins) that convert to aldehydes by oxidation, becoming coloured when exposed to the periodic-acid Schiff reaction.
Mucosa: Epithelial cover of internal cavities.
Mucus: A viscous secretion covering mucosal surfaces. It is formed by large, highly glycosylated proteins (mucins), which bind water and may contain inorganic ions, further reactive proteins such as antimicrobial peptides, various enzymes and immunoglobulins. It is produced by goblet cells or cells found in mucous glands.
Multicellular: Many-celled.
Mutagen: A physical or chemical agent that may induce alterations to the genetic material of an organism. A mutagen has the potential to raise the frequency of mutations above the level found naturally.

Mycelium: The network of the hyphae, the filamentous cells of a fungus, oomycete or a bacterium as a whole.
Mycobacteriosis: A chronic systemic bacterial disease of fish caused by *Mycobacterium* spp.
Myocarditis: Inflammation of the heart muscle.
Myofibril: Basic organelle of a muscle fibre, which allows the contraction of the muscle. Myofibrils are composed of long filamentous proteins, including actin, myosin and other proteins. During contraction, myosin and actin filaments are sliding along each other.
Myoid: Resembling or composed of muscle fibres.
Myomere: One of the vertically marked segments that comprise the lateral muscle mass.
Myopathy: A wasting condition affecting the muscle tissue.
Myxospore: Developmental stage in the life cycle of myxozoan parasites, which develop in the vertebrate host and infect the invertebrate host.

Naïve: Without any prior exposure or experience (e.g. immunologically naïve to a parasite or pathogen).
Native: A species that is a member of the natural biotic community.
Necrosis: Dead or dying cells or tissues within the living body.
Nematocyst: A specialised cell found in all cnidarians, including myxozoa. Nematocysts contain a coiled hollow filament, which is ejected upon an external trigger, penetrates the target organism and injects a toxic compound. In myxozoa, the everted filament presumably attaches the sore to the host's epidermis during the infection process.
Neoplasm: An uncontrolled growth of cells.
Nephritis: Inflammation of the kidneys.
Nephrocalcinosis: A condition of renal insufficiency due to the precipitation of calcium salts in the tubules of the kidney.
Neural: Relating to the nervous system.
Neurofibroma: Tumour of peripheral nerves due to abnormal proliferation of Schwann cells.
Neurolemmoma: Tumour of a peripheral nerve sheath.
Neutralisation: Inactivation of a pathogen, usually a virus complexing with a specific antibody.

Neutrophil: A leucocyte having no affinity for acid or basophilic dyes, but stainable in neutral dyes.
Nodule; Node: A small lump, knot or node. This is an embracing term which can include any small tumour, whether it be neoplastic or not.
Non-ionised (ammonia): Also known as un-ionised. In the case of ammonia this is the highly toxic NH_3 form.
Non-infectious: Not capable of being transmitted from one organism to another.
Norm: An accepted standard.
Notifiable disease: A disease which, when first identified, must be reported to the appropriate authority. Notifiable diseases are listed by various countries, and they are the more serious and may need some form of containment or eradication.
Nucleus: An organelle in eukaryotic cells that contains the majority of the genetic information of the cell.

Oedema: Excessive accumulation of fluid in the tissue spaces due to increased exudation of the fluid from the capillaries resulting from an inflammatory process. (US sp.: edema)
Oedematisation: In pathology, refers to the accumulation of liquid in a tissue.
Oesteoma: A tumour of bony (osseous) tissue.
Oestrogenic hormone: Hormone responsible for the development and the regulation of the female reproduction system. In American, spelling is 'estrogenic'.
Omnivorous: An animal that eats both plant and animal matter.
Oocyst: A developmental stage in the life cycle of sporozoan parasites (Apicomplexa). It denotes a zygote surrounded by a thick-walled envelope, which is resistant to adverse environmental conditions. In the oocyst the zygote undergoes sporogony to produce infective sporozoites.
Oogonium: Structure in which non-motile round female gametes (oosphere) develop, characteristic for oomycetes. It may contain one to several oospheres.
Oospore: A thick-walled spore that developed from a fertilised oosphere. Oospores are found in some algae, fungi and oomycetes.

Operculum: A flap-like protective gill covering, present in cyprinids and other fish.
Opportunist agent: A pathogen or parasite that infects an abnormal or weakened host.
Oral: Pertaining to the mouth.
Osmoregulation: The regulation of the water and ion balance of an organism/living cells.
Osmotrophic: Uptake of dissolved organic compounds from the extracellular space, mainly by diffusion across a cell membrane, for nutrition.
Osseomucoids: Mucoids of bony tissue.
Osseous: Bony tissue.
Ossification: Formation of bone.
Ossified: Converted to bony tissue.
Osteosarcoma: Malignant tumour of bony tissue.
Outbreak: Occurrence of a disease within a site, region, or country during a limited period of time.
Overt disease: A disease apparent or obvious by gross inspection; a disease state exhibiting clinical signs.
Ovicidal: Acting against eggs, e.g. killing parasite eggs.
Ovulation: The release of mature eggs from ovarial tissue.
Oxygen saturation: The solubility of oxygen in water is very poor and is related to external pressure and temperature. The maximum amount of oxygen that can be dissolved in water at a given temperature and a given pressure is referred to as oxygen saturation.

Palintomic division: Mitotic divisions in a parasitic protist, in which no growth of the daughter cells is observed between subsequent divisions.
Pansporoblast: In the life cycle of some myxozoan parasites, a sporoblast that gives rise to more than one spore.
Papilloma: A benign tumour of epithelial tissue, which grows projecting outwardly.
Paralysis: In zoology, the loss of (voluntary) muscle movement.
Paraphyletic: In evolutionary biology, a group of organisms of a given systematic rank with a common ancestor, but some descendants of this ancestor were not included in this group.

Parasite: An organism that lives at the expense of, in or on another organism (the host) and which depends on the host for its metabolism; often has a higher reproductive potential than the host. By definition, a parasite will cause serious harm to the host when it is present in large numbers.
Parasitism: The mode of life in which one species, the parasite, lives in or on another species, the host. The former gains benefit from the association usually at the expense of the latter.
Parasitoid: Living in close association with a host organism, at the host's expense and finally killing the host. Thus, this form of parasitism is close to predation.
Parasitology: The study of parasites.
Parasitophorous vacuole: The vacuole that encloses intracellular stages of apicomplexan parasites in the cells of their host.
Paratenic host: The host of a parasite, which is not needed for the developmental cycle of the parasite and in which the parasite does not develop further. It does, however, act as a reservoir to transmit the parasite to its definitive host, in which the parasite then reaches maturity.
Parenchyma: The functional specific tissue of an organ.
Pathogen: Any organism which, in living on or within another organism (the host), causes disease to the host.
Pathogenesis: The origin or development of disease.
Pathogenic: Producing disease or pathological changes.
Pathological condition: A deviation from normal of known or unknown origin.
Pathology: The study of disease by scientific methods.
Peracute: Relating to a very severe and very acute condition.
Pericarditis: Inflammation of the pericardium, the sac enclosing the heart.
Pericardium: The connective tissue surrounding the heart.
Pericyte: An elongate contractile cell surrounding precapillary arterioles outside the basement membrane.
Periostial tissue: Connective tissue surrounding bone.
Peristomal ciliates: Arrangement of cilia in the area of the mouth opening of a ciliate.
Peritonitis: Inflammation of the peritoneum.

Peritrichous flagella: An organism that has flagella which are distributed uniformly.
Pesticide: A chemical substance that is used to kill unwanted organisms, like insects, molluscs or wild plants.
Petechiae: Small red spots formed by the escape of blood
Petechial: Pertaining to small red spot(s) caused by the escape of blood from blood vessels.
Phagocyte: A cell capable of ingesting bacteria, foreign particles and other cells.
Phagocytosis: The uptake of particulate matter into eukaryotic cells by the formation of membrane-bound vesicles that bud into the cell.
Photosynthetic: Referring to the process in which green plants generate organic compounds from inorganic substances by using energy absorbed from light. In this process, light energy is converted to chemical energy and stored in organic compounds.
Pinocytosis: The uptake of liquids into eukaryotic cells by formation of small membrane-bound vesicles.
Pituitary: In vertebrates, an endocrine gland located at the floor of the diencephalon as a part of the forebrain. It regulates various physiological processes.
Plankton: Aquatic organisms that float in the open water column and are unable to swim against the current or the wind.
Plasma cell: Antibody-secreting cell of the immune system which differentiates from the B lymphocyte after recognition of an antigen.
Plasmotomy: A type of reproduction in multi-nucleated organisms which gives rise to multi-nucleated daughter individuals. In these organisms, the division of nuclei occurs independently from the division process of the cell. Plasmotomy is observed in some myxozoa, in *Opalina,* a multi-nucleated gut parasite of amphibia, and in some foraminifera.
Poikilothermic: Of an animal whose temperature varies with that of the surrounding medium. Known commonly as 'cold-blooded'.
Polar capsule: Structure found in the spores of myxozoan parasites. It is a specialised cell, which forms and contains the coiled polar filament.

Polar tube; Polar filament: A structure found in the spores of microsporidian and myxozoan organisms. In myxozoans, polar filaments are found coiled in polar capsules. It presumably serves to temporarily attach the spore to the host's epidermis during the infection process.
Pollutant: Substance or effect that causes pollution.
Pollution: The introduction by human activities, directly or indirectly, of substances or energy into the environment resulting in deleterious effects on living organisms.
Polychlorinated biphenyls (PCBs): A group of toxic synthetic lipid-soluble hydrocarbons that are used in various industrial processes.
Polyculture: The culture or rearing of more than one species in a system.
Polyphyletic: In evolutionary biology, a taxon that contains organisms with similar features but which are of different evolutionary origin.
Polyplanetism: Several swarming periods with different zoospores occurring in the developmental cycle of a fungus or oomycete, separated by non-motile (aplanosporic) stages.
Potamodromous: In fish biology, migratory fish that migrate for spawning reasons in freshwater and do not enter marine waters. They swim a relatively short distance upstream and spawn in upstream waters of the main river system.
Prevalence: The total number of disease cases in a population. It can be represented as a percentage of a population.
Proboscis: A trunk-like extension from the head that in some parasites (e.g. acanthocephalans) comprises teeth and is used for attachment to the host.
Pronephros: The area of the kidney that develops first in the embryonic or larval life and in fish loses its excretory function and is a major immune organ.
Prophylactic: An agent that prevents disease.
Protozoal: Belonging to the single-celled animals (Protozoa).
Pseudofaeces: In fish medicine, this term refers to white mucus clumps hanging from the anus of diseased fish and appearing like faeces deposits.

Pycnosis (pyknosis): In necrotic cells with abundant chromatin, dense masses are formed that break up (karyorrhexis) to form granules inside the nuclear membrane or throughout the cytoplasm.
Regeneration: New growth or repair of structures or tissues lost by disease or injury.
Reinfection: A second infection by the same type of microorganisms.
Resistance: The ability, natural or acquired, of an organism to overcome the process of a disease.
Reticular: Pertaining to a reticulum; net-like.
Rhabdomyoma: A tumour originating from striated muscle cells.
Rhizocysts: In dinoflagellates, rod-like organelles that penetrate into epithelial cells and thus firmly attach the parasitic dinoflagellate to the host cell.
Rhizoids: Filamentous outgrowth.

Saline: Solution of sodium chloride (salt) at a physiological concentration.
Saprophytic: Nutrition by absorbing dissolved organic material from dead or decaying organic matter such as dead or decaying animals or plants.
Sarcoma: A malignant tumour whose parenchyma is composed of anapaestic cells resembling those supportive tissues in the body.
School: In fish biology, a school (or shoal) describes a social group of fish that stay together, swimming in the same direction in a coordinated manner. Fish benefit from schooling, for instance by a better defence against an attack of predators or by enhanced foraging success.
Scoliosis: Lateral curvature of the vertebral column.
Secretion: The production of a substance that differs chemically and physically from the body producing it.
Section: To divide by cutting; a slice (e.g. histological section).
Septicaemia: A clinical syndrome characterised by a severe bacteraemia infection, generally involving the significant invasion of the bloodstream by microorganisms.
Serodiagnostic test: A test that incorporates immune sera (i.e. serum containing antibodies).

Serology: The study of immune sera and the use of antisera to characterise pathogens, antigens, cells etc.
Serosa: A serous membrane.
Serous membrane: A membrane that lines a cavity that has no communication with the external environment.
Serum: The fluid part of the blood after removal of clotted cells (pl. sera).
Slough: A piece of dead (necropsied) tissue that has separated from a wound or inflamed surface.
Somatic ciliature: Arrangement of cilia on the cell surface of a ciliate opposite the ciliature around the mouth opening (oral ciliature).
Sporangiophore: In fungi, a hypha that carries a sporangium, a structure in which spores develop and mature.
Sporangium: In plants, algae, fungi or oomycetes, a structure where spores are formed.
Sporocyst: A developmental stage in the life cycle of some sporozoan parasites. The sporocyst develops during sporogony and protects sporozoites by a thick resistant cyst wall from adverse environmental conditions. In many apicomplexan species, several sporocysts are enclosed in an oocyst.
Sporogony: A stage in the life cycle of sporozoan parasites in which an encysted zygote undergoes several asexual reproduction cycles and forms sporozoites that are encysted in a spore.
Squamous: Simple epithelium of nucleated cells.
Sterile: Free from microorganisms; infertile.
Stieda body: A structure in the polar region of the sporocyst in some species of coccidia, gut-dwelling apicomplexan parasites of vertebrates. A knob-like structure that occludes a hole in the sporocyst wall like a stopper. The degradation of the Stieda body allows the excystation of sporozoites through this opening.
Stratum spongiosum: A layer in the skin of vertebrates, located in the dermis, beneath the outermost layer, the epidermis, and characterised by loose connective tissue. Also referred to as stratum laxum.
Stress: The sum of non-specific reactions of an organism to an adverse stimulus, thus affecting the functional processes of that organism, possibly leading to pathological changes.

Stressor: Any adverse stimulus (e.g. of physical or chemical nature) that leads to stress.
Subcutaneous: Located under the skin.
Sublethal: Term applied to a dose that is not quite fatal.
Submucosa: The tissue lying beneath the mucous membrane.
Sulcus: Longitudinal cell groove in dinoflagellates.
Suppuration: Formation or discharge of pus
Swab: A mounted piece of cotton-wool used for obtaining specimens (of a secretion, etc.) for laboratory examination.
Swelling: The abnormal enlargement of an organ.
Syndrome: A group of signs which, when considered together, characterise a disease.
Systemic: Throughout the body, involving the whole body.

Teratogen: A substance that may induce malformations during the development of an embryo.
Teratoma: Abnormal growth of an oocyte or testes germ cell.
Thallus: The multi-cellular body of a plant or a fungus, which is not organised in horizontal stem, root and leaves as found in vascular plants.
Theca, thecal plates: In dinoflagellates, a structure to reinforce the cell membrane complex.
Therapeutic: Any procedure for the treatment of a disease.
Therapy: Any type of treatment of a disease, e.g. vaccine therapy, chemotherapy.
Theront: In some parasitic ciliates with a complex life cycle, a motile, elongate stage that infects the host.
Thrombocyte: Spindle-shaped nucleated cells involved in blood-clotting in fish.
Tomite: In some parasitic protists, the stage that results from a series of mitosis in the encysted tomont developmental stage. The tomites subsequently differentiate into free-swimming therons, which colonise a suitable host.
Tomont: Adult stage of life cycle in some parasitic protists from aquatic organisms. After phase of feeding and growing, the parasite leaves the host, attaches to any submerged object and undergoes a series of mitoses to become a tomont.

Torpor: In zoology, a stage of reduced metabolic activity, during which the animal is inactive and barely responds to external stimuli. This stage enables the animal to survive a period of water or food shortage.
Transmission: The transfer of a disease agent from one individual to another.
Trauma: A wound or other external injury.
Treatment: The medical management of a disease or disorder as a whole.
Trophont: In protists with a complex life cycle, the stage in which the protist feeds and grows.
Tubercle: A rounded, solid, raised lump on the skin or surface of an organ.
Tumorigenesis: Pathological process involving the transformation of normal cells to a neoplastic form, usually leading to cell proliferation.
Tumour: Literally means any swelling, but by common consent the term is not held to include passing swellings caused by inflammation etc. It should be restricted to neoplasms.

Ulcer: A break in the skin or mucous membrane with loss of surface tissue, disintegration and necrosis.
Unsegmented: Lacking segments.

Vaccine: a biological preparation that induces a response of the adaptive immune system in vertebrates that protects the treated individual from a particular infectious or malignant disease.
Ventral: In anatomy, the front or lower side of the body. In vertebrates, the side of the body where the stomach is located.
Vertical transmission: Transmission of a pathogen, usually a virus, from parent to offspring via the reproductive tract or gametes
Virology: The study of viruses
Viriome: The total viral content in a habitat or environment.
Virulence: The relative capacity of a pathogen to produce disease.
Vitamins: A group of unrelated organic substances that in small amounts are necessary for normal metabolic functioning of the body.

Vitellaria: Yolk glands.
Vitellogenesis: Yolk formation.
Vitellogenin: Protein that is converted into yolk.
Viviparous: Producing young alive rather than laying eggs.

Water hardness: The content of divalent metal ions, namely calcium (Ca^{2+}) and magnesium (Mg^{2+}), in water.
Western blot: A technique in which proteins/peptides are transferred after electrophoresis from a gel medium to nitrocellulose paper. Aids further analysis of the protein.

Yellow bodies: Aggregates of enterocytes infected with sporulated coccidian oocysts and macrophages enriched with melanin. The aggregates have a yellow colouration and are found in the intestine of carp infested with enteritic coccidia.

Zooid: A single animal as a part of a colony of animals that live in a close association with each other and are connected by tissue.
Zoonosis: A disease or infection naturally transmitted between a vertebrate animal and humans.
Zooplankton: Aquatic eukaryotic microorganisms from the planktonic community with heterotrophic nutrition. Planktonic organisms float in the water column and cannot swim effectively against water currents.
Zoospore: A flagellated spore that moves in water by the aid of one or several flagella.
Zygospore: A thick-walled, multi-nucleated spore formed from a zygote (after the fusion of two equal-sized flagellated gametes). A zygospore occurs as a dormant stage in the life cycle of some fungi and algae.

Chapter 1

World Carp Production: Economic and Societal Importance

Carp have long been recognised as the most important group of fish in the global market and their production contributes significantly, not only to world finfish production for sustainable food, but also within the recreational activities and ornamental sectors. FAO data published in 2021 indicated that carp represent in tonnage over 80% of world production of fish, and representatives of the carp group occupy six of the top seven positions within world aquaculture. Over the past decade the production of individual carp species has increased dramatically and percentage increases range from 3.7% for *Labeo rohita* (rohu) to 10.5% for *Ctenopharyngodon idella* (grass carp). Carp species therefore make a significant contribution to the economic basis of several countries, regional areas and individuals. Prominent carp species include those that occur globally, such as *Cyprinus carpio* (common carp) and *Carassius* species, as well as carp species that predominate in Europe and Asia such as grass carp, *Hypophthalmichthys molitrix* (silver carp), *Hypophthalmichthys nobilis* (bighead carp) and more specialised production of, for example, *Gibelion catla* (catla) and rohu in the Indian subcontinent (Fig. 1.1).

Carp production is recognised as a major contribution in addressing world food limitations, providing a relatively cheap and sustainable source of protein and essential fatty acids for human health. Compared with other finfish aquaculture, for example trout and eel, carp species production utilises far less compound feed, comprising only 0.4% wild fishmeal. Only 57% of carp are reared

 Carps: A Practical Approach to Disease Control and Health Management (eds David Hoole *et al.*)
DOI: 10.1079/9781836990024.0001

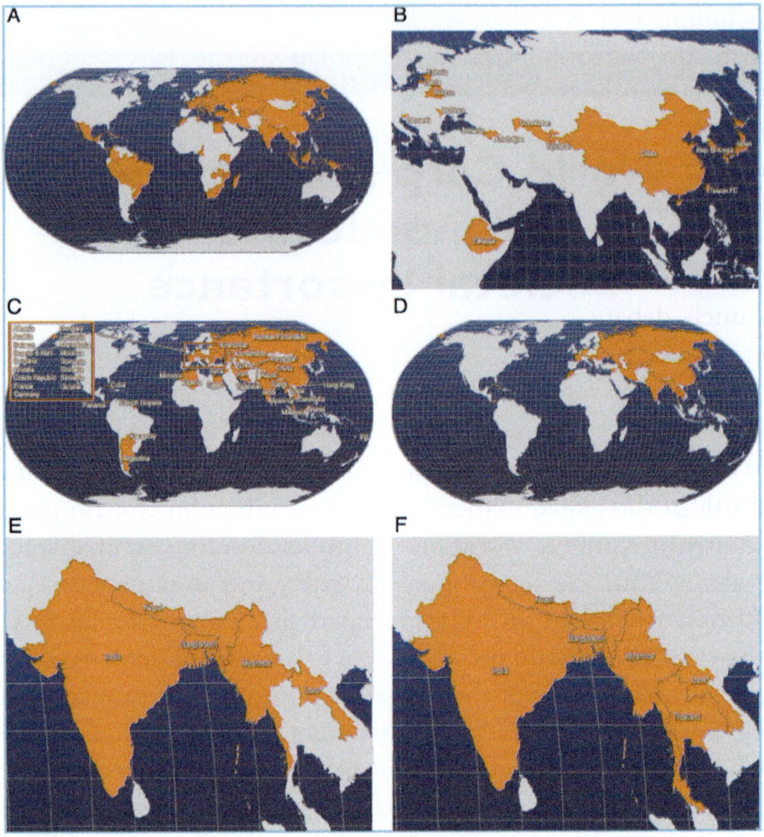

Figure 1.1 Global distribution of the major carp species. Main producer countries for (A) *Cyprinus carpio* (common carp)[b], (B) *Carassius carassius* (crucian carp)[b], (C) *Ctenopharyngodon idella* (grass carp)[a], (D) *Hypophthalmichthys molitrix* (silver carp)[b], (E) *Gibelion catla* (catla)[b] and (F) Indian major carp species *Cirrhinus mrigala* (mrigal)[b] and *Labeo rohita* (rohu)[b]. Credit: FAO Fishery Statistics, [a]2002, [b]2006. Courtesy of the Food and Agricultural Organization of the United Nations (FAO).

on such formulated feed; the rest utilise a range of pond-derived food sources.

The historic nature of the selective breeding of the various carp species has produced a thriving worldwide ornamental trade, which in 2015 was estimated to be worth US$3 billion annually, although this is considered to be a gross underestimation, as Japan exported 295 tonnes of koi carp in 2016, estimated to be worth

3.5 billion yen (approximately US$31.3 million). There is much debate on the origins of the ornamental (koi) carp production, which currently comprises over 100 different varieties. Both wild and feral carp species are considered to be the sources, whilst other authorities consider that koi carp were developed from common carp in Niigata Prefecture in Japan over 100 years ago. Goldfish production is thought to have originated in China in the Jin dynasty (AD 266–420) where coloured versions of carp were noted. There is much debate on the taxonomic origin of goldfish and recent genetic analysis suggests that they originated from the domestication of *Carassius auratus*, rather than *Carassius carassius* (crucian carp) or *Carassius gibelio* (gibel carp), although hybridisation can occur. There are now thought to be over 200 varieties of goldfish exhibiting a range of colours and body form, which means that the goldfish is perhaps the most favoured pet fish in aquaria and pond facilities.

A major component of the production of carp worldwide is associated with capture and leisure fisheries. It is estimated that in England angling makes a £1.4 billion annual contribution to the country's economy and supports over 27,000 fulltime equivalent jobs. Over 19 million days are spent by anglers per year, with carp being the most sought-after fish, making up 37% of this commitment.

Carp, whether produced in aquaculture, the ornamental trade or in angling, therefore make a significant contribution to the economic growth of several countries and regions therein, and may provide an important source of income to families where other types of employment are not readily available. In addition, the variety of production units, either natural sources or manmade, provide habitats for a range of threatened species and are beneficial to regional water balance and hydro-environmental sustainability. Diseases, however, represent a serious threat to the growth of carp production worldwide.

Selected References for Further Reading

FAO (2023) Fisheries and Aquaculture: Cultured Aquatic Species Information Programme. Available at: https://www.fao.org/fishery/en/aqspecies/ (accessed 13 March 2023).

Chapter 2

Range of Species and Production Units

2.1 Introduction

Carp are members of the teleost family Cyprinidae and are widely distributed in freshwater in Eurasia, Africa and North America. Cyprinids are naturally absent from South America, Australia and Madagascar, but some, like the common carp (*Cyprinus carpio*), have been introduced into waters of these regions for economic reasons. The family Cyprinidae comprises a large number of species found in stagnant or slow-flowing waters. The highest species diversity occurs in Chinese and South-east Asian waters, and the lowest in Africa and North America. Many cyprinid species tolerate fluctuations in temperature and in concentrations of dissolved oxygen and pH. This is thought to be an adaptation to differences in ambient temperature that occur between summer and winter, and to changes in water conditions, which are frequently found in still waters due to photosynthetic activity of plants during summer. Adult common carp and *Carassius* species, for instance, survive at temperatures ranging from 1 to 35°C, while best growth is achieved in a temperature range around 25–30°C. These two carp species also survive for a few hours in waters with a range of dissolved oxygen concentrations from supersaturation to almost anoxia, which can occur in summer or under the ice in winter. However, in order to ensure good health and optimal growth, all cyprinids require oxygen-saturated water at a neutral pH and of moderate calcium hardness.

Most cyprinids are omnivorous opportunist feeders and vary their diet according to life stage, season and abundance of food items. Many species are zooplankton feeders, but also accept benthic

DOI: 10.1079/9781836990024.0002

organisms, detritus, filamentous algae and phytoplankton. Some species filter phyto- or zooplankton from the water column or ingest plant materials. At optimal water temperature, in particular in tropical or subtropical waters, cyprinids may grow to several kilograms in weight per year, while in temperate regions growth is much less and fish may reach 1–2 kg after 2–3 years.

In global aquaculture, cyprinids play a major role that cannot be overestimated. According to FAO statistics, cyprinid fish account for a quarter of global aquaculture production in freshwater in quantity and value. The four carp species grass carp, silver carp, common carp and bighead carp are among the top ten fish species in freshwater aquaculture. These carp are mainly cultivated in ponds (Fig. 2.1) or reservoirs in many Asian and Continental European countries.

Many species occupy a low trophic level in the food chain and, in particular in pond aquaculture, are not dependent on external feed input, but ingest food items that develop in the pond ecosystem. In many countries, carp species are considered as 'every man's food', as they provide components of traditional dishes and are sold locally at a relatively low price. While households from the middle social class consume carp less, these fish play an important role for food security and alleviate hunger and poverty in rural areas, particularly in many Asian countries.

2.2 Bighead Carp

Bighead carp (*Hypophthalmichthys nobilis*) (Fig. 2.2) are currently ranked 5th among cultured freshwater fish by the FAO. They originate from freshwater habitats in China with a broad distribution from the Pearl River drainage to the Heilongjiang (Amur) river catchment in Northern China. Because of this wide geographical distribution, bighead carp tolerate water temperatures over a wide range of 0.5–38°C. They live in lakes, rivers and reservoirs in the upper layer of the water column and feed on zooplankton, but also accept artificial feed containing products from grain processing and organic detritus. Bighead carp grow rapidly and may reach up to 40 kg in weight.

Figure 2.1 (A) Aerial view of a carp farm with ponds for on-growing fish and large production ponds in southern Poland. (B) Carp ponds for on-growing fish in Central Europe.

Bighead carp are mainly farmed in polyculture systems together with omnivorous and herbivorous carp species (except silver carp) and tilapia. In these ponds production costs can be very low through the use of organic fertilisation of water with manure or plant waste. The aquaculture of bighead carp can also be well integrated into crop farming or livestock production in order to improve the

Figure 2.2 Bighead carp, *Hypophthalmichthys nobilis*. Courtesy of the Food and Agricultural Organization of the United Nations (FAO).

utilisation of natural resources. This integration of land and water food production is therefore recommended by FAO for freshwater aquaculture in developing countries. Most important for fish production is the supply of seed for aquaculture (i.e. juvenile stages from hatchling, spawn and fry to fingerling). This can be affected by the slow gonadal development in adult fish, the production of fingerlings for stocking of ponds and the slow growth of the fish in the nursery stage. While initially seed can be collected from natural stocks, artificial propagation has been developed to ensure a sufficient production of eggs and fry. However, care should be taken to ensure the genetic quality of the seed for hatchery populations. Artificially bred fish may escape from aquaculture and mate with riverine populations, which could have a negative impact on the germplasm of natural populations.

2.3 Common Carp

The aquaculture of common carp (*Cyprinus carpio*) (Fig. 2.3) has a long tradition both in China and in Europe. Initially, common carp was considered to be a single species (*Cyprinus carpio*) with a wide distribution in Asia and Europe. More recently, authorities have separated the European and West Asian populations from those that are native in East Asian waters from the Amur River to

Figure 2.3 Common carp, *Cyprinus carpio*. Courtesy of the Food and Agricultural Organization of the United Nations (FAO).

the Red River basin and have named the latter *Cyprinus rubofuscus*. In addition, other *Cyprinus* species have a more restricted distribution and are mainly found in the waters of the Yunnan Province in south-west China. In China, carp were kept in undrainable ponds which were regularly stocked with fry from rivers. In these ponds common carp were raised in polyculture with other cyprinids and several semi-domesticated carp races were developed. In Europe, common carp were raised in fish ponds in monoculture, where the largest individuals were selected as broodfish. In China, the broodstock is derived from local and exotic varieties of *C. carpio* and *C. rubofuscus* and includes various local genetic varieties of these species as well as local *Cyprinus* species.

Controlled semi-natural breeding and rearing in drainable ponds (Figs. 2.4 and 2.5) was developed in the 19th century in Europe. During this period, several domesticated common carp strains were developed. Currently, the species has an outstanding importance in freshwater aquaculture and coloured varieties are highly popular ornamental fish. The natural habitats of wild common carp are middle and lower streams, flooded areas and shallow confined waters such as ponds, lakes or water reservoirs. In these situations, common carp are bottom dwellers but feed at all levels of the water column. They are omnivorous and ingest benthic insect larvae, worms or molluscs, but also plant material. Fry and juvenile stages feed on zooplankton. Common carp ponds are usually shallow eutrophic ponds with a muddy bottom, some underwater vegetation and a

Figure 2.4 Drainable semi-natural pond for on-growing carp in Central Europe.

Figure 2.5 Outlet of a drainable pond (monk structure).

high development of zooplankton. Common carp tolerate a water temperature between 1°C and 35°C, and grow best between 23°C and 30°C, when the daily growth rate can be up to 2–4% of body weight. In such situations, common carp may reach a body weight of up to 1 kg within one season, but growth in temperate zones is much slower and fish may reach only 1–2 kg body weight after three to four rearing seasons. Domesticated common carp release mature eggs within a limited period of a few hours, while fish from natural populations deposit their eggs in portions over a period of several days (Fig. 2.6).

The eggshell, which becomes sticky in water, enables the eggs to attach to submerged vegetation. Larvae hatch from fertilised eggs after 60–80 degree-days, attach to submerged substrates, and after another developmental period of 60–80 degree-days the larvae swim to the water surface, ingest air to fill their swim bladder, and subsequently start feeding on small plankton, mainly rotifers. Therefore, small shallow ponds without aquatic weeds are most suitable as a common carp nursery; they provide a high abundance of zooplankton as food, initially rotifers and subsequently cladocerans, such as *Daphnia* and *Moina* species. The production of fingerlings and market-sized fish usually takes place in fertilised semi-intensive ponds, which enhances the generation of natural food. Cereals or pellets can be used as supplementary feed to enhance production. In addition to a pond production in monoculture, common carp are farmed in ponds in polyculture together with Chinese carp or Indian major carp, tilapia or mullets. They are also farmed in cages, associated with irrigation reservoirs or in ponds with running waters in intensive monoculture using artificial feeds. In Europe, common carp are consumed regionally as components of traditional dishes. In Asia, common carp are sold fresh on local markets at a relatively low price.

2.4 Crucian Carp

The aquaculture of crucian carp (*Carassius carassius*) (Fig. 2.7) dates back in China to the East Han Dynasty (AD 25–189), but for a long time it was limited to small-scale production in China and

Figure 2.6 Artificial reproduction of carp: obtaining eggs from carp. (A) Typical method of 'stripping' female carp for eggs. (B) Culturing of stripped carp eggs.

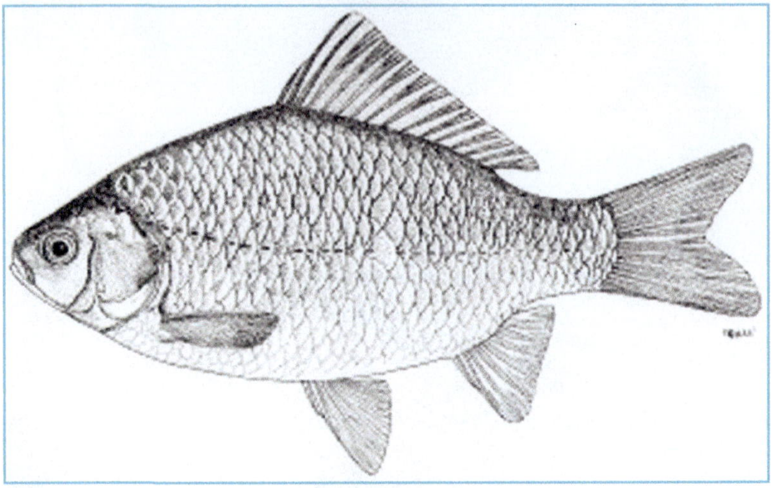

Figure 2.7 Crucian carp, *Carassius carassius*. Courtesy of the Food and Agricultural Organization of the United Nations (FAO).

Japan. Since the mid-1960s, production has expanded to other regions in East Asia and Eastern Europe and currently the annual production reaches almost 3 million tonnes. The natural habitat of crucian carp is the bottom layer of the water column in lakes, rivers and reservoirs. Crucian carp, like common carp, can tolerate a wide range of temperature and have a remarkable ability to survive in hypoxic conditions, even without dissolved oxygen, such as occurs in water for days in summer or months under the ice covering shallow lakes in winter. This exceptional tolerance to anoxia is attributed to their ability to generate energy anaerobically by glycolysis, with ethanol as the major end product. In their natural habitat, crucian carp are omnivorous and feed on small benthic organisms, detritus, filamentous algae and pieces of plants. Fry or fingerlings feed on zooplankton. In aquaculture, crucian carp also accept artificial feed containing by-products of grain processing, oil extraction or pelleted feed. The diploid crucian carp, which possesses 100 chromosomes, reproduces sexually, grows moderately and can reach a weight of about 1.5 kg. The morphology of crucian carp is very plastic, and the ornamental goldfish (*Carassius auratus*, syn. *C. auratus auratus*) (Fig. 2.8) has been cultured throughout history in

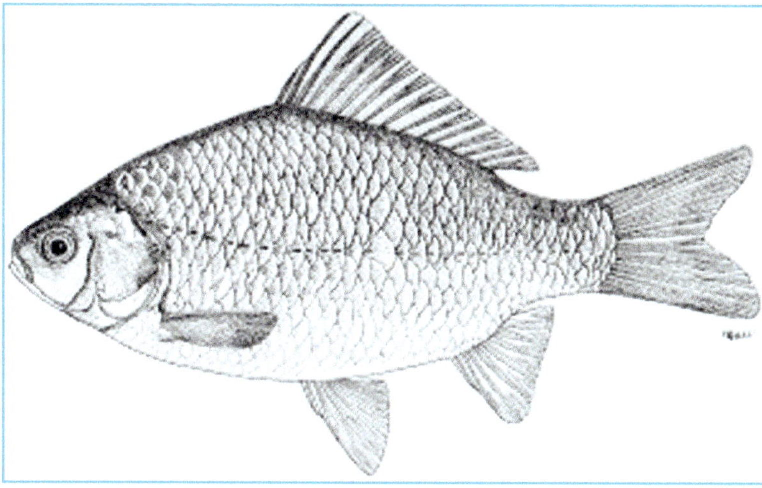

Figure 2.8 Goldfish, *Carassius auratus*. Courtesy of the Food and Agricultural Organization of the United Nations (FAO).

a large number of different morphological varieties. In aquaculture, mainly naturally occurring polyploid gibel carp (*Carassius gibelio*, syn. *C. auratus gibelio*), with more than 150 chromosomes, were cultivated. In these carp, heterologous sperm (from other cyprinids) activates embryo development (termed heterologous gynogenesis, or allogynogenesis). These allogynogenetic gibel carp grow faster than the parent fish, and thus enable a much higher yield than other varieties of the *C. carassius/C. auratus* genetic complex. Although fish can spawn naturally, currently spawning is mainly induced artificially by injection of pituitary gland extracts or hormones. It is important to provide substrata, such as grass bundles, fine tree branches or plastic weeds, for egg attachment. Ovulated eggs can be inseminated with sperm from common carp to activate gynogenesis and are subsequently moved into a spawning pond or incubated in hatching jars. Fingerlings and market-sized fish may be produced in monoculture in ponds with natural feeds, which are supplemented with commercial feed based on soybean cake, rapeseed cake or wheat bran. Currently, the majority of *Carassius* spp. production results from polyploid gibel carp. Most commonly, crucian carp and gibel carp are stocked together with other carp species in polyculture in ponds/pens or rice paddies. Crucian carp/gibel carp is a locally consumed product and preferred in most parts of China because of its meat quality and high nutritional value. However, it possesses a large number of fine inter-muscular bones, which limits its acceptance outside China and by other consumers.

2.5 Grass Carp

Grass carp (*Ctenopharyngodon idella*) (Fig. 2.9), or white amur, is naturally distributed in rivers, lakes and water reservoirs of the catchment area of the Pearl River in Southern China to the Heilongjian (Amur) river in Northern China. Grass carp dwell in mid–lower layers of the water column, are basically herbivorous and feed on aquatic weeds. Fry and larvae feed on zooplankton. In aquaculture, grass carp accept terrestrial grasses, vegetable oil extraction meals and pelleted feed. The fish prefers relatively clear water, is semi-migratory and, in natural populations, mature broodstock move upstream to propagate. Essential stimuli for propagation are flowing water and changes in water level. In aquaculture ponds, fish can reach sexual maturity, but cannot spawn naturally because flowing water is an essential stimulus for this to occur. Aquaculture of this species began in China during the Tang Dynasty (AD 618–904), but was dependent on a natural supply of seed. When induced breeding technology by hormone injection was developed, this ensured a sufficient supply of seed, and aquaculture production rose rapidly. Now

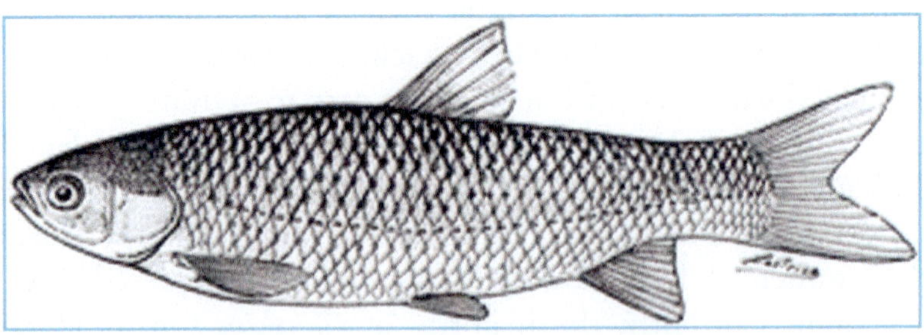

Figure 2.9 Grass carp, *Ctenopharyngodon idella*. Courtesy of the Food and Agricultural Organization of the United Nations (FAO).

the culture of grass carp accounts for more than 15% of global freshwater aquaculture production. Grass carp are grown in more than 40 countries in various regions in Asia, Europe, Northern Africa and South America. In aquaculture, grass carp are grown in monoculture in the nursery stage, but for the production of fingerlings and market-sized fish, grass carp frequently are kept in polyculture with other carp species in fertilised ponds. For intensive culture, grass carp are stocked together with additional fish in cages and fed with aquatic weeds/terrestrial grasses, by-products from vegetable oil extraction and grain processing, and pelleted feed. In nature grass carp grow rapidly, but the production of market-sized fish of <1500 g requires at least 2 years. Grass carp are sold at a low price; they are therefore affordable for buyers with a low income, and are mainly sold locally.

2.6 Silver Carp

Silver carp (*Hypophthalmichthys molitrix*) (Fig. 2.10) is an important species for aquaculture in China, because it is herbivorous and can be raised in polyculture with other cyprinids. Its herbivorous feeding habit has also meant that silver carp have been introduced into European and Israeli waters for algal control. Natural

Figure 2.10 Silver carp, *Hypophthalmichthys molitrix*. Courtesy of the Food and Agricultural Organization of the United Nations (FAO).

populations of silver carp live in backwaters and slow-flowing areas of large rivers in the temperate zone in Asia. These natural populations are potamodromous, migrate upstream to breed, and eggs and larvae float downstream to floodplain areas where they populate stagnant or slow-flowing waters. Silver carp are active, swim just below the water surface and jump out of the water when disturbed. As a typical planktivorous fish, silver carp filter diatoms, dinoflagellates, planktonic green algae and cyanobacteria from the water column by means of their gill rakers. In addition, silver carp ingest detritus, rotifers and small crustaceans. Spawning occurs at warm water temperature, optimal at 22–28°C, in late spring/early summer. The eggs are non-adhesive and in stagnant water stay at the bottom or, in flowing water, float in mid-water until the fry hatch. In aquaculture, spawning is induced by injection of oestrogenic agents, including luteinizing release hormone or fish pituitary gland extracts. Released eggs require good water quality and high oxygen concentrations for undisturbed embryonic development and hatching. Because hatched larvae are delicate and are frequently victims of various predators, such as wild fish, tadpoles or larger crustaceans, nursery ponds have to be carefully cleaned and sterilised, for instance with quicklime (CaO), before stocking with eggs or larvae. In order to produce natural food, ponds are fertilised by applying fermented manure or compost a few days before stocking with fry. The production of fingerlings and market-sized fish is mainly done in polyculture together with additional carp species or with tilapia. Silver carp are consumed locally and fresh. The market price is relatively low.

2.7 Indian Major Carp

Species from the cyprinid subfamily Labeoninae are widely distributed in tropical Asia and Africa. The northern Indian river system of Indus, Ganges and adjoining waters are populated with various labeonid fish, from which, in particular, catla (*Labeo catla*), mrigal (*Cirrhinus mrigala*) and rohu (*Labeo rohita*) are cultivated in polyculture in India and other South-east Asian countries.

2.7.1 Catla

Catla (*Gibelion catla*, also known as *Catla catla*, *Labeo catla* or *Cyprinus catla*) (Fig. 2.11) is endemic in waters from the river system in Northern India, Pakistan, Bangladesh, Nepal and Myanmar. Catla dwell in surface to mid-water layers of rivers, floodplains, ditches and canals. The fry feed on plankton and adults prefer zooplankton (rotifers and crustaceans), insects and protozoa, but also ingest algae and plant material. In aquaculture, catla also accept rice bran or oil cake. Catla is highly preferred by consumers in some countries, and therefore is an important component in the carp polyculture systems in South and South-east Asia. It is the fastest growing species among the three Indian major carp. In natural habitats during the monsoon season, catla undergoes a spawning migration to the upper stretches of rivers and the resulting seed floats downstream to other areas. In aquaculture ponds, although catla reaches maturity, natural breeding does not occur in the pond environment because it is a riverine species. Reproduction therefore requires artificial induction of spawning by hormone injection, although this may give poor results compared with its use in Chinese carp and other Indian major carp. The spawning period is short and precise environmental conditions

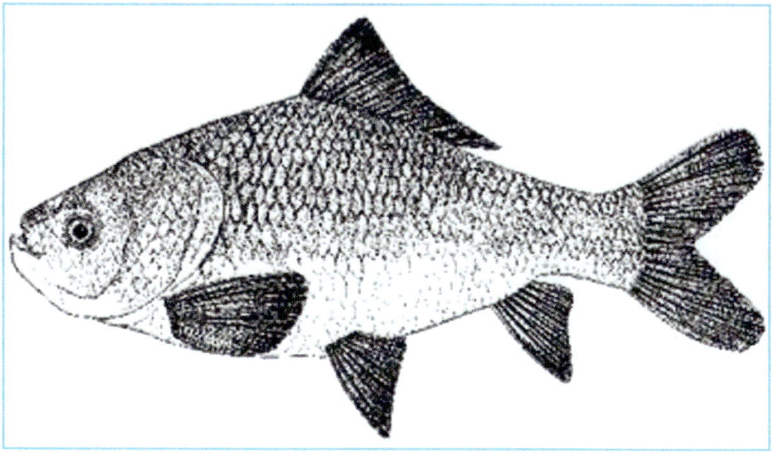

Figure 2.11 Catla (*Gibelion catla, Catla catla, Labeo catla*). Courtesy of the Food and Agricultural Organization of the United Nations (FAO).

are required for successful spawning. Hatched larvae are stocked in nursery ponds, which are cleared of predatory fish by liming, and fertilised by organic or inorganic fertilisers a few days before stocking of the fry. Fingerlings and on-growing catla are further reared together with rohu and mrigal or additional carp species in polyculture in fertilised ponds. For supplementary feeding conventionally, a mixture of rice bran and oilcake is provided. Catla is marketed fresh at a size of 1–2 kg in local markets.

2.7.2 Mrigal

Mrigal (*Cirrhinus mrigala*) (Fig. 2.12) is endemic to waters of the Indus–Ganges river system, where it dwells in mid-lower layers of the water column and adults move to the bottom. It mainly feeds on detritus and decaying vegetation from the bottom and thus acts as a scavenger, although fish can also ingest phytoplankton and zooplankton from the water column. In aquaculture, a mixture of rice bran and oil cake is provided as supplementary feed. Mrigal is eurythermal and tolerates a minimal water temperature of about 14°C. In natural populations, spawning depends on the onset and duration of the south-west monsoon, which in India, Bangladesh and Pakistan extends from May to September. In pond culture, mrigal reaches maturity but propagation is not achieved, because it requires riverine conditions for spawning. Breeding in captivity

Figure 2.12 Mrigal (*Cirrhinus mrigala*). Courtesy of the Food and Agricultural Organization of the United Nations (FAO).

relies on the injection of pituitary extracts or oestrogenic hormones. Hatched larvae are stocked in nursery ponds from which predatory fish have been removed and the water is fertilised with organic manure or inorganic fertilisers. Additional feed such as rice bran and oil cake can also be provided. Fingerlings and on-growing fish are raised in polyculture with other carp species in fertilised ponds. Mrigal reaches a market size of 600–700 g after a grow-out period of about 1 year. *Cirrhinus* is the least preferred among the three Indian major carps and is sold at a lower price than rohu or catla.

2.7.3 Rohu

Rohu (*Labeo rohita*) (Fig. 2.13) is the most important of the three Indian major carp species raised in carp polyculture. Natural populations are found in rivers of northern and central India, Pakistan, Bangladesh and Myanmar. The species was also introduced into waters in various regions in East Asia and Africa. Fry and fingerlings feed on zooplankton, mainly rotifers and crustaceans, but also ingest phytoplankton, which most likely are filtered from the water column by means of fine, hair-like gill rakers. Fry and juveniles form shoals for feeding. Adult rohu filter phytoplankton from the water column, but also ingest decaying organic matter and soft plant material from the bottom of the ponds. Rohu is eurythermal and does not grow at a temperature below 14°C. It does, however, have a fast growth rate and may reach 700–800 g within 1 year of cultivation. Fish reach

Figure 2.13 Rohu (*Labeo rohita*). Courtesy of the Food and Agricultural Organization of the United Nations (FAO).

sexual maturity after 2 years, but complete fecundity is reached only after 4–5 years. Natural populations spawn in shallow marginal areas of flooded rivers during the south-west monsoon in April–September at a temperature of 22–31°C. In pond aquaculture, rohu reaches maturity after 2 years, but as a riverine species, it does not spawn in stagnant waters and propagation relies on the injection of pituitary extracts or oestrogenic hormones. Fecundity is high and ranges between 200,000 to 300,000 eggs/kg body weight, depending on fish size. Fry are stocked in small nursery ponds, which have to be prepared prior to stocking by the removal of aquatic weeds, predatory fish and water insects by liming, and fertilised with organic manure or inorganic fertilisers to enhance the production of zooplankton as feed. As supplementary feed, rice bran and oil cake are offered. Fingerlings and on-growing rohu are frequently cultivated in polyculture with other carp species in fertilised ponds with rice bran/oil cake as supplementary feed. In these polycultures, rohu is stocked as the major species at a proportion of 30–40% of the total fish. Rohu, a carp highly preferred by consumers, is marketed mainly at a size of 1–1.5 kg, which is reached within a culture period of about 18 months. Rohu is marketed fresh in local markets or shipped on ice to urban markets, where it is sold at a relatively high price.

2.8 Selected References for Further Reading

Bais, B. (2018) Fish scenario in India with emphasis on Indian major carps. *International Journal of Avian and Wildlife Biology* 3, 409–411.

Balon, E.K. (1995) Origin and domestication of the wild carp, *Cyprinus carpio*: From Roman gourmets to the swimming flowers. *Aquaculture* 129, 3–48.

Balon, E.K. (2004) About the oldest domesticates among fishes. *Journal of Fish Biology* 65(Supplement A), 1–27.

Billard, R. (1999) *Carp: Biology and Culture.* Springer-Praxis Series in Aquaculture and Fisheries. Praxis Publishing, Chichester, and Springer, London.

FAO (2023) *Carassius auratus.* Cultured Aquatic Species Information Programme. Text by Weimin, M. Fisheries and Aquaculture Division,

FAO. Available at: https://www.fao.org/fishery/en/culturedspecies/carassius_carassius/en (accessed 13 March 2023).

FAO (2023) *Catla catla*. Cultured Aquatic Species Information Programme. Text by Jena, J.K. Fisheries and Aquaculture Division, FAO, Rome. Available at: https://www.fao.org/fishery/en/culturedspecies/catla_catla/en (accessed 13 March 2023).

FAO (2023) *Cirrhinus mrigala*. Cultured Aquatic Species Information Programme. Text by Weimin, M. Fisheries and Aquaculture Division, FAO, Rome. Available at: https://www.fao.org/fishery/en/culturedspecies/cirrhinus_mrigala/en (accessed 13 March 2023).

FAO (2023) *Ctenopharyngodon idellus*. Cultured Aquatic Species Information Programme, Text by Weimin, M. Fisheries and Aquaculture Division, FAO. Available at: https://www.fao.org/fishery/en/culturedspecies/ctenopharyngodon_idella/en (accessed 13 March 2023).

FAO (2023) *Cyprinus carpio*. Cultured Aquatic Species Information Programme. Text by Peteri, A. Fisheries and Aquaculture Division, FAO, Rome. Available at: https://www.fao.org/fishery/en/culturedspecies/cyprinus_carpio/en (accessed 13 March 2023).

FAO (2023) *Hypophthalmichthys molitrix*. Cultured Aquatic Species Information Programme. Text by Yang, N., Fisheries and Aquaculture Division, FAO, Rome. Available at: https://www.fao.org/fishery/en/culturedspecies/hypophthalmichthys_molitrix/en (accessed 13 March 2023).

FAO (2023) *Hypophthalmichthys nobilis*. Cultured Aquatic Species Information Programme. Text by Weimin, M. Fisheries and Aquaculture Division, FAO, Rome. Available at: https://www.fao.org/fishery/en/culturedspecies/hypophthalmichthys_nobilis/en (accessed 13 March 2023).

FAO (2023) *Labeo rohita*. Cultured Aquatic Species Information Programme. Text by Jena, J.K. Fisheries and Aquaculture Division, FAO, Rome. Available at: https://www.fao.org/fishery/en/culturedspecies/labeo_rohita/en (accessed 13 March 2023).

Gui, J.-F., Tang, Q., Li, Z., Liu, J. and De Silva, S.S. (2018) *Aquaculture in China: Success Stories and Modern Trends*. Wiley, Hoboken, New Jersey.

Jhingran, V.G. and Pullin, R.S.V. (1985) *A Hatchery Manual for the Common, Chinese and Indian Major Carps*. Asian Development Bank, International Center for Living Aquatic Resources Management, Manila, Philippines.

Kohlmann, K. and Kersten, P. (2013) Deeper insight into the origin and spread of European common carp (*Cyprinus carpio carpio*) based on mitochondrial D-loop sequence polymorphisms. *Aquaculture* 376–379, 97–104.

Kottelat, M. (2006) *Fishes of Mongolia. A checklist of the fishes to occur in Mongolia with comments on systematics and nomenclature*. The World Bank, Washington, DC.

Nelson, J.S. (2006) *Fishes of the World*, 4th edn. Wiley, Hoboken, New Jersey.

Sharma, J.G., Singh, S.P., Mittal, P. and Chakrabarti, R. (2016) Impact of temperature gradient on the Indian major carp *Catla catla* larvae. *Proceedings of the National Academy of Sciences India, Section B: Biological Sciences* 86, 269–273.

Stout, C.C., Tan, M., Lemmon, A.R., Lemmon, E.M. and Armbruster, J.W. (2016) Resolving cypriniformes relationships using an anchored enrichment approach. *BMC Evolutionary Biology* 16, 44.

Takada, M., Tachihara, K., Kon, T., Yamamoto, G., Iguchi, K. *et al.* (2010) Biogeography and evolution of the *Carassus auratus*-complex in East Asia. *BMC Evolutionary Biology* 10, 7.

Yang, L., Sado, T., Hirt, M.V., Pasco-Viel, E., Arunachalam, M. *et al.* (2015) Phylogeny and polyploidy: Resolving the classification of cyprinine fishes (Teleostei: Cypriniformes). *Molecular Phylogenetics and Evolution* 85, 97–116.

Chapter 3

Getting Started: The Norms in Biological Structure and Diagnostics

3.1 Parameters of a Healthy Carp

The popularity of carp in aquaculture is due to several factors, such as good growth capacities, high adaptive flexibility and high resilience to suboptimal environmental conditions, ability to utilise natural resources of ponds, and relative lack of predatory/cannibalistic interactions. Carp are also perceived to be relatively resistant to diseases, but health problems can occur and recognising them is essential to counteract effectively any health challenges.

Healthy fish have intact skin and fins without any bleedings or other pathological changes (Fig. 3.1). Eyes and skin have a shiny, clear colouration and do not appear cloudy or milky or of dull colour. Healthy fish also often swim freely in loose schools in the tank/pond and do not aggregate at the water inlet or separate at the water surface or pond banks or at the bottom of tanks. In addition, some playful behaviour with the water current is typically normal. Healthy carp respond quickly to external stimuli with energetic swimming behaviour. When adapted to artificial feeding they can shoal, jump and readily take feed by hand.

The gills of healthy fish should be vibrant red, with a limited amount of mucus and clearly visible primary lamellae. Within the body cavity are organs that can be subject to a range of diseases or nutritional disorders. When dissected from the lateral aspect and the muscle fillet removed, the most conspicuous organs are the liver, gut, swim bladder and reproductive organs. In healthy fish, the

 Carps: A Practical Approach to Disease Control and Health Management (eds David Hoole et al.)
DOI: 10.1079/9781836990024.0003

Figure 3.1 Healthy common carp fingerling prepared for blood sampling.

liver should appear red/brown and occupy a larger portion of the body cavity. In addition, an intense dark red, elongated spleen can on occasions be scattered into fractions within the body cavity. The gut, which extends from the oral cavity to the cloaca, can appear in several folds and has a milky brown colour unless containing food. The reproductive organs, i.e. ovaries and testes, are located dorsally and their appearance changes along with maturation and the seasonal reproductive cycle. The two-chambered swim bladder appears as an extensive white structure located ventrally to the spinal column. Between the swim bladder chambers is the 'saddle' of the red-coloured kidney, which extends along the dorsal aspect of the body cavity.

The internal body cavity, when opened during necropsies, should be free of excessive amounts of liquid and should contain well-defined organs without adhesions. It is the alteration of these internal and external organs that forms the source of material for a range of diagnostic methods (Fig. 3.2).

3.2 Introduction of Diagnostic Procedures

Effective and early observation of behavioural abnormalities can be essential for detecting the first signs of the infection. For example,

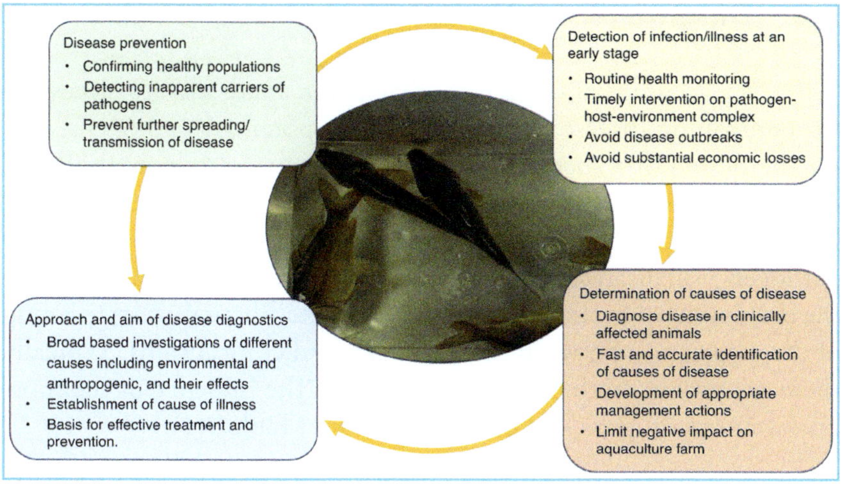

Figure 3.2 Role and aim in aquaculture health management.

the normal uptake of the food is one of the important indicators of a fish's wellbeing. In the earthen ponds that often house carp, the turbidity of the water can be an indicator of normal feeding behaviour, whilst clear water may suggest that the fish have stopped feeding. With the development of automatised surveillance of the fish, the observation of natural behaviour could be taken over by machine learning incorporating algorithms of artificial intelligence. Until this is done the owner is the best source of surveillance, noting any changes in fish behaviour.

Observation of individual fish is also essential for spotting health and welfare problems. Intact mucosal barriers are essential for fish to survive in osmotically and microbiologically hostile aquatic environments. Therefore, intact skin covered by clear mucus, red vibrant gills and non-inflamed intestine are essential for the wellbeing and survival of the fish. Detection of any gross pathological changes, for example discolouration, bleedings, ulcers, or fin erosions, should be treated seriously, as they can be the harbinger of bigger problems ahead. In this sense, diagnosing the causative agent of a disease or disorder is essential for further prevention, mitigation or treatment strategies.

3.3 Diagnostic Methods

Sensitive, accurate and decisive diagnostic methods are necessary for controlling disease and the spread of potential pathogens, and for the design of a successful treatment plan. Multiple methods are constantly being developed for spotting causes of the health problems of fish and the detection of pathogens. Quantitative methods, pathogen visualisation within the lesion, and metagenomics enable the evaluation of pathogen load and facilitate a better understanding of the association between the pathogen and clinical signs. Typically there are several methods that can confirm the presence of the most important pathogens and causes of disease (Table 3.1).

3.3.1 Testing types

There are generally two types of testing that can be performed to detect a pathogen or an environmental problem challenging the fish. Good interaction between point-of-care testing (POCT) and laboratory testing provides the most meaningful results (Fig. 3.3).

Point-of-care testing (POCT)

Point-of-care testing (POCT) or pond-side testing is the basis for the future of fish diagnostics. POCT is defined as diagnostic testing at or near the fish-keeping facilities. The development and use of diagnostic tools or devices designed to provide rapid diagnosis of fish health issues at the pond/tank side are likely to increase and will provide fish owners and breeders with nearly instantaneous diagnosis (Fig. 3.4). This development of pond-side testing will also allow a faster reaction time to health problems and ultimately limit morbidity and mortality in fish.

Laboratory testing

Despite the dynamic development of POCT, the sensitivity of these procedures may not always allow the detection of certain pathogens. Furthermore, according to good diagnostic practices

Table 3.1. List of diagnostic methods indicating their use, advantages and disadvantages

Method name	Use	Advantages	Disadvantages	Remarks
Manual behaviour / appearance observation	Detection of a problem by detecting changes in behaviour and appearance	Cheap, fast	Requires experienced personnel Does not give a clear diagnosis	
Artificial intelligence (machine learning) behaviour / appearance observation	Detection of a problem by detecting changes in behaviour and appearance	Fast	Requires development of algorithms for each of the fish species Requires installation of monitoring system (cameras) Expensive	Highly dependent on clarity of the water
Light microscopy wet mounts	Detection and visualisation of parasites and bacteria	Cheap, fast Allows visual evaluation of pathology and parasites associated with the lesions	Requires experienced personnel	Can be automated in future by use of artificial intelligence
Histology	Visualisation and characterisation of pathological changes and parasites/bacteria cells	Allow morphological characterisation of uncultivable pathogens	Requires experienced personnel and expensive equipment	Can be automated in the future by use of artificial intelligence
Electron microscopy	Visualisation and morphology of pathogens	Allows morphological characterisation of uncultivable pathogens	Requires experienced personnel and expensive equipment	Can be automated in the future by use of artificial intelligence

Continued

Table 3.1 Continued

Method name	Use	Advantages	Disadvantages	Remarks
Cell culturing (virus)	Detection and isolation of infective virus	Allows isolation of the viral pathogen	Slow detection method (up to 14 days) Not all viruses can be cultured	
Bacteria culturing	Culturing of bacteria from a lesion	Allows isolation of the pathogen	Relatively slow detection method Only 10% of bacteria can be cultured Some bacteria require special media	
Biochemical bacteria identification	Allows identification of bacteria using enzymatic profiles	Fast and cheap	Sometimes lacks precision	
MALDI–TOF	Mass spectrometry-based identification of bacteria	Fast and cheap	Requires expensive equipment Database still needs more information about fish pathogens	Usability will be increased with the growth of databases
PCR and RT-PCR	Allows detection of nucleic acids of pathogens	Fast and cheap	Could detect non-infective remains of pathogen Prone to contamination	

Quantitative PCR (quantitative RT-PCR)	Allows detection and quantification of nucleic acids of pathogens	Very fast and relatively cheap Quantitative results	Requires specialised equipment and sample preparation	Can be multiplexed to increase the throughput Currently the gold standard
Droplet digital PCR	Allows detection and quantification of nucleic acids of pathogens	Very fast and potentially cheap Quantitative results Superior sensitivity	Requires specialised equipment and sample preparation	Can potentially replace the classical and quantitative PCR
In situ PCR	Allows visualising infected cells in the tissue	Relatively cheap Provides information about how many cells are infected	Requires specialised equipment and sample preparation	
In situ hybridisation	Allows visualising the infected cells in the tissue	Relatively cheap Provides information about how many cells are infected	Requires specialised equipment and sample preparation	
Antigen detection	Allows detection of proteins of pathogens	Fast and cheap	Could detect non-infective remains of pathogen Generally less sensitive than PCR-based methods	
LAMP	Allows detection of nucleic acids of pathogens	Fast and cheap	Generally less sensitive than PCR-based methods	

Continued

Table 3.1 Continued

Method name	Use	Advantages	Disadvantages	Remarks
Immunohistochemistry	Allows visualisation of infected cells in the tissue	Allows confirmation of pathogen presence within the pathological change	Highly dependent on antibodies Requires specialised equipment and sample preparation	
Metagenomics (pathogenomics)	Detection of nucleic acids of all pathogens and symbionts in the sample	Allows detection of multiple pathogens	Requires extensive bioinformatics Relatively expensive Requires specialised equipment and sample preparation	Possibility of detection of pathogens in environment (eDNA/eRNA)
Blood morphology	Allows cell blood counts	Fast method to confirm the infection with high white blood cell counts or anaemia when red blood cell counts are low	Requires specialised equipment or personnel Does not give the clear diagnosis	
Pathophysiological blood analysis	Detection of changes in ions, proteins, pH or other parameters in blood	Allows confirmation of physiologically compromised fish	Does not give the clear diagnosis Requires specialised equipment	Can be useful for treating the clinical signs (salt treatment when disruption of osmoregulation is present)

ELISA	Detection of antibodies against a pathogen or detection of the pathogen	Allows indication that a fish had been in contact with the pathogen	The fish may have cleared the pathogen Highly dependent on availability of antibodies	Can be used for evaluation of vaccines
SNT	Detection of neutralising antibodies against a pathogen	Allows indication that the fish had been in contact with the pathogen and how high is the protection based on antibodies	Slow method, the fish may have cleared the pathogen	
Medical imaging (X-ray, ultrasound, computed tomography, magnetic resonance, positron emission tomography etc)	Detection of internal changes and abnormalities by specialised imaging	Detailed information on internal structures obtained *intra vitam*	Limited availability of specialised instrumentation Expensive equipment and staff required	Good in diagnosing tumours and gross abnormalities
Water quality measures	Allows detection of environmental problems	Fast and relatively simple	Does not give a clear diagnosis Often requires specialised equipment	

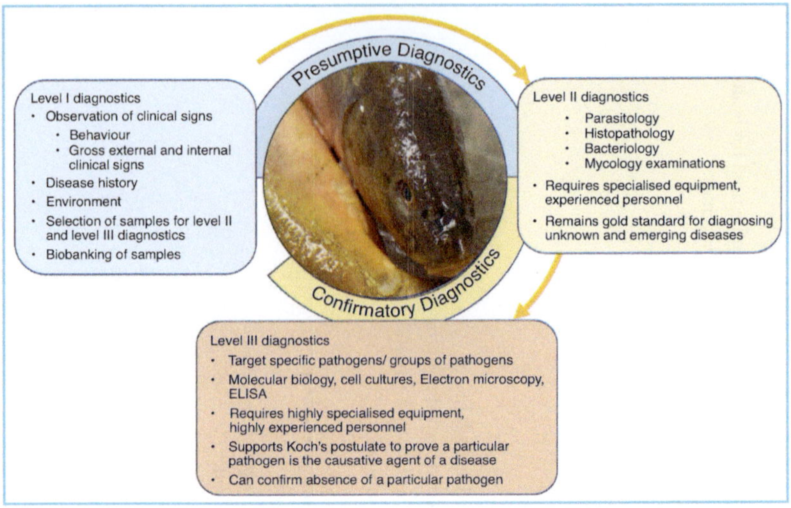

Figure 3.3 Levels of diagnosis.

Figure 3.4 Tools for water analysis at the pond side as an example of point-of-care testing.

any positive POCT result should be confirmed by laboratory methods. In principle, laboratory testing is more advanced and requires well-trained and experienced staff who follow exact protocols and are aware of the science that lies behind the laboratory test (Fig. 3.5). The use of more sophisticated diagnostic processes

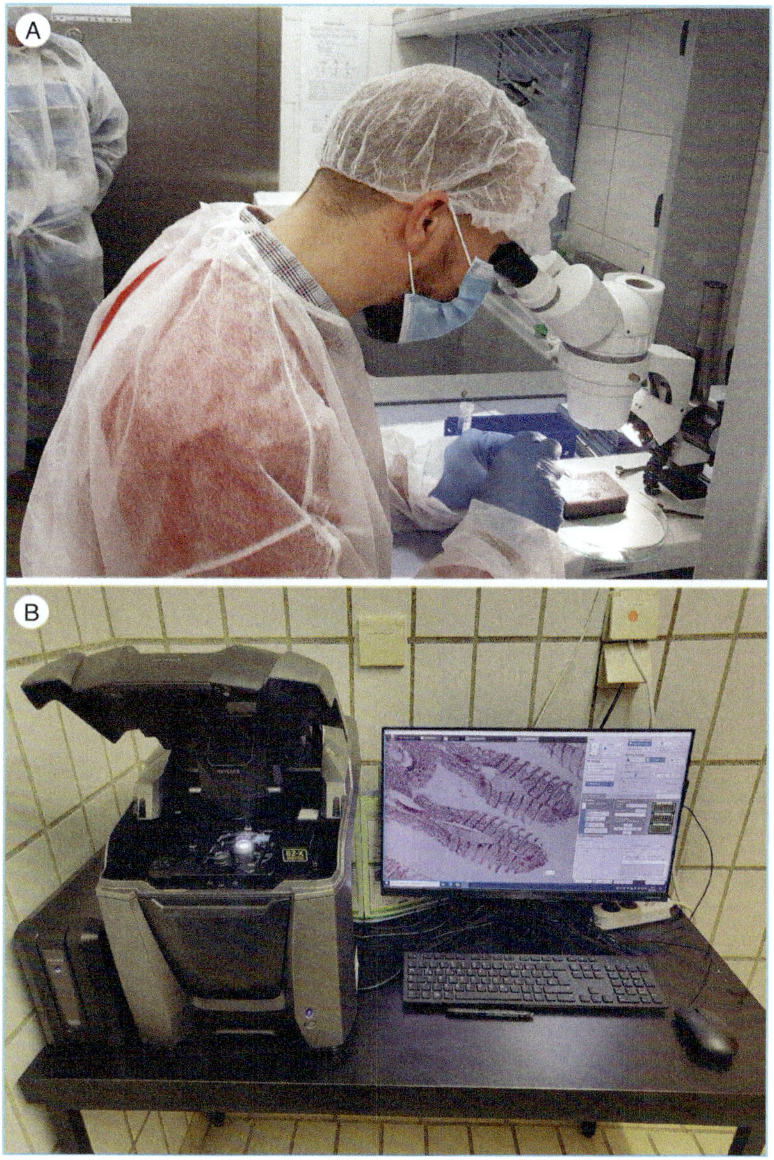

Figure 3.5 Examples of laboratory testing. (A) Binocular microscope-assisted collection of internal organs from fingerling as an example of laboratory-level testing. (B) Microscopic evaluation of a histological slide for observing pathological changes in gills related to disease.

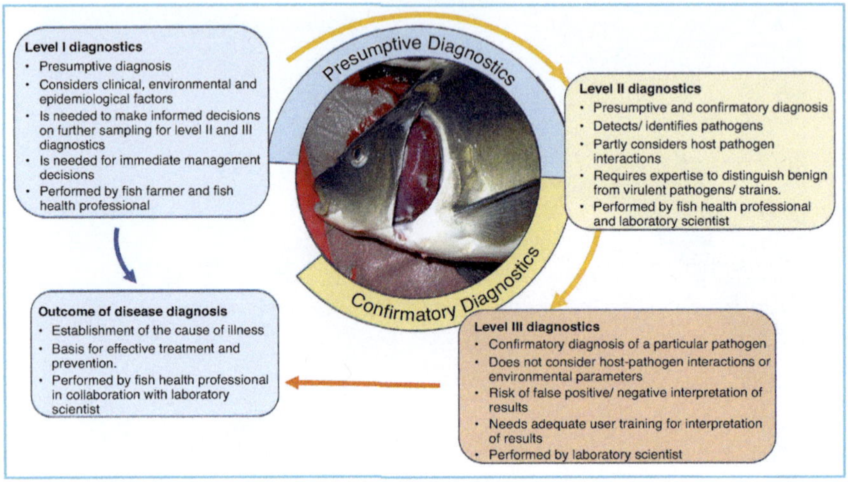

Figure 3.6 Requirements of diagnosis.

will ensure greater accuracy, analysis, validation and recording of results. Laboratory testing may be required to evaluate properly the causative relationship between the pathogen and clinical signs, for example in a lesion (Fig. 3.6).

3.3.2 Types of diagnostic procedures

Behaviour or (patho)morphology

Behavioural observation of the fish in the pond or natural environment is essential for spotting the primary signs of a health problem. During the diagnostic check-up or necropsies the observations should be extended to closer examinations of the external appearance of the fish, and the appearance and location of internal organs.

Fresh mounts light microscopy

In order to examine skin and gills for the presence of small ectoparasites or an increased amount of bacteria, fresh tissue samples

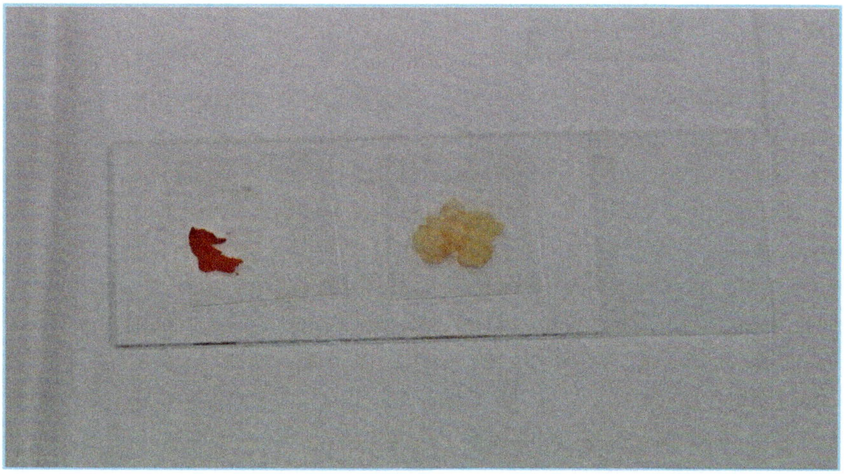

Figure 3.7 Fresh mount of internal organs (spleen and liver) of common carp, ready for light microscopy.

including samples of skin scrapes, gill biopsies and gut scrapings can be collected for microscopic examination. From gill biopsies a structure analysis can be performed. Fresh mounts from internal organs can be observed microscopically and evaluated for any pathological changes such as melanomacrophage centres, granulomas and necrotic abnormalities (Fig. 3.7).

Histology

Histology is a fundamental tool in veterinary science for diagnosing diseases at the microscopic level. It is the study of the microscopic structure of tissues, which plays a crucial role in the visualisation and characterisation of pathological changes within the body, as well as in the identification of parasites and bacterial cells. This detailed analysis is essential in understanding diseases at the cellular level (Fig. 3.5B), allowing for accurate diagnoses and informing treatment strategies. It may require experienced personnel and expensive equipment. In the future it may become automated by use of artificial intelligence.

Pathogen cultivation

Certain pathogens (e.g. viral, bacterial) can be cultivated to aid their identification. *In vitro* cell cultures from fish tissues can be used for virus replication (Fig. 3.8). For this, the cells have to fulfil certain characteristics, including susceptibility for the virus infection and replication. Some viruses, for example rhabdoviruses, replicate in a very wide number of cell lines, while herpes and poxviruses have a narrower supportive viral invitrome and require a specific cell type or differentiation stage of the cells, which most likely has to be established from the host fish species. Bacterial culture media must also be adjusted to the species of bacteria that are suspected to cause the pathology. Some bacterial species require specialised media to allow their growth and limit the growth of other more opportunistic bacteria (Fig. 3.8). Isolated and cultured pathogens can be subjected to further analysis, allowing their identification.

Antibody-based pathogen identification: ELISA and immunocytochemistry/immunohistochemistry

Enzyme-linked immunosorbent assay (ELISA) is a rapid method of detecting the presence of antibodies against certain pathogens, as well as the presence of a pathogen in the sample (Fig. 3.9). It can be used for high-throughput population screenings revealing the presence of the pathogen in the population. The detection of the pathogen is highly dependent on the availability of antibodies specific for the pathogen. The detection of the presence of antibodies in the blood of a fish reacting against a specific pathogen can indicate that the fish has been in contact with the pathogen concerned. However, the characteristic of the antibody response in fish is relatively slow and low titre levels are produced. This brings certain challenges in detecting chronic, as well as healed infections.

Immunocytochemistry/immunohistochemistry uses pathogen-specific antibodies to locate and visualise the presence of a specific antigen in cells/tissues. It is highly dependent on the availability of antibodies and requires optimisation before it can be employed for pathogen characterisation (Fig. 3.10).

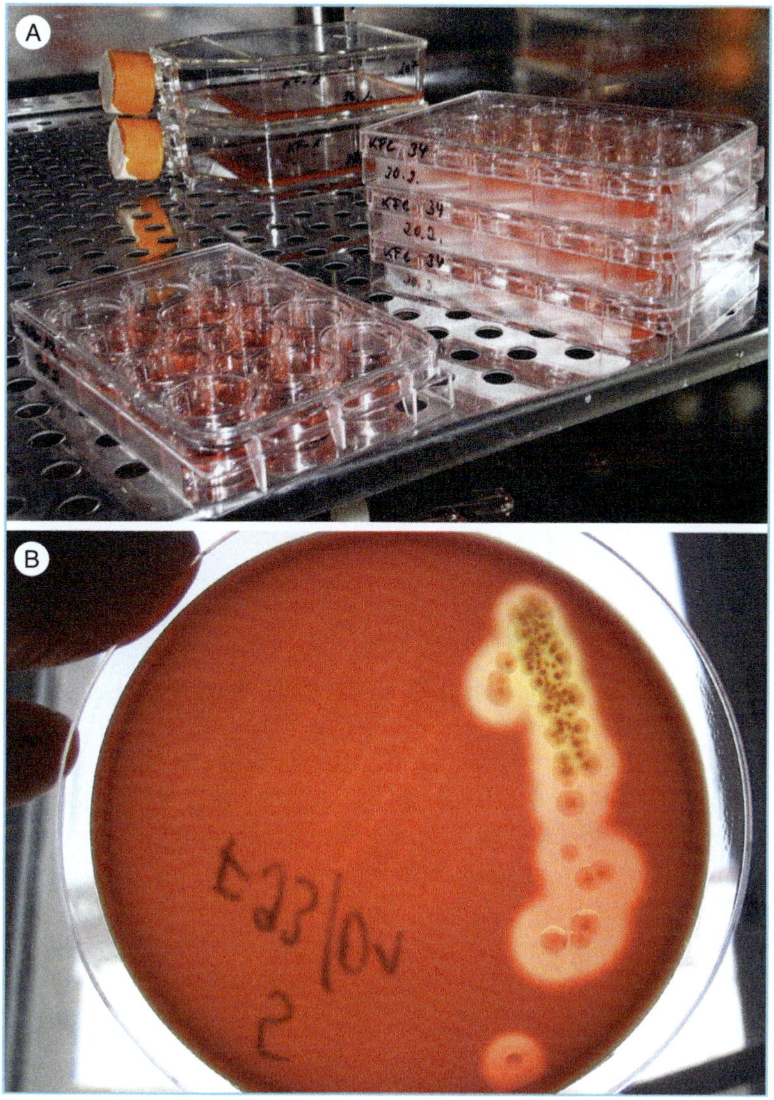

Figure 3.8 Methods for cultivation of pathogens. (A) Culture vessels with cell monolayers for cultivation of a virus. (B) Bacterial colonies growing on Columbia agar with sheep blood. Note haemolytic activity of the bacteria making the agar transparent in the vicinity of the bacteria colonies.

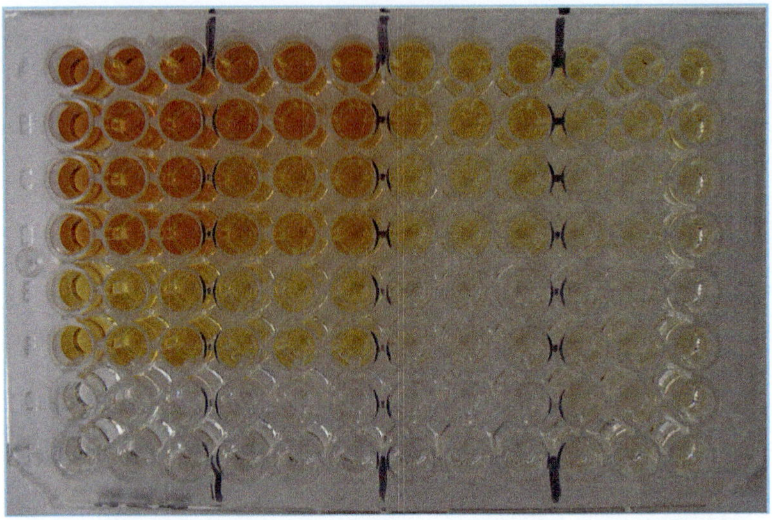

Figure 3.9 Enzyme-linked immunosorbent assay (ELISA) plate with colorimetric test for the presence of antibodies against a pathogen.

Mass spectrometry (MALDI–TOF) for bacterial analysis

Matrix-assisted laser desorption/ionization–time of flight mass spectrometry (MALDI–TOF MS) is a broadly used diagnostic tool for microbial (e.g. bacteria) identification. The main advantages are speed of analysis, sensitivity and relative low cost, which allows multiple samples to be identified. The limitation of the technology is its dependence on databases with pathogen-specific spectra, which may be lacking for certain bacterial species. The identification is highly dependent on the resolution of the data in the database.

Biochemical and enzymatic profiles for bacterial analysis

The biochemical identification of bacteria uses the fact that bacteria can use various substrates for active growth/multiplication. It is the detection of certain patterns of enzymatic activity by biochemical

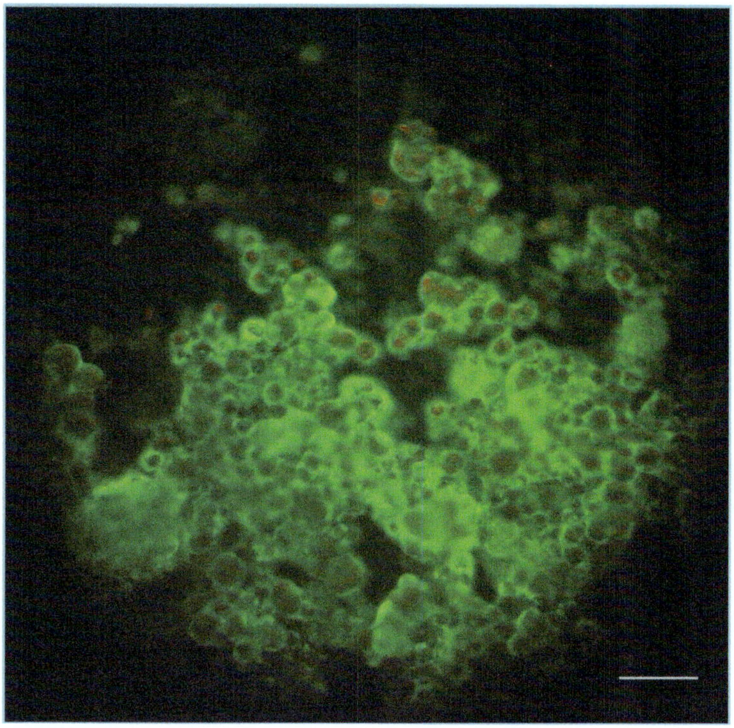

Figure 3.10 Immunocytochemical analysis confirming *Cyvirus cyprinidallo 3* (Cyprinid herpesvirus 3) infection (green fluorescence) in common carp brain cells. Indicative scale bar 30 μm.

methods, which can facilitate identification of the genus of the bacteria. It is also possible that the profile of enzymes produced by a bacterium can allow species classification, especially when the enzymatic profiles are characteristic of a given bacterium (Fig. 3.11). Although the method allows fast identification of bacteria, it does not always work reliably for the aquatic bacteria.

Nucleic acid analysis

The detection of nucleic acid sequences specific for a particular pathogen provides a quick method to assess the presence of the pathogen in the environment and in fish, pathologically changed organs or even single cells.

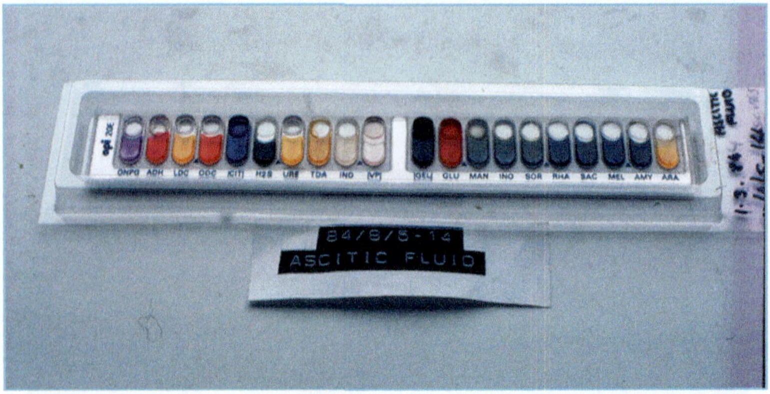

Figure 3.11 Biochemical identification of bacteria. Courtesy of the Environment Agency, National Fisheries Laboratory, UK.

PCR. Various PCR-based methods for the detection of pathogen-specific nucleic acids (RNA or DNA) are available for the detection of known viruses, bacteria and parasites. There are several protocols available (Fig. 3.12) with different degrees of complication; however, the quantitative double-labelled probe-based PCR (qPCR) is often treated as the gold standard. It also allows the assessment of pathogen load and thus enables a better detection of the association of the pathogen with the disease. The new droplet digital PCR (ddPCR) technology provides unparalleled precision, low detection limit and the ability of absolute quantification, which most likely will make the ddPCR methods a replacement for classical end-point as well real-time quantitative PCR.

IN SITU HYBRIDIZATION. A labelled complementary DNA or RNA *in situ* hybridization can be used to show the location of specific nucleic acid sequences of pathogens in the tissues. This can help in understanding the visible pathology and allows the association of certain pathogens with observed lesions.

IN SITU PCR. Using a labelled deoxynucleotide triphosphate (dNTP) *in situ* PCR can show the location of rare or single-copy number nucleic acid sequences of pathogens in the tissues. Similar to the *in situ* hybridisation technique, this can help in understanding

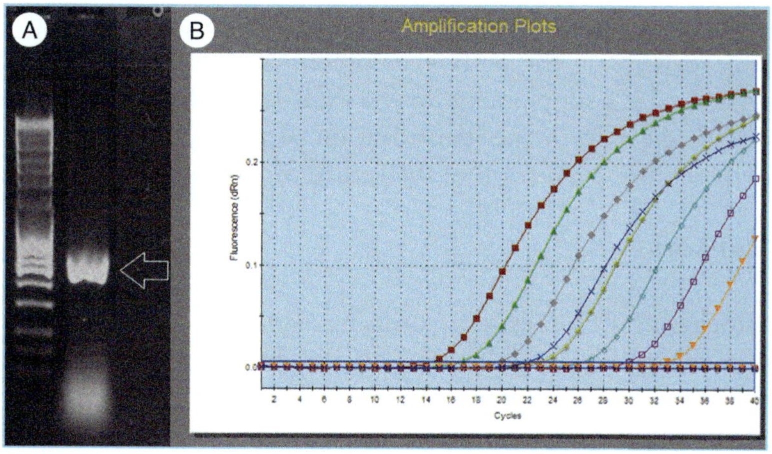

Figure 3.12 End-point and quantitative PCR results. (A) Agarose gel showing DNA marker ladder (left) and amplified DNA fragment (arrow) stained with intercalating dye visible under UV light. (B) Quantitative PCR amplification curves showing the increase in DNA fluorescence indicating the accumulation of the target PCR product.

the visible pathology and allow the linking of a certain pathogen to an observed lesion. This method generally has a lower detection threshold and therefore it can be used also to detect asymptomatic or latent carriers of the pathogens.

Additional methods for nucleic acids detection

Loop-mediated isothermal amplification (LAMP) assay, recombinase polymerase amplification (RPA), or gold nanoparticle-based hybridization assay can provide a quick detection of viral DNA/RNA in the POCT setting.

3.3.3 New technologies

Barcoding/sequencing (identification)

PCR-dependent or independent sequencing of the nucleic acids using any generation of sequencing procedures (Sanger first generation, short-sequence high-throughput second generation or long-sequence high-throughput third generation sequencing) can help in identifying the pathogens present in the sample or lesion. This method is a supplement for the PCR-based detection of pathogens and can help in the identification of pathogens, as well as in epidemiological evaluation of the spread of pathogens.

Biosensors

Electrochemical biosensors utilising antibodies, aptamers or imprinted polymers can be used for the detection of pathogens in a fish sample or in the environment. Although this method is in an initial phase of development, it can be envisaged that a constant monitoring of water for the presence of specific pathogens can be achieved.

Metagenomics/metaproteomics

The diversity of the microbial world within the environment or the pathologically changed tissue can be highly complex. High-throughput methods, such as metagenomics or metaproteomics (also known as community omics, or environmental omics) can be used to unravel all pathogens that could be detected in the sample. This is most important in the case of co-infections or superinfections, which lead to a multipathogenic responsibility for the disease, and where a pathobiome analysis has to be performed. This is becoming more feasible thanks to the employment of high-throughput metagenomics approaches, which in a single analysis allow the identification of all pathogens present in the affected tissue.

MEDICAL IMAGING. Detection of internal changes and abnormalities in the body of fish can be carried out using specialised imaging techniques, such as radiography (X-ray), ultrasonography, computed tomography, magnetic resonance, positron emission tomography, etc. These methods provide detailed information on internal structures that can be obtained *intra vitam* or at post-mortem (Fig. 3.13). Due to the need for expensive equipment and well-trained specialist staff, these methods may be of limited availability for

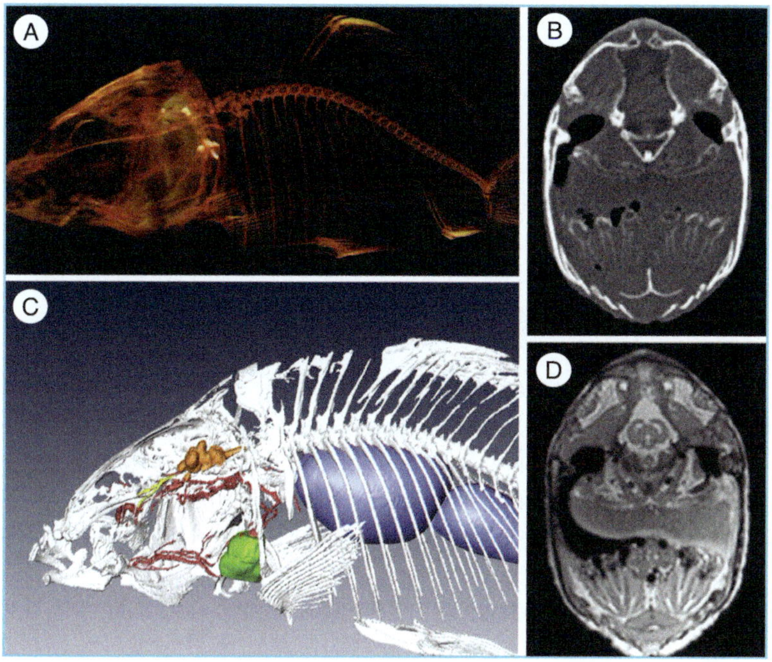

Figure 3.13 Results from medical imaging. (A) Computer model of skeletal structure of common carp based on computed tomography. (B) Computed tomography cross-section of head of common carp. (C) Computer model of skeleton (white), brain (orange), optical nerves (yellow), main blood vessels in the head (red), heart (green) and swim bladder (blue) based on computed tomography and magnetic resonance tomography. (D) Magnetic resonance tomography of cross-section of head of common carp.

day-to-day diagnostic work. As in veterinary and human medicine, these methods are particularly useful in the diagnosis of tumours, gross abnormalities and injuries.

3.3.4 Limitations of diagnostic procedures

There are several limitations that can prevent the use of a certain diagnostic procedure, equipment or methods. The diagnostic approaches should be tailored to which methods are available and what diagnostic outcome it is hoped to achieve.

Physical limitations

Certain methods are not designed to provide diagnostically useful outcomes when used beyond their physical limits. For example, there is no ability to detect small particles such as viruses using the naked eye or even the light microscope; they can only be visualised using an electron microscope.

Usability limitations

Certain methods are not designed to answer given diagnostic questions. For example, a bacteriological examination can be performed from live or freshly killed fish, while in a dead fish the results may be compromised by contaminations from the bacterial flora that develops post-mortem.

Equipment limitations

Certain methods and diagnostic approaches require access to specialised and sometimes very expensive equipment and software. Furthermore, personnel trained and experienced in the use of the equipment and the software could be required.

3.3.5 Sample types

Collection of correct samples allowing the identification of the causative agent is the key element of diagnosis. Several types of samples can be collected.

Environmental

With the development of novel methods, environmental samples are increasing their usability with development of the high-throughput methods, such as high-throughput sequencing (next-generation sequencing) of environmental DNA or RNA (eDNA/eRNA). Targeted PCR-based methods can be also be used. Furthermore, novel sensors are being developed to detect the pathogens in the environment without killing the fish. The main advantage of these methods is the lack of the need to kill the fish for diagnostic purposes. This action can limit the financial losses related to sacrificing a significant number of fish just for routine diagnostics or monitoring for presence of pathogens. These methods are less useful in the case of the internal lesions caused by pathogens, as well as in fish where only one pathogen is causing the change.

Environmental analysis can be carried out on a range of media, as follows.

WATER. Samples are easy to collect in the aquatic environment. Due to the likely high dilution of the pathogen, the sample has to be prepared before being used in diagnostics. For example, the water has to be filtered to extract bacteria and larger organisms, or concentrated by centrifugation to pellet bacteria or viruses, which achieves a higher probability for pathogen detection. Note that it confirms the presence of a pathogen, but not a causative relation with any visible pathology.

BIOFILMS. Biofilms are very good environmental diagnostic samples for monitoring the presence of pathogenic bacteria in the ecosystem. Biofilms provide a sample that can be cultured or processed for molecular methods. As with any environmental sample, it can confirm only the presence of a pathogen, not a causative relation with any visible pathology.

POND GROUND. Environmental samples can be also collected from the pond ground. As this is a complex environmental sample, it can have high concentration of inhibitors, which can limit the detection of the pathogens. It confirms the presence of a pathogen, not a causative relation with any visible pathology.

PLANKTONIC WATER FILTRATORS. The free-floating microorganisms found in the water column, which are feeding on detritus, bacteria and algae, can be good intermediate samples for detecting some pathogens of fish.

BOTTOM FILTRATORS (E.G. MOLLUSCS). Benthic organisms are becoming excellent sources of samples for detecting fish pathogens in the vicinity of fish farms. These organisms are constantly filtering the water; therefore, certain parasites, bacteria and viruses are taken up by them and can be detected by molecular methods. However, they cannot be used in the case of closed systems without contact with natural waters.

SENTINEL FISH. Placing additional fish into the tanks can help to detect pathogens in aquaculture facilities. These sentinel fish have to be specific pathogen-free and susceptible to any pathogen that can infect or should be detected in the diagnosed fish. The use of sentinel fish is useful if the whole population requires evaluation, for example tanks, cages or recirculation systems. The epidemiologically relevant number of sentinel fish has to be used, and these fish have to cohabit with the tested fish for a period of time, allowing the sentinels to be infected with the given pathogen.

Fish samples

The samples collected directly from the fish have an advantage over environmental samples as they can provide tangible evidence that a certain pathogen is responsible for the observed pathology.

INTRA VITAM. Mucus, blood, reproduction products, faeces and tissue biopsies can be collected from the fish without the necessity of killing the animal, and are therefore termed *intra vitam* – 'during life' diagnosis. Depending on the pathogen's biology, there is no need to kill the specimen to detect the pathogen. Skin or gill parasites can be detected with the naked eye or in fresh microscopic mounts utilising, for example, scrapings of the tissues. Infectious diseases can also be diagnosed by analysing blood samples for the presence of pathogens, antibodies to pathogens, or changes in physiological or metabolomic parameters that may indicate a non-infectious disorder (Fig. 3.14). The sample collection can be performed at the point of care, for example at the pond side or in quarantine facilities, and the fish can be returned to their original tanks whilst the samples can be processed at the pond side or in a specialised laboratory.

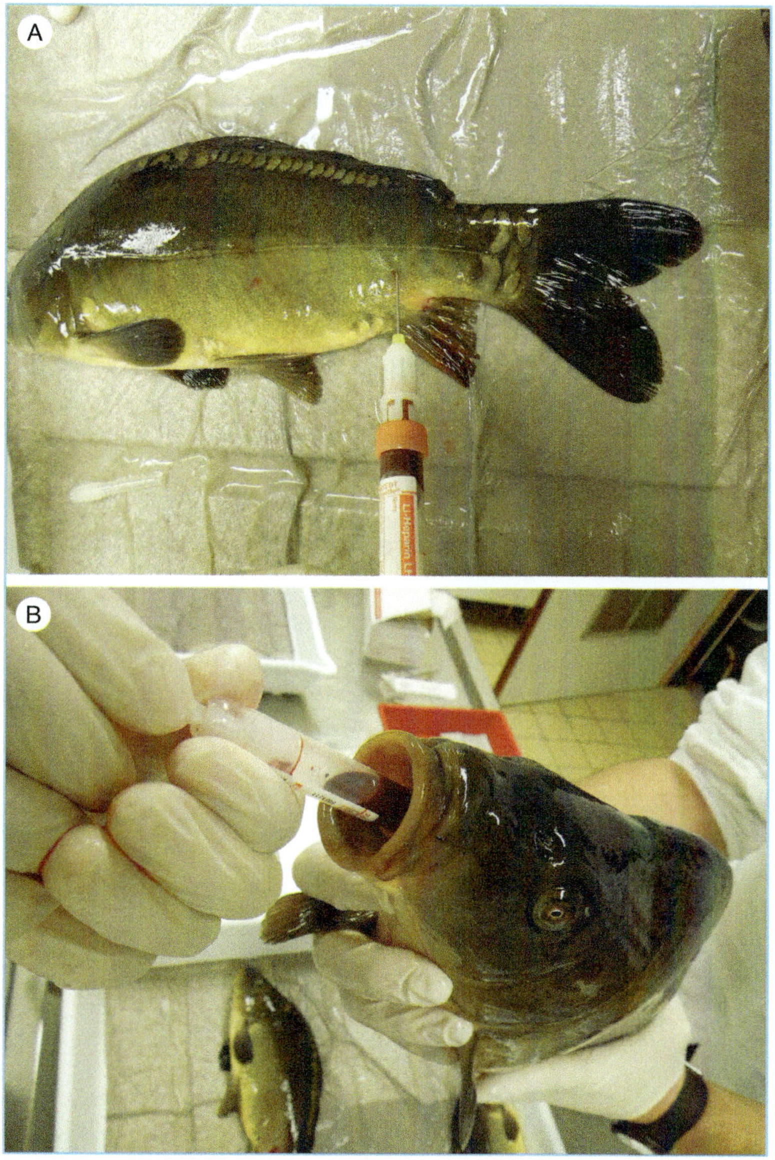

Figure 3.14 Blood collection techniques. (A) Blood collection from caudal vein. (B) Blood collection from the dorsal aorta.

POST-MORTEM. Sample collections from freshly killed or dead fish during necropsy are the most classical and very often the most useful source of diagnostic samples. Mucus, blood, reproduction products and tissues can be collected from individuals with the most prominent pathological and/or behavioural changes. The fish that succumbed from the disease or showed clinical signs of the disease will most likely harbour the highest pathogen load or toxin concentration. For certain diseases only internal organs such as brain, spleen, kidney or pancreas should be collected to ensure that the pathogen is detected. Then the fish have to be killed before collection of the samples (Fig. 3.15). Bacteriological examinations are more usually carried out on freshly killed fish, as dead fish can be contaminated by the bacterial flora developing post-mortem.

3.3.6 Diagnostic procedures based on pathogen type

Diagnosis of the infection with different types of pathogens requires utilisation of divergent diagnostic approaches and methods. Diagnostic procedures that could be used for detection of pathogenic viruses, bacteria or parasites are presented here.

Figure 3.15 Necropsy of apparently healthy common carp. Note open body cavity and swab sampling of internal organs for bacteriological analysis.

Viral infection

If there is a suspicion of viral infection, the following diagnostic procedures can be performed to confirm and characterise the pathogen.

POCT TESTING. Viruses cannot be seen by the naked eye, therefore a method detecting their nucleic acids or protein has to be applied. Loop-mediated isothermal amplification (LAMP) assay, lateral flow devices based on PCR, recombinase polymerase amplification (RPA), or gold nanoparticle-based hybridization assay can provide a quick detection of viral DNA/RNA even at the pond site. Although an antigen test can be utilised to detect viral proteins, care should be taken during the interpretation of the results, due to sensitivity issues, and positive results should be confirmed by laboratory methods.

LABORATORY TESTING. Multiple cell lines can be used for isolation of viruses from carp. For example, *C. idella* kidney (CIK) cells or grass carp swim bladder (GSB) cells are susceptible to infection with grass carp reovirus (GRSV); Koi fin 1 (KF-1), koi fin cells (KFC), epithelioma papulosum cyprini (EPC), or fathead minnow cells (FHM) are susceptible to *Cyvirus cyprinidallo1* (CyHV-1) infection, where the virus causes a cytopathic effect (CPE). KF-1 and common carp brain (CCB) cells are susceptible to *Cyvirus cyprinidallo3* (CyHV-3), whilst goldfish fin (GFF), gibel carp brain (GiCB), gibel caudal fin (GiCF), fantail goldfish fin (FtGF), fantail goldfish gill (FtGG), fantail goldfish liver (FtGL) and fantail goldfish brain (FtGB) cells have been shown to be useful in isolating and culturing of *Cyvirus cyprinidallo2* (CyHV-2). The development of morphological changes in cells, such as focal lysis or syncytia formation, can be observed by days 3–14 post-infection, which makes the culturing methods relatively slow (Fig. 3.16). Immunocytochemistry can also be performed in order to label the infected cells with virus-specific antibodies, or PCR can be used to identify the virus. Electron microscopy can be employed to visualise the viral particles in infected cells.

Faster detection methods exist, such as PCR-based techniques that detect viral nucleic acids (RNA or DNA). There are several

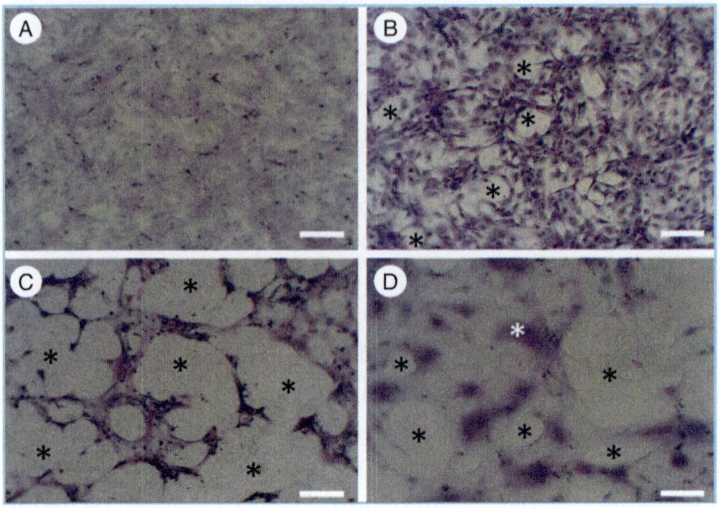

Figure 3.16 Common carp brain (CCB) cell monolayers stained with crystal violet showing different extent of cytopathic effects. (A) Non-infected cells showing continuous monolayer. (B) *Cyvirus cyprinidallo 3* (Cyprinid herpesvirus 3) -infected cells showing focal degeneration (*). (C) Common carp iridovirus-infected cells showing extensive degradation of the monolayer with multiple large spots lacking cells (*). (D) Common carp paramyxovirus-infected cells showing syncytium formation (white*) and multiple large spots lacking cells (black*). Indicative scale bar 30 μm.

protocols available with different degrees of complication, but quantitative double-labelled probe-based PCR is often treated as the gold standard, and allows the assessment of pathogen load and the association of a pathogen with the disease. Tests based on antibody detection, such as ELISA or serum neutralization assay, can be used for detecting the virus-specific antibodies or virus-neutralizing antibodies. Results for the detection of virus can be inconclusive, in which case metagenomic (high-throughput DNA/RNA sequencing) can be used to identify the virus.

Bacterial infection

In cases of suspicion of bacterial infection, the following diagnostic procedures can be performed to confirm and characterise the pathogen.

POCT TESTING. Bacteria are difficult to identify by POCT and they cannot be seen by the naked eye. However, methods detecting their nucleic acids or protein can be applied in theory. Loop-mediated isothermal amplification (LAMP) assay, lateral flow devices based on PCR, recombinase polymerase amplification (RPA), or gold nanoparticle-based hybridization assay can provide quick detection of bacterial DNA/RNA even at the pond side. However, as bacteria are ubiquitous in the aquatic environment and the sensitivity of these procedures may not always allow the detection and correct identification of the bacterial pathogen, any results should be confirmed by laboratory methods.

LABORATORY TESTING. Bacterial cultures have to be adjusted in line with the bacterial group or species suspected of causing the pathology. Some bacteria require special media to allow their growth and limit the growth of most opportunistic bacteria. After the bacteria have been isolated, several methods such as biochemical analysis, MALDI–TOF, 16S sequencing can be used to identify the bacteria species. PCR-based methods detect specific bacterial nucleic acids (RNA or DNA). There are several protocols available with different degrees of complication. A quantitative double-labelled probe-based PCR is often treated as the gold standard and allows the assessment of bacterial load, thus allowing a better association of the pathogen with the disease. Antibody detection using ELISA or serum neutralisation assay can be used for detecting bacterium-specific antibodies. Identification of bacteria to species level can be challenging, in which case methods such as PCR, biochemical/spectrometrical analysis or metagenomic (high-throughput DNA/RNA sequencing) can be utilised.

Parasitic infection

The diagnosis of a parasitic infection is still largely based on visual inspection and morphological characterisation of the pathogen, which can support rapid treatment of the infection.

POCT TESTING. Fresh mounts or scrapings of skin and gill mucosa, faeces etc. can be observed in light microscopy and the parasite can be morphologically identified. Although some large metazoan parasites can be tentatively identified at the pond side, further laboratory tests should be carried out as confirmation.

LABORATORY TESTING. A range of laboratory-based methods are available for the detection of parasitic infections. Initial inspections utilise microscopical analysis, which can be undertaken at the pond side as long as facilities are available. Fresh mounts or scrapings of, for example, skin and gill mucosa, or faeces can be observed using a light microscope to aid the morphological identification of the parasite. Histology can also be performed for the detection of parasites residing in internal organs. In addition, electron microscopy can be employed to visualise parasite cells in affected tissues, but this entails an extensive and time-consuming process for preparation of the material. PCR-based methods detecting the nucleic acids (DNA or RNA) are available for detecting and identification of some parasite species. There are several protocols available for identification and barcoding of parasites, and a range of quantitative PCR have been developed against some parasitic species, which allows assessment of pathogen load. ELISA or serum neutralisation assays can also be used for detecting parasite-specific antibodies or parasite-neutralising antibodies, and more recently metagenomics (high-throughput DNA/RNA sequencing) has been developed to provide *de novo* genome sequences aiding parasite systematics.

3.4 Selected References for Further Reading

Austin, B. (2019) Methods for the diagnosis of bacterial fish diseases. *Marine Life Science & Technology* 1, 41–49.

Cunningham, C.O. (2002) Molecular diagnosis of fish and shellfish diseases: Present status and potential use in disease control. *Aquaculture* 206, 19–55.

Kovas, K., Hatzilygeroudis, I., Dimitropoulos, K., Spiliopoulos, G., Poulos, K. *et al.* (2023) Using level-based multiple reasoning in a web-based intelligent system for the diagnosis of farmed fish diseases. *Applied Sciences* 13, 13059.

Li, D., Li, X., Wang, Q. and Hao, Y. (2022) Advanced techniques for the intelligent diagnosis of fish diseases: A review. *Animals (Basel)* 12(21), 2938.

Noga, E.J. (2011) *Fish Disease: Diagnosis and Treatment*, 2nd edn. Wiley-Blackwell, Chichester, UK.

Smith, S.A. (ed.) (2019) *Fish Diseases and Medicine*, 1st edn. CRC Press, Boca Raton, Florida.

Vergneau-Grosset, C., Nadeau, M.E. and Groff, J.M. (2017) Fish oncology: Diseases, diagnostics, and therapeutics. *Veterinary Clinics of North America: Exotic Animal Practice* 20, 21–56.

Chapter 4

Examination Procedure, Pathological Terms and Clinical Signs

4.1 Examination Procedure

When examining a diseased carp, for example to advise on an appropriate treatment, or any other control measures, which can lead to a proper diagnosis, careful examination of the affected fish is required. Many clinical manifested diseases in carp are caused by multiple factors and therefore in many cases more than one diagnosis is required. Specific pathogenic agents might be some of these factors, but the environment, especially the water quality, and other husbandry factors play a very important role in the emergence of carp diseases. The examination of the environmental conditions is of prime importance. A systematic stepwise approach is recommended to come to a correct disease diagnosis. The following should be addressed.

4.1.1 *Anamnesis*

A detailed anamnesis should be recorded. Questions on the disease history should be asked first. It is important to know when the clinical signs are occurring, what kind of clinical signs are seen, how many fish are affected, which fish species are affected and the age of the affected fish (Fig. 4.1). The water temperature at the time of the first recognised disease signs can be important and should be ascertained. It should be noted if mortalities have occurred and when. For a possible treatment of the fish, it is also important to know if the

 Carps: A Practical Approach to Disease Control and Health Management (eds David Hoole *et al.*)
DOI: 10.1079/9781836990024.0004

Figure 4.1 Anamnesis of ornamental carp aquarium presenting an interaction between the veterinary professional and the fish owner.

fish have been pretreated with therapeutics. If pretreatments were performed then the substances that were used, their concentrations and which treatment protocols were applied should be determined. Regarding fish kept in production systems, the fish species, the age of the fish and the number of fish should be ascertained. Also, it should be noted when the last new fish were introduced into the system and the source of that introduction.

After answers relating to the acute disease problems observed, more general questions on the keeping conditions of the animals should be discussed. Questions should refer to the management of the system, such as the water source, water exchange rates and water flow, and the amount of water available for the system. It must be noted if the fish are kept in special summer or winter ponds, and when the fish were introduced into their current location. All technical devices used in the system should also be noted and the feeding regime evaluated, including the storage, type, amount and frequency of food given. As water parameters are of great importance, it should be ascertained if and what parameters are measured regularly on farm. Additionally, the current water temperature should be noted.

4.1.2 Examination of the environment

Water samples should be optimally taken from the holding water and in- and outflows, and the oxygen content, the water temperature and the pH should be measured directly using a multimeter. It is always recommended that amounts of the components of the nitric cycle, for example ammonia/ammoniac, nitrite and nitrate, as well as carbonate and total hardness, are measured. All these parameters should be recorded on-site using a portable photometer with appropriate test kits or by commercially available test systems, which usually assess levels using a change in colour. If the chemical water parameters are to be measured in a laboratory, at least 100 ml of water should be taken in clean glass bottles. It is important to fill the bottles completely with water so that no air is left in the bottles (Fig. 4.2A). Water samples should be taken to the laboratory as soon as possible and analysed using a range of photometer test kits, which can be obtained from different manufacturers (Fig. 4.2B). In addition to the aforementioned water parameters, and dependent on the clinical signs and the anamnesis report, it might be useful to examine the water samples for other parameters, such as zinc, copper, manganese, phosphorus and iron. If a bacterial infection is suspected or if the hygienic conditions in the water are macroscopically suboptimal, an examination of a water sample for the number of bacteria per millilitre and the most abundant bacterial species is recommended. The examination of the microbiological water quality is only possible in specific laboratories and whether a laboratory is able to perform such examinations should be ascertained before sending samples. Water samples for microbiological examination should be cooled and dispatched directly after sampling.

4.1.3 Observation of fish in the keeping system

Observation of the fish in their keeping system and the construction of the keeping system itself can give additional hints

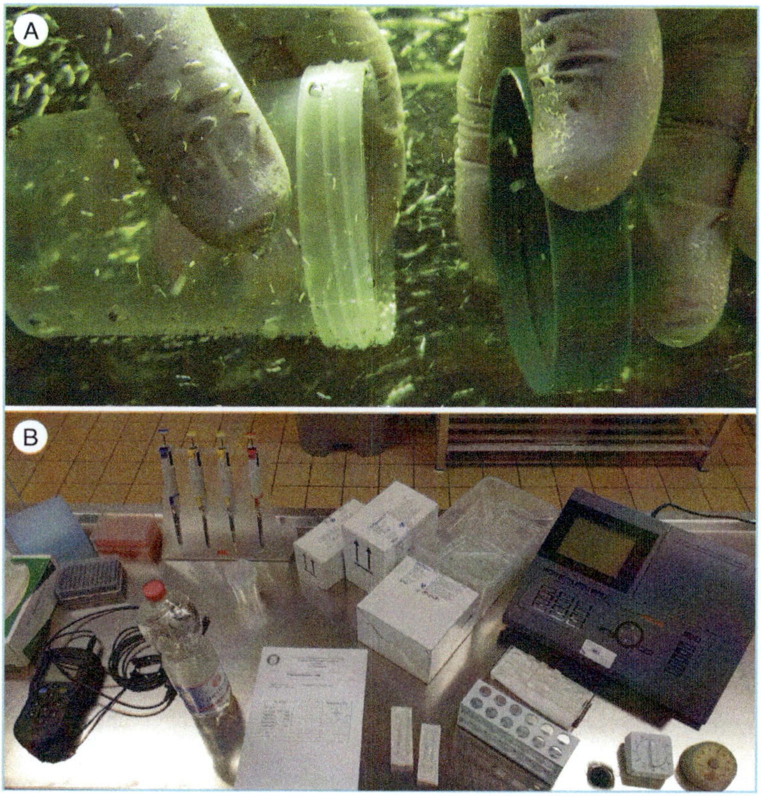

Figure 4.2 Water analysis. (A) Collection of the water sample. (B) Equipment and tools used for water analysis.

regarding the pathogenesis of the disease. Therefore, whenever possible a visit on-site should be planned and the behaviour of the fish observed (Fig. 4.3A-D). It is also advisable that the owner should be present and should feed the fish so that the feeding behaviour can be observed.

It is recommended that answers to the following questions be obtained.

- Are the fish showing swarming behaviour or are some fish remaining solitary?
- Where are the fish positioned in the water column: at the surface or at the bottom?

Figure 4.3 Observation of the fish in the keeping system. (A) Observation of behaviour of fish and fish-eating birds at large earthen pond used for growing common carp. (B) Observation of ornamental fish fingerlings in large concrete tank. (C) Monitoring of fertilised carp eggs in outdoor incubators. (D) Monitoring of fertilised carp eggs in indoor hatchery.

- Are the fish in lateral or in dorsal recumbence, or are they in an upright position?
- Are the movements of the fins normal and regular, or are some fish clamping one or more fins to the body?
- Are some fish showing erratic movements or shooting through the water?
- Are the fish rubbing their skin, especially on the head and the opercula, on structures in the keeping system?
- Are the breathing frequency and intensity higher in some fish?
- Are the fish collecting in regions with high oxygen levels, like the water inlets, the water surface or near an aerator?
- Are the fish gasping for air at the water surface?
- How is the colouration of the fish and are some fish darker than others?

- Are the fish showing any lesions of the skin, fins, eyes, or any other abnormalities?
- Are all fish reacting to the feed or are some fish not interested in feeding?

The keeping system itself should be carefully scrutinised. The number of fish in the tanks, the turbidity of the water, the material of the tank walls and bottoms, and the hygienic situation should be evaluated. Is there still feed seen in the water or are there large amounts of organic waste on the bottom of the tanks or ponds? All technical devices that are installed should be checked, such as biological filters and devices for water hygienisation, such as UV-C reactors and ozone generators. The change interval of the UV-lamps should be determined and the possibility that free ozone might enter the fish-keeping tanks should be excluded.

4.1.4 Clinical examination of fish

In most disease outbreaks not all the fish in one keeping system will show signs of the disease. To obtain a fast and correct diagnosis it is very important to examine those fish that are showing behavioural or external alterations. A meaningful number of fish, for example ten diseased fish, should be caught from the keeping system and examined. Particular attention should be paid to the fish that are separated from the shoal and are exhibiting behavioural alterations or visible disease changes. The fish that are the first on the feed are not usually the best individuals for the clinical examination! In cases where the fish are dying acutely and no clinical signs are seen before death, the number of examined fish must be higher. If a visit on-site is not possible, the fish owner has to be informed which fish should be selected for a clinical examination (Fig. 4.4). The examination of already-dead fish might give some information, but their usefulness is limited due to rapid autolysis, and the examination of live fish is preferred (Fig. 4.5).

The fish can be killed prior to examination. This should be done humanely and the relevant national legislation on stunning and killing of fish should be consulted prior to any action being taken. Killing can be carried out using an overdose of anaesthetics, such

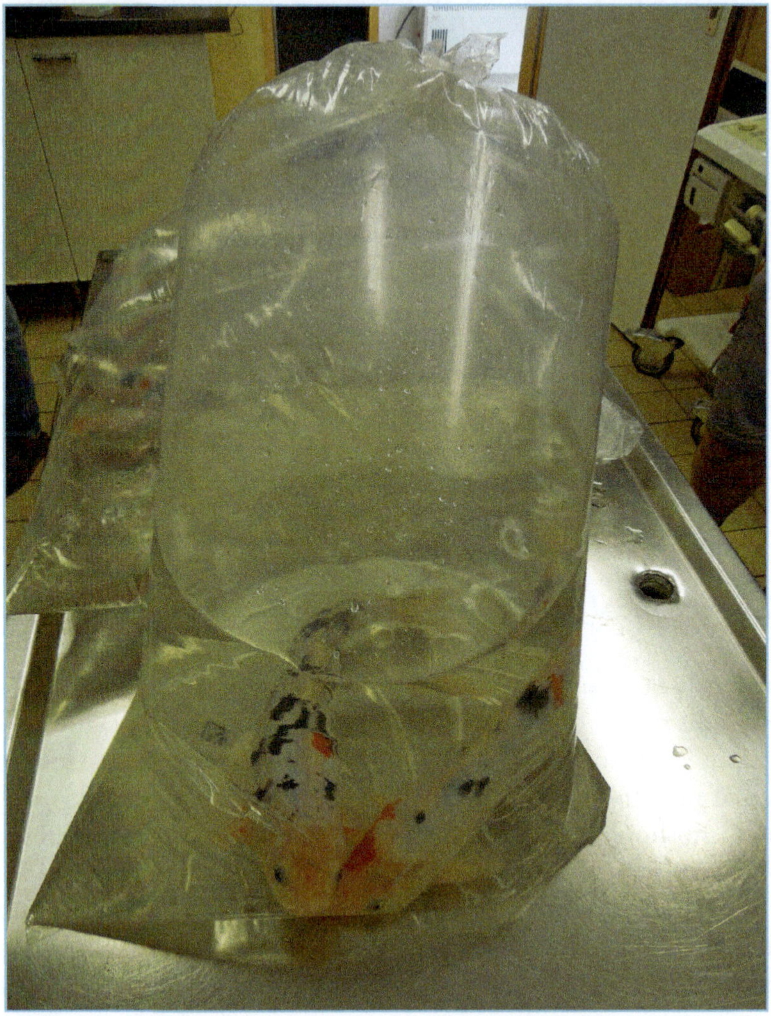

Figure 4.4 Transport of live fish in plastic bags filled with water and oxygen, preferably in a 1:2 ratio.

as tricaine (MS-222), eugenol, clove oil, 2-phenoxyethanol etc., or mechanically by a blow on the head and subsequent exsanguination, or by severing the vertebral spine (Fig. 4.6). If ornamental fish are examined that should survive the examination, the fish should be anaesthetised. More information on this is found in Chapter 20, An Ornamental Perspective.

Figure 4.5 Observation of sickness behaviour in a group of fish after transport to the laboratory.

The external features of the fish should be examined prior to any internal examination. For the skin (Fig. 4.7) a range of questions should be considered. For example, is the colouration normal or are there any areas of depigmentation or hyperpigmentation? Are there any lesions, ulcers, tumours or haemorrhages? It must be noted that some fish species can change their colour quite rapidly when stressed. Are there scales missing, or are some scales covered with algae or fungal growth? Are the scale pockets oedematous and do the scales protrude out of the scale pockets? Are the fins damaged? The mouth should also be examined and any lesions or bleedings around or inside the mouth should be noted. Also, are the barbels damaged and is the area around the nostrils damaged? The eyes should be examined for any signs of eno- or exophthalmos, bleedings in the eye chamber or around the eyes, lesions of the cornea and prolapse or clouding of the lens. The opercula on both sides of the head should be deflected, and the gills should be examined macroscopically (Fig. 4.8). The colour of the gills should be bright red and the mucous layer on the gills should be thin (Fig. 4.8A). Any pigmentations, depigmentation

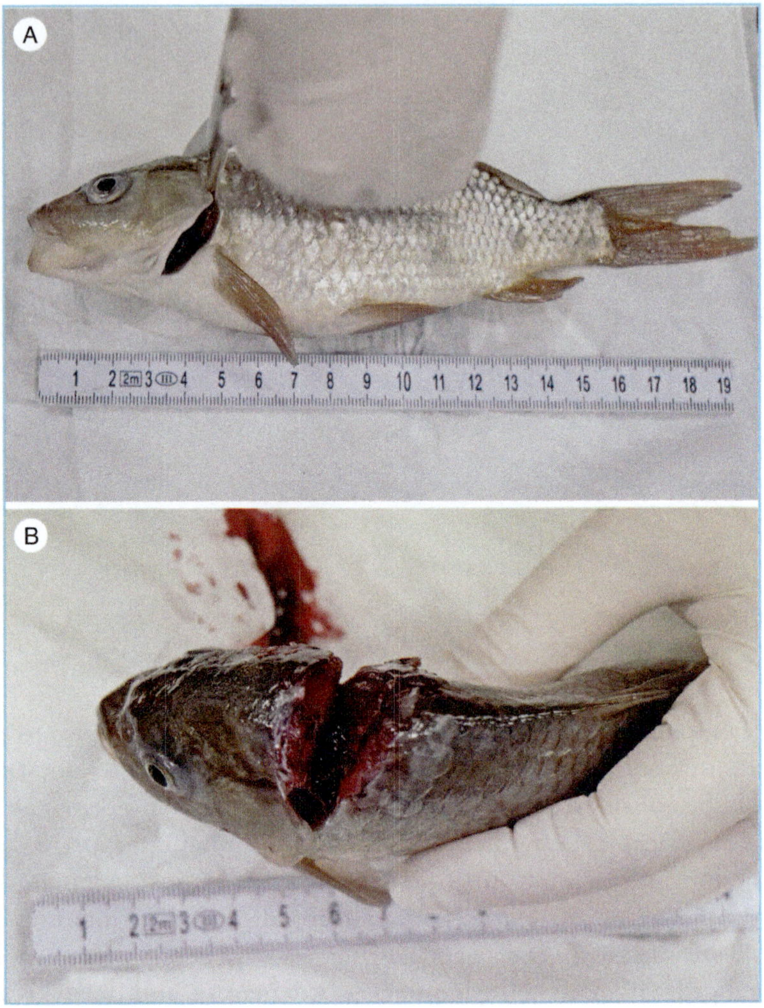

Figure 4.6 Killing of small carp by cutting the vertebral spine. (A) Position of the cut in the 'neck' area. (B) Completed cut.

and colour changes, like very pale gills (Fig. 4.8B) or brown gills, should be noted. Also, the gills should be examined macroscopically for parasites, the presence of large amounts of mucus, necrosis (Fig. 4.8C), oedema and the presence of algae or fungus. The area around the vent should be scrutinised for lesions and haemorrhages, and for any signs of inflammation, protrusion, and ulceration of the vent itself, which should be noted.

Figure 4.7 External appearance of healthy and sick fish. (A) Intact skin and fins of a healthy carp. (B) Fin and skin erosion due to virus infection in carp. (C) Damaged skin and fins in carp due to predatory bird attack.

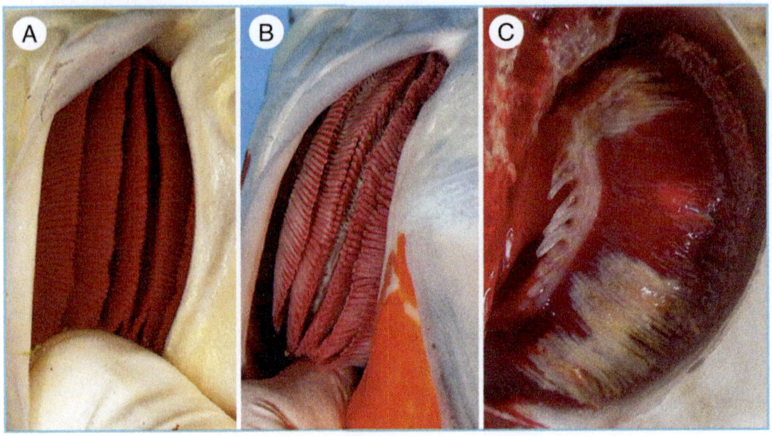

Figure 4.8 External appearance of healthy and infected gills. (A) Healthy gills. (B) Gills with increased mucous cover leading to pale appearance. (C) Gills with extensive necrosis.

Some ectoparasites can be seen by the naked eye, but for an examination of smaller ectoparasites samples should be taken from the mucus of the skin and of the gills for microscopic examination. Skin scrapes can be taken with a blunt scalpel blade or the edge of a glass coverslip (Fig. 4.9). It is recommended that these samples be obtained from different regions of the body, such as mucus from the skin in the dorsal and ventral region of the body, and from the region dorsal to the pectoral fins. Mucus should also be removed from the gills; and small parts of the gill tissue, ideally from both sides of the fish, should be transferred to a glass slide and covered by a coverslip. All samples should be examined as soon as possible, preferably within a few minutes, using a microscope. Bacterial and fungal infections of the skin and the gills can be detected by this method, and ectoparasites are particularly noticeable by their movements. Samples for parasitological examinations can also be taken from the intestinal tract. In living fish, samples can be obtained by flushing the pseudogaster and the hindgut with a small amount of water, using a syringe with a curved irrigation cannula (pseudogaster sampling) (Fig. 4.10), or the plastic inside of an IV catheter (gut sampling) (Fig. 4.11). Approximately 0.1 ml of water is applied to the pseudogaster of the gut and after a short massage of the abdomen

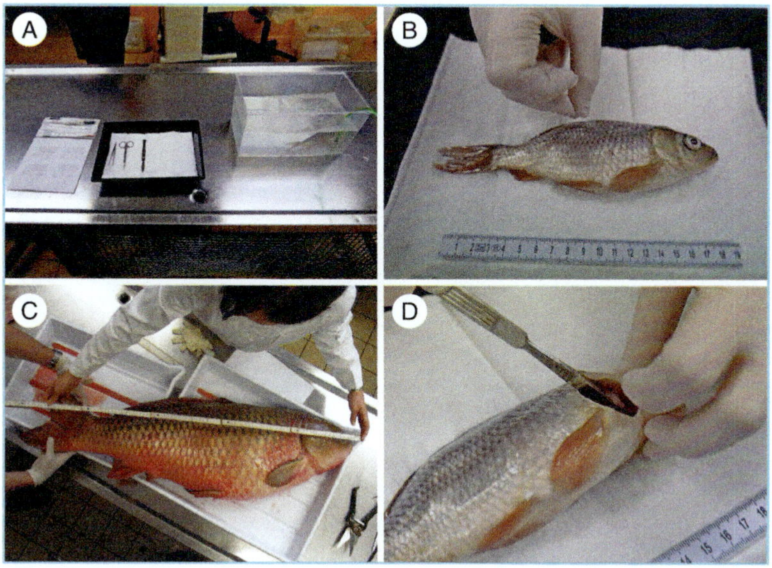

Figure 4.9 Clinical examination of fish. (A) Workplace prepared for clinical examination of fish. (B) Collection of a skin mucus sample using a coverslip. (C) Recording the body size of a large carp. (D) Collection of a gill biopsy using a scalpel.

of the fish, the liquid is aspirated with the syringe. The sample should be analysed microscopically as soon as possible. Protozoan parasites, proglottids of cestodes and eggs of nematodes can be recognised if present. Blood samples are collected from anaesthetised fish from the caudal vein, dorsal aorta (Fig. 4.12) or heart puncture, and used for diagnostic purposes, such as blood cell counts, blood chemistry, or blood parasite detection. In ornamental fish, imaging techniques like X-ray, ultrasound or endoscopy can be used to examine a diseased fish (Fig. 4.13).

4.1.5 Dissection of moribund fish

In most cases of disease outbreaks in food fish, a dissection of an adequate number of diseased animals should be performed. After killing the fish by euthanasia or by stunning followed by blood withdrawal, the examination of skin and gills should be performed as described earlier. The dead fish is opened laterally, and the internal organs are

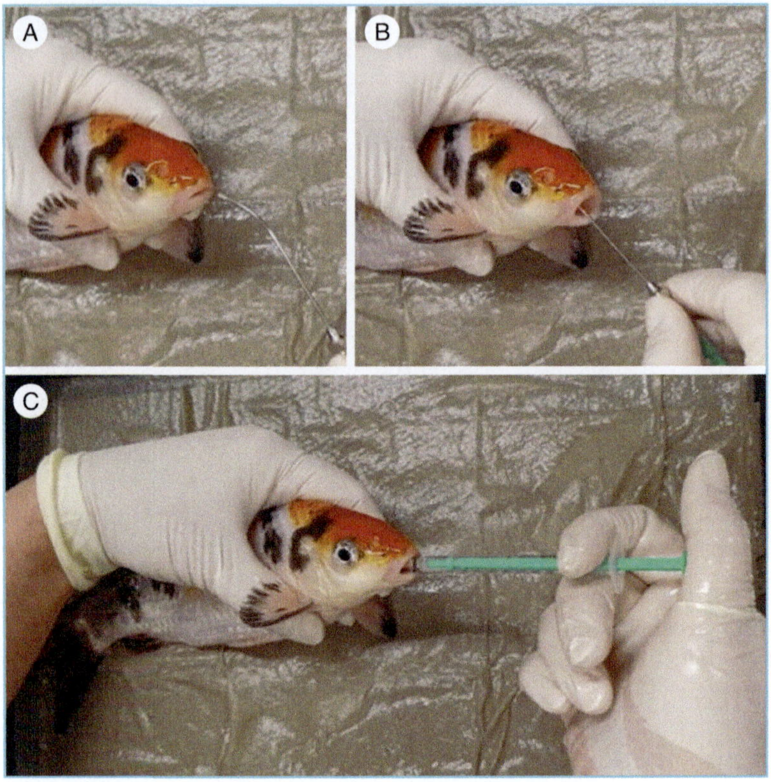

Figure 4.10 Oral gavage procedure. (A) Curved irrigation cannula fitted to the syringe used for pseudogaster sampling. (B) Insertion of the cannula. (C) Application of liquid.

examined macroscopically (Fig. 4.14). The presence of liquid or large amounts of free fat inside the body cavity should be ascertained. The position, the size, the colour and any alterations of the organs should be noted. The swim bladder is examined for the filling ratio and for any liquids inside the lumen, which should be recorded. It should be checked if both chambers of the swim bladder, for example in common carp (*Cyprinus carpio*), are filled with air. Liver and spleen are examined for the sharpness of their edges and for bleedings, and checked additionally for a congestion of bile. The kidneys and the head kidneys require examination for their size, which should be equal on both sides of the fish's body. Also, possible necrotic regions, haemorrhages and other alterations should be noted. The intestinal

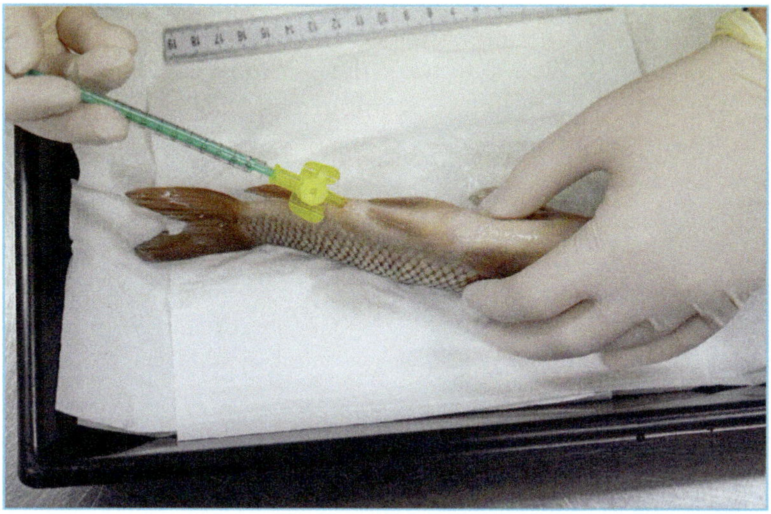

Figure 4.11 Hindgut sampling with plastic syringe attached to an IV catheter.

tract is examined externally and afterwards cut open laterally, and checked for bleedings and parasites, and the amount of food is noted. The heart is examined for liquid inside the pericardial cavity, fat storage around the heart muscle, and the size and thickness of the walls of ventricle and atrium. The skeletal muscle is cut in the region of the 'neck' of the fish and examined for haemorrhages.

Small pieces of liver, spleen and kidney should be placed on a glass slide and squashed with a cover glass for microscopic examination. Alterations such as granulomas, the presence of large numbers of macrophages and red blood cells can be seen in fresh tissue preparations without staining, and thus the first signs of the pathogenesis of the disease obtained. Swabs from the internal epithelium of the intestinal tract should be taken with a cover glass for microscopic examination. Protozoan parasites, eggs of nematodes and proglottids of cestodes can often be detected if present.

4.1.6 Sampling for laboratory examinations

Samples of the examined fish should be further analysed histopathologically for infections with bacteria or viruses, and it is always

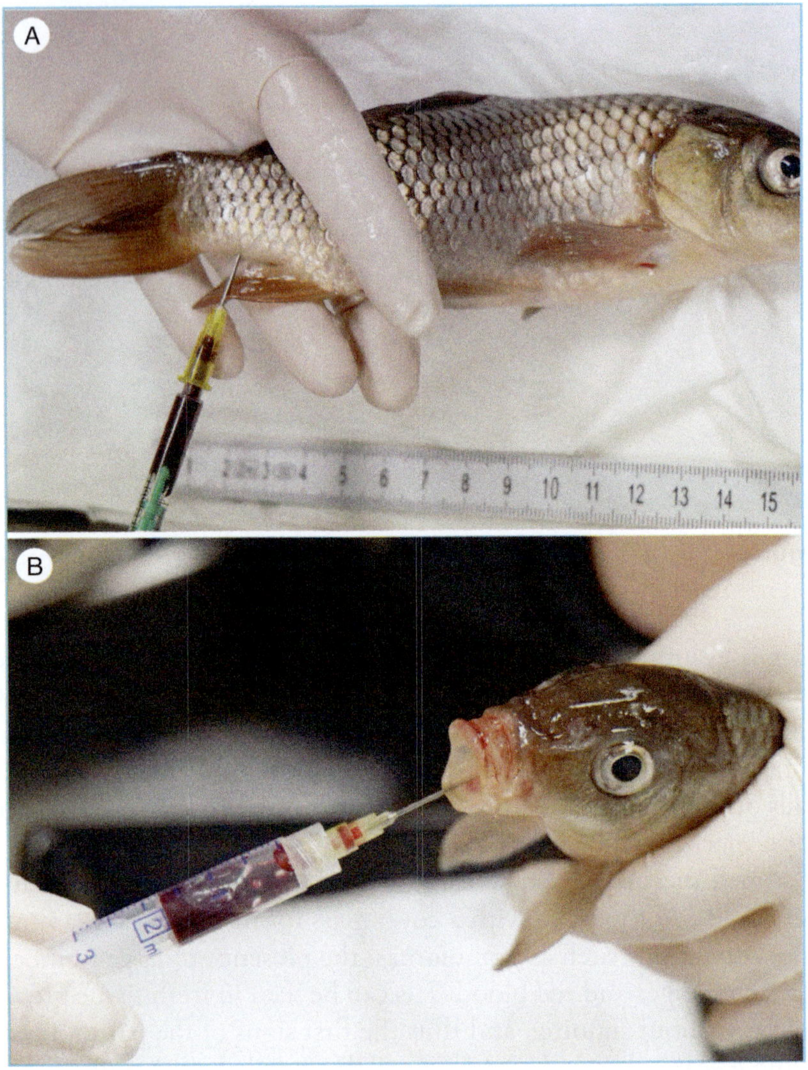

Figure 4.12 Blood sampling. (A) Collection of blood from the caudal vein. (B) Blood collection from the dorsal aorta.

recommended that a specialised laboratory is contacted prior to sampling. The best methods of sampling, fixing, storing and transporting the samples can be clarified in this way to achieve the most reliable and meaningful results.

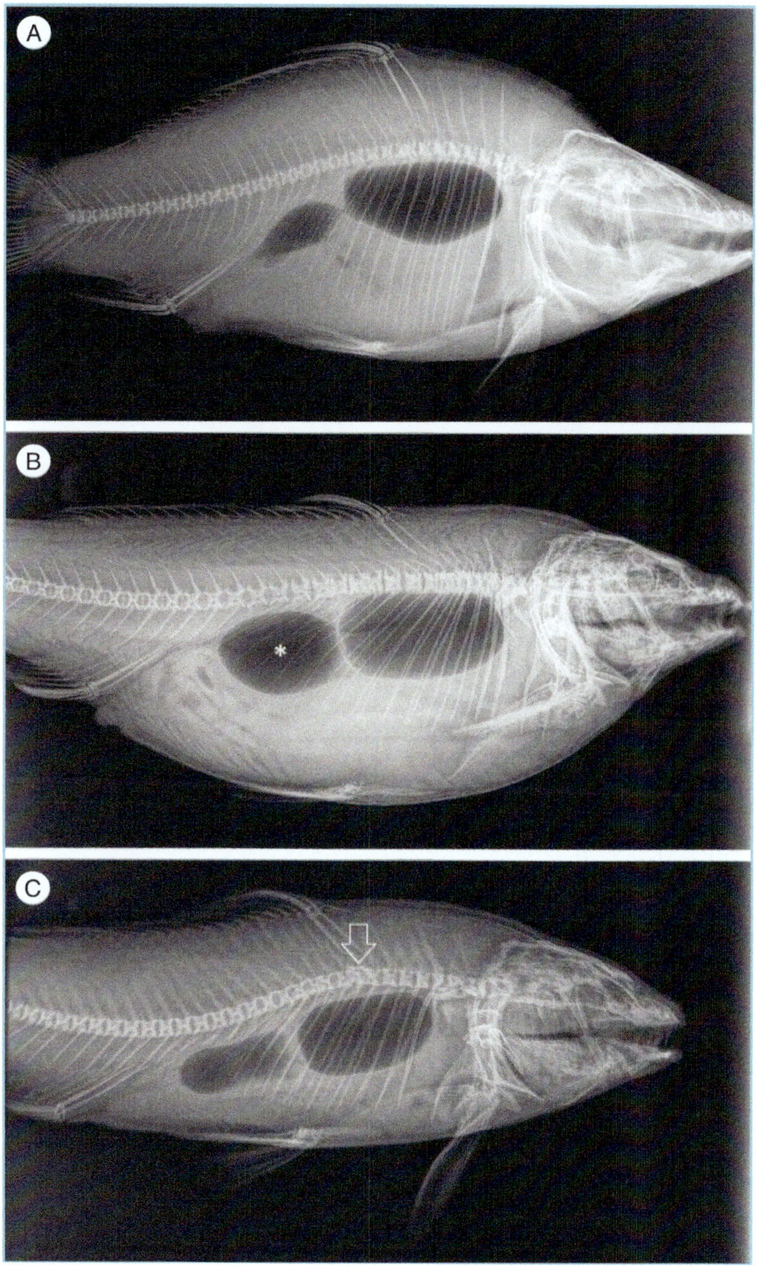

Figure 4.13 X-ray images of healthy and diseased carp. (A) X-ray of a healthy carp. (B) Swim bladder disorder with enlarged second chamber (*). (C) Broken vertebral column (arrow).

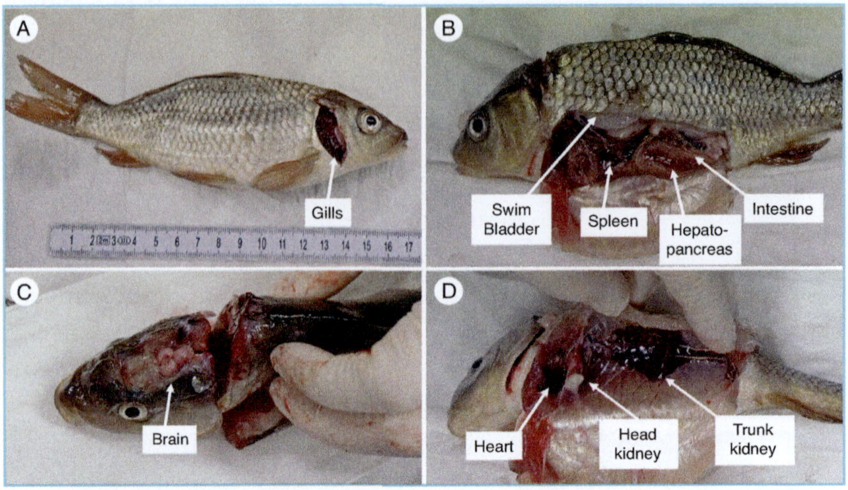

Figure 4.14 Internal organs of common carp. (A) Carp with operculum removed, exposing gills. (B) Open body cavity showing internal organs such as swim bladder, spleen, hepatopancreas and intestine. (C) Open skull showing carp brain. (D) Body cavity with most of internal organs removed, showing heart, head and trunk kidney.

Sampling for bacteriological examinations

When sampling for bacteriological examinations and especially when interpreting the results from those examinations, it must be considered that fish are always in direct contact with many bacteria from the physiological microbiome in the surrounding water. If samples are taken from the skin or the gill, bacteria will always be detected. To get a meaningful result from those examinations a correct sampling procedure is necessary (Fig. 4.15). If the fish are showing any lesions of the skin or the gills, the sample should be taken either directly with a sterile loop and placed onto an agar plate, or with a sterile swab that can be stored in a nutrition medium until arrival in a laboratory. In both cases, the samples must be taken at the very edge of the lesions and never from the centre. The bacteria responsible for the alterations are proliferating from the diseased to the healthy tissue, and thus can be found at this transition area.

If a dissection is performed and samples for bacteriology analysis are required, samples should be taken from the internal

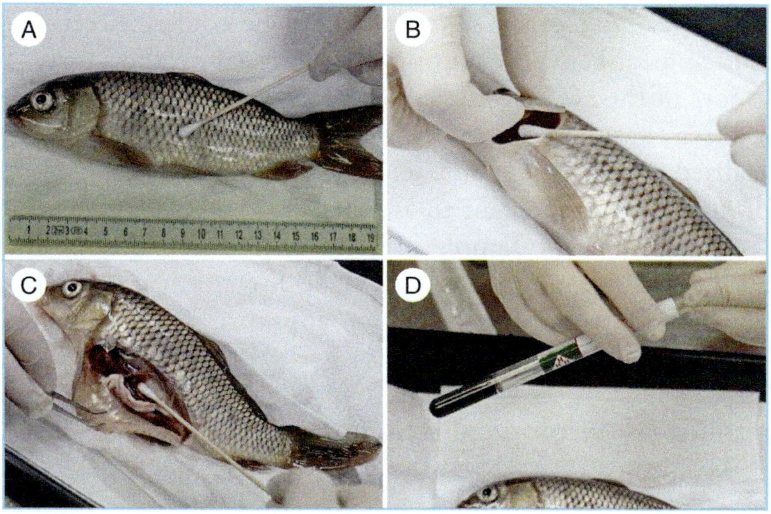

Figure 4.15 Examples of sampling for bacteriological examinations using swabs. (A) Collection of a skin sample. (B) Collection of a gill sample. (C) Collection of a sample from internal organs. (D) Insertion of the swab sample into a tube containing transportation medium in order to prevent desiccation of the sample.

organs, such as small parts of kidney, liver and spleen. Samples should be obtained using sterile scissors and placed into a sterile tube or into a sterile nutrition medium. Whilst the examination of single organs is more expensive, it does however give the most detailed results. The examination of the kidney, in particular, gives normally a very good overview on a possible septicaemia in the fish. As this organ is highly vascular and has a phagocytic function, any organism causing a bacteriaemia is likely to be recovered from the kidney. Examination of a pooled sample from the internal organs can be performed, but the information value is limited if macroscopic and microscopic examinations of alterations are not carried out during dissection. If alterations are seen during dissection in other internal organs, such as the gut and the swim bladder, it is recommended that a sample from these organs is obtained. It is important to note that, during sampling, contamination of the sample by touching the skin of the fish or by cutting the intestinal tract must be avoided.

All swabs and organ samples should be cooled directly after sampling and kept cool until arrival in the laboratory. The transport of the swab to the laboratory should be as soon as possible after sampling. As not all bacteriological laboratories are specialised in analysing fish bacteria or do not have the necessary equipment, it is important to ascertain that the receiving laboratory can examine the sample prior to dispatching. An antibiotic sensitivity test should always be made for any bacterial species detected in the samples, to select an appropriate antibiotic agent for treatment.

Sampling for virological examinations

In cases of a suspicion of a viral infection, or to exclude viral infections in control examinations, samples from an adequate number of fish (ten fish in cases of disease outbreaks, 30 fish for control examination) must be taken using recognised sampling and shipping techniques (Fig. 4.16A-C). For examinations for CEV, samples of the gills are sufficient that can be taken from living animals under anaesthesia. For examination of CyHV-3, samples of the gills or blood from living fish can be examined as well, but in cases of latent infections the reliability of a negative examination result is somewhat limited. More reliable is the sampling of internal organs, such as kidney, spleen, brain, gut and gills from recently killed fish. Samples of individual organs can be pooled for examination. But note: tissues from no more than five fish that are smaller than 10 cm, or two fish that are larger than 10 cm, should be pooled together. For the examination of an infection with SVCV, samples from kidney, spleen, gill and brain should be taken. Only specialised laboratories can analyse samples for virological examinations, and therefore, as with the laboratory examination of bacteriological samples, it is advisable to contact a laboratory to assess their ability to analyse the material and discuss the best method for sampling and transportation of the sample. The samples should be sent as soon as possible, according to the requirements given by individual laboratories.

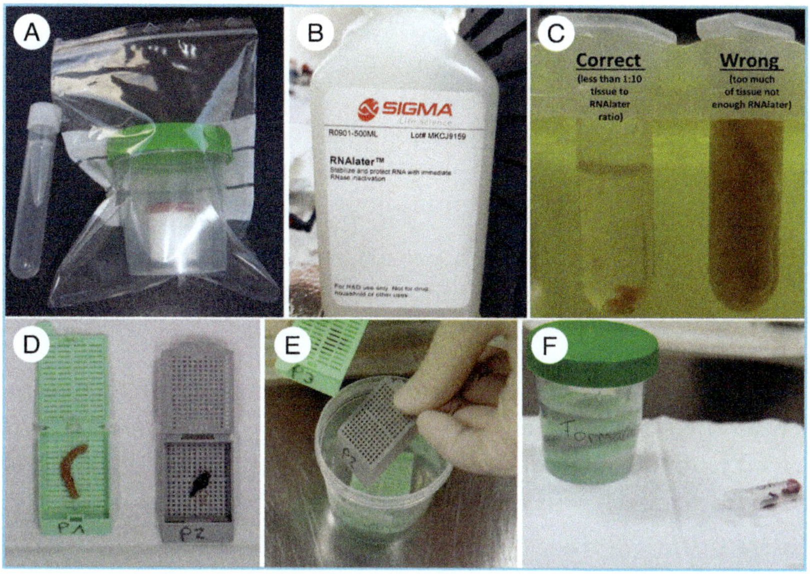

Figure 4.16 Sampling for virological and histological examinations. (A) Transport of virological samples in leakage-proof double packaging. (B) Example of nucleic acids preserving liquid. (C) Correct and incorrect ratio of the sample to the preserving liquid. (D) Placement of internal organs in the histological cassettes before fixation. (E) Placing cassettes with organ samples into fixation liquid. (F) Samples for histological analysis in fixation liquid (left) and tissues for virological examination in a reaction tube (right).

Sampling for histological examinations

In a histological examination, the tissues of the organs are cut in very thin slices and stained with specific dyes to make the distinction of individual cell components easier. Histology allows the alterations of the organs that cannot be seen macroscopically to be detected. For histological examinations it is most important to take samples as soon as possible after the death of the fish. Fish that already show signs of autolysis and frozen fish or tissues are not suitable for histological examinations. The samples of organs or tissues to be analysed are taken and placed in, for example, 4% buffered formalin or another commercially available fixing solution (Fig. 4.16D-F). The fixed tissues, appropriately labelled, should be sent in the fixation solution to a specialised laboratory.

4.2 Pathological Terms

When examining diseased carp, it is essential to describe the diagnosed alterations as exactly as possible. Therefore, the common pathological terms for those alterations should be used. In the following, the most common terminology is described in alphabetical order.

4.2.1 *Abscess*

An abscess is a localised accumulation of pus in any part of the body, such as the skin or internal organs. The pus is found in a cavity formed from the affected tissue. The tissue is damaged by an infection with bacteria, when they gain access by means of small tissue damages or wounds. Toxins are released by multiplying bacteria that destroy host cells. Pus is formed when the immune system is fighting the invading bacteria, as leucocytes absorb the bacteria by phagocytoses and breaking down dead cells. The broken-down tissue together with the leucocytes, dead bacteria and extracellular fluid form the yellowish pus. A capsule is formed by the healthy tissue around the pus that shields the damaged area from the surrounding tissue. Cell damage results in inflammation leading to swelling, redness, heat, and pain in the area around the abscess. When the abscess is ruptured naturally or by surgical intervention the pus is released, and the signs of the inflammation are revealed.

4.2.2 *Ascites*

Ascites is an abnormal accumulation of fluid within the abdominal cavity (Fig. 4.17). In fish, ascites is often noticed in cases of kidney diseases or dysfunction of kidney and/or liver. Ascites can also occur in cases of general bacterial infections.

4.2.3 *Deformation*

A deformation (also called deformity) is an alteration in shape, size, or structure of a part of the body (Fig. 4.18). A deformation can be

Figure 4.17 Carp showing an extensive expansion of its body resulting from an accumulation of ascites fluid and an abdominal tumour. Courtesy of the Environment Agency, National Fisheries Laboratory, UK.

Figure 4.18 Koi carp showing deformation of abdomen.

Figure 4.19 Carp showing sunken eyeball, typical of enophthalmia. Courtesy of the Environment Agency, National Fisheries Laboratory, UK.

present at birth and is then called congenital, or can be acquired after birth. Acquired deformations can be due to mechanical forces that distort a previously normally formed structure.

4.2.4 Enophthalmia

Enophthalmia is the sinking of an eyeball into the orbital cavity (Fig. 4.19) that can occur uni- or bilaterally and is often caused by a degradation of the fat body behind the eye. In carp, enophthalmia can occur when the fish are heavily infected with ectoparasites, especially *Ichthyobodo necator*, or due to viral infections, such as CyHV-3 or CEV.

4.2.5 Erosion

An erosion is the loss of cells on the surface of a tissue by physical or chemical toxic substances (Fig. 4.20). A skin erosion, for example, is characterised by the partial loss of the epithelium, with the basement membrane left intact.

Figure 4.20 Carp showing extensive skin and fin erosion.

4.2.6 Exophthalmia

Exophthalmia (also called exophthalmus or exophthalmos) is the protrusion of the eyeball from the orbital cavity (Fig. 4.21). Exophthalmia, which can occur uni- or bilaterally, can be congenital or caused by different diseases. In carp, exophthalmia can be caused by viral infections, such as SVCV, and is then normally seen bilaterally. A retro-orbital tumour or an injury of the eye can also lead to a unilateral exophthalmia.

4.2.7 Haemorrhage

A haemorrhage is a bleeding that is defined as an escape of blood from one or more ruptured blood vessels. Haemorrhages can be external, like bleeding from the skin, or internal, like bleeding within internal organs (Fig. 4.22). Usually, a large quantity of blood is lost in a short time. Blood loss can be in the form of isolated petechiae which can occur, for example, within muscle or the swim bladder (Fig. 4.23).

Figure 4.21 Koi carp showing protrusion of eyeballs. Courtesy of the Environment Agency, National Fisheries Laboratory, UK.

4.2.8 Hyperplasia

Hyperplasia is the enlargement of a tissue or an organ caused by an increased number of cells. The multiplication rate of the cells is

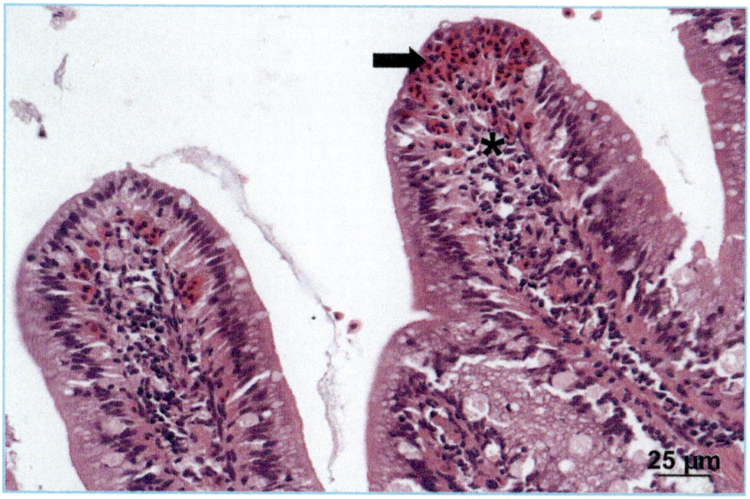

Figure 4.22 Mild haemorrhage (arrow) originating from capillaries of the intestinal lamina propria characterised by the presence of erythrocytes (*) located within the mucosal epithelium. Sample positive for *Cyvirus cyprinidallo 3* (Cyprinid herpesvirus 3) via molecular confirmation. Scale bar 25 µm. H&E stain. Images and interpretation provided by John Bignell of the Centre for Environment, Fisheries and Aquaculture Science (Cefas), UK. Crown copyright, Cefas 2022.

Figure 4.23 Petechiae located in the wall of a swim bladder of a carp infected with SVC. Courtesy of the Environment Agency, National Fisheries Laboratory, UK.

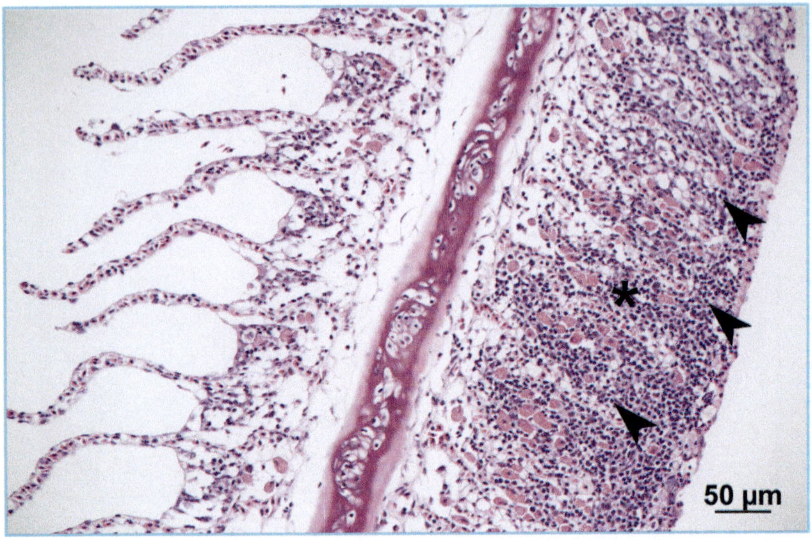

Figure 4.24 Diffuse inflammation (branchitis) of gill characterised by moderate lymphocytic infiltration of hyperplastic epithelium (*) located between secondary lamellae. Note the presence of relatively more basophilic lymphocytes (arrowhead) in comparison with epithelial cells. Sample positive for herpes-like virus via molecular confirmation. Scale bar 50 µm. H&E stain. Images and interpretation provided by John Bignell of the Centre for Environment, Fisheries and Aquaculture Science (Cefas) UK. Crown copyright, Cefas 2022.

increased, and more cells are then usually produced in a short period of time. The size of the produced cells is normal. Hyperplasia can occur to compensate for damages in tissues, but also due to inflammatory responses (Fig. 4.24).

4.2.9 Hypertrophy

Hypertrophy is the enlargement of a tissue or an organ caused by an increased size of cells. The number of cells stays normal in hypertrophic tissues. It can occur physiologically if specific cells, such as muscle cells, have an increased workload. Pathological hypertrophy can occur in tissues to compensate for dysfunctions of organs or tissues.

4.2.10 Inflammation

An inflammation is a response of the body triggered by any damages to tissues. It is marked by redness, pain, swelling, heat and loss of function. Inflammations occur as a response and a defence mechanism of the body to harmful stimulants, such as pathogens, toxic substances or other cell damages, such as injuries. The immune system is activated by those stimulants, and inflammatory cells and cytokines are released. The permeability of the blood vessels is increased, and fluid, proteins and white blood cells can invade the site of tissue damage. Inflammations can be acute or chronic (Fig. 4.24). Acute inflammations last only a few days, for example after injuries of the skin. Chronic inflammations are of longer duration.

4.2.11 Lesion

A lesion is any region in an organ or tissue that was damaged, for example through injury or disease. A lesion can be a wound, an ulcer, an abscess or a tumour. Lesions can be external (Fig. 4.25) or internal.

4.2.12 Necrosis

Necrosis is a cell injury that results in the permanent death of cells or tissues in an organ, and results from disease, an injury, or the failure of blood supply. Necrosis can be caused by pathogens, such as necrosis of gill cells in infections of carp with CyHV-3, or by chemicals or injuries. Necrosis can occur in single cells (single-cell necrosis), many cells, or indeed, all cells within a tissue or organ. In cells that undergo necrosis the cell organelles swell, the plasma membrane is ruptured, the cell is lysed and the intracellular content is released into the surrounding tissue (Fig. 4.26). Usually, necrosis is associated with inflammatory processes.

4.2.13 Oedema

Oedema, also called edema or dropsy, is the accumulation of fluids in cells or tissues. The affected tissues are swollen (Fig. 4.27). An

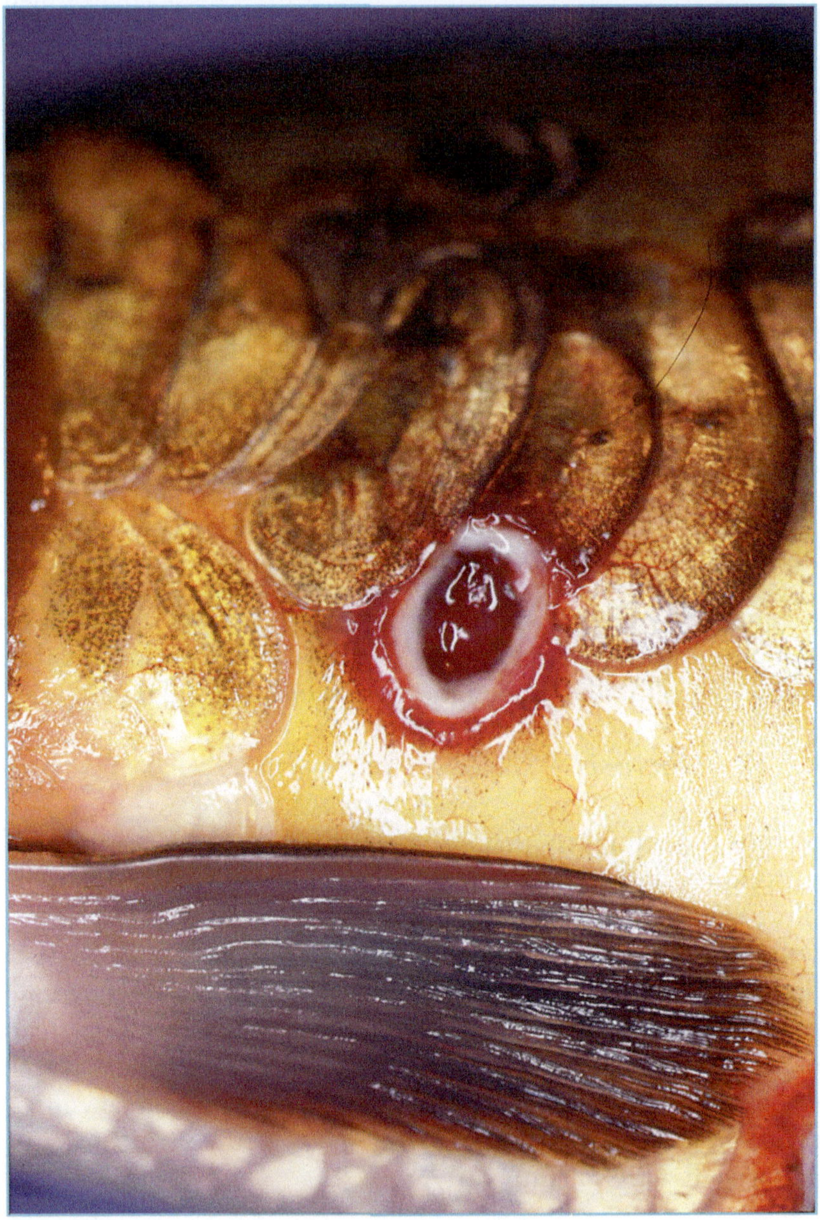

Figure 4.25 Lateral surface of a carp with a typical lesion caused by carp erythrodermatitis. Courtesy of the Environment Agency, National Fisheries Laboratory, UK.

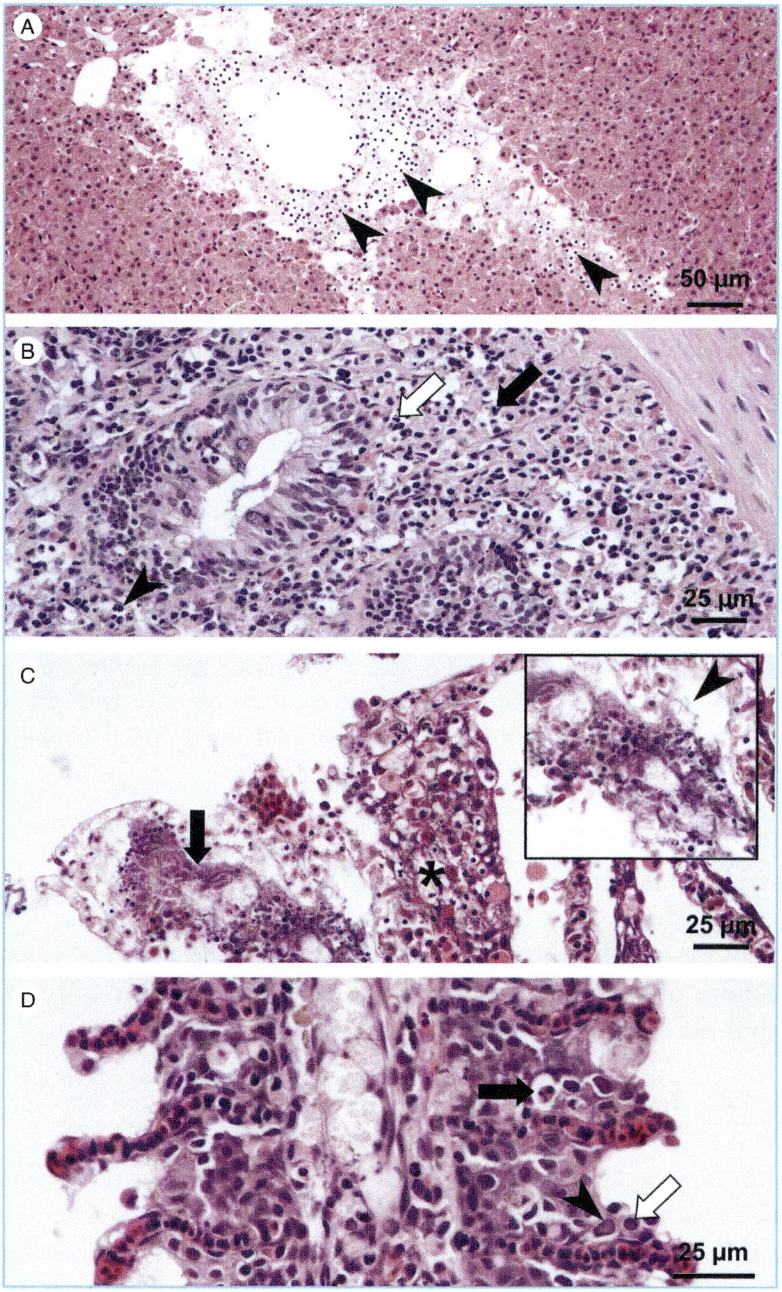

Figure 4.26 Images showing characteristic examples of necrosis in various organs on carp.

Figure 4.26 Continued

(A) Hepatocellular necrosis evidenced by numerous examples of pyknotic nuclei (arrowhead) surrounded by degenerative cytoplasmic material. Sample positive for *Cyvirus cyprinidallo3* (CyHV-3) via molecular confirmation. Scale bar 50 μm. H&E stain. (B) Diffuse necrosis of the intestinal lamina propria with numerous necrotic cells exhibiting nuclear pyknosis (black arrow), karyorrhexis (white arrow) and margination of chromatin (arrowhead). Sample positive for CyHV-3 via molecular confirmation. Scale bar 25 μm. H&E stain. (C) Fusion and necrosis (*) of two individual secondary gill lamellae leading to secondary infection with filamentous bacteria corresponding to regions of relatively pale basophilia. A higher magnification inset of area indicated by an arrow reveals the presence of individual filamentous bacteria (arrowhead). Sample positive for CyHV-3 via molecular confirmation. Scale bar 25 μm. H&E stain. (D) Necrosis of gill interlamellar cell mass. Note the presence of nuclei exhibiting karyorrhexis (black arrow) and margination of chromatin with eosinophilic intranuclear inclusion eliciting a signet-rink-like appearance (white arrow). Pale eosinophilic intranuclear inclusions can also be seen in some surrounding nuclei (arrowhead). Sample positive for CyHV-3 and carp oedema virus (CEV) via molecular confirmation. Scale bar 25 μm. H&E stain. Images and interpretation provided by John Bignell of the Centre for Environment, Fisheries and Aquaculture Science (Cefas), UK. Crown copyright, Cefas 2022.

oedema can be limited, for example to single cells or body tissues, or can occur throughout the body (Fig. 4.28). In carp, oedema of the skin and especially the scale pockets can often be seen in bacterial infections of the skin. Oedema of gill epithelial cells can occur due to viral infections, such as CEV.

4.2.14 Tumour

A tumour is a swelling of an organ or tissue, or a part of it, that is caused by an abnormal mass of cells (Fig. 4.29). The cells either grow more than usual or have a reduced death rate. In general, an inflammation is not seen in tumorous tissues. Tumours can be malign or benign. Benign tumours, which are defined as non-cancerous, can grow large, but do not invade other tissues. Malign tumours are

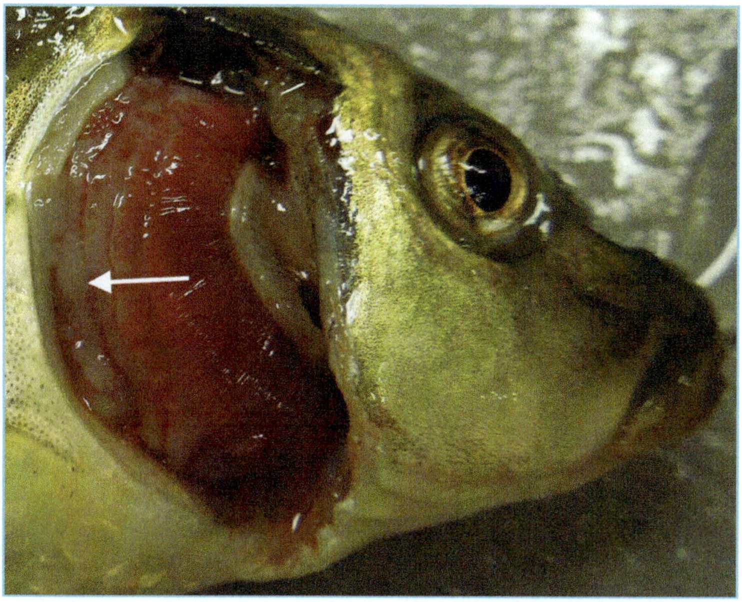

Figure 4.27 Operculum removed from a carp showing oedema of gill as swollen white patches.

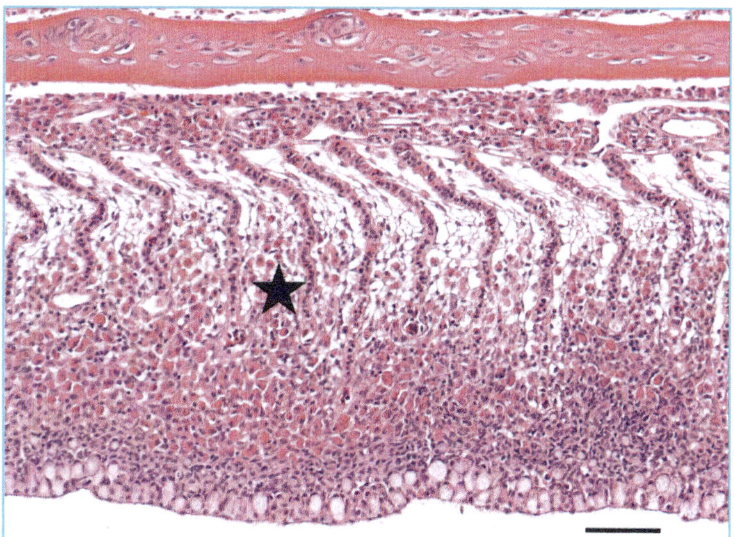

Figure 4.28 Histological section of a gill infected with carp oedema virus showing extensive oedema within the tissue (*). Indicative scale bar 40 μm. Courtesy of the Environment Agency, National Fisheries Laboratory, UK.

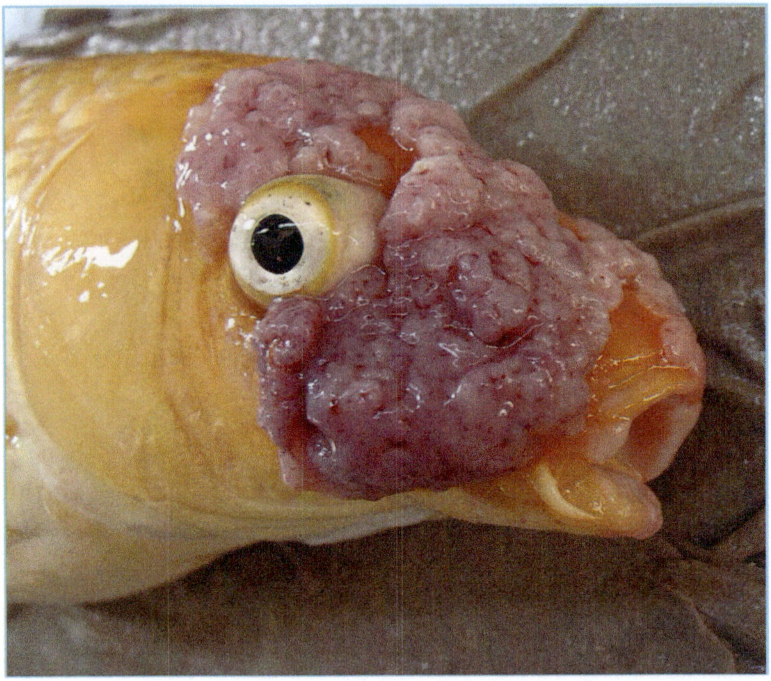

Figure 4.29 Extensive epithelial skin tumour around the eye of a koi carp.

defined as cancer and can grow and infiltrate into surrounding tissues or organs. Abnormal cells spread into other parts of the body through the blood system. Tumours can occur in carp, especially in ornamental koi carp, where malign tumours of the ovaries are seen regularly.

4.2.15 Ulcer

An ulcer is a sore area with loss of tissue on an external or internal surface of the body. Ulcers are caused by a break in the skin or a mucous membrane that is not healing or is only slowly healing. The surface cells are inflamed and die. Ulcers are often associated with cell necrosis and pus. Ulcers are associated with a range of infections, particularly caused by bacteria. In carp, ulcers of the skin and sometimes also the underlying muscle tissue are often seen (Fig. 4.30). Ulceration is the development of an ulcer.

Figure 4.30 Extensive skin ulcer on the lateral aspect of a goldfish.

4.3 Selected References for Further Reading

Ferguson, H.W. (2006) *Systemic Pathology of Fish: A Text and Atlas of Normal Tissues in Teleosts and Their Responses in Disease*. Scotian Press, London.

Noga, E.J. (2010) *Fish Disease. Diagnosis and Treatment*, 2nd edn. Wiley-Blackwell, Ames, Iowa.

Roberts, R.J. and Rodger, H.D. (2012) The pathophysiology and systematic pathology of teleosts. In: Roberts, R.J. (ed.). *Fish Pathology*. Blackwell Publishing, Hoboken, New Jersey, pp. 62–143.

Stoskopf, M.K. (1993) *Fish Medicine*. Saunders, Philadelphia.

Wildgoose, W.H. (2001) *BSAVA Manual of Ornamental Fish*, 2nd edn. British Small Animal Veterinary Association, Quedgeley, Gloucester, UK.

Chapter 5

Management Practices, Therapeutics and Vaccines

5.1 Therapeutics and Vaccines

As is shown throughout this book, many diseases in fish are caused by multiple factors. Often suboptimal environmental conditions in combination with different pathogens are leading to disease signs in fish and may induce high morbidity and mortality. Therefore, the examination of the diseased fish as well as the environment, especially the water quality, is most important to establish a disease diagnosis. Having a proper diagnosis is crucial for any treatments and management strategies.

In many cases different factors that may contribute to the clinical disease signs will be found and more than one diagnosis will have to be made. To achieve a long-term improvement of the health situation in a fish population, all these different diagnoses must be considered, and different pathogens might have to be treated. As this is normally not possible at the same time, the factors that are leading to the clinical signs must be prioritised. Acute health-threatening factors, such as very critical environmental conditions, must be addressed first. In the case of problems with water quality, the fish must be moved immediately to uncontaminated water, or water changes must be made immediately. High numbers of pathogens, such as protozoan parasites and flavobacteria, on the gills are a severe health problem for fish and must also be treated immediately. Other parasitic or fungal infections are less severe and, although they need to be diagnosed promptly, they can be treated after the acute problems have been solved. In many cases, improvement of the environmental conditions and treatment of possible parasitic infection will also

DOI: 10.1079/9781836990024.0005

improve the situation. Most bacterial infections are secondary infections and must be treated therapeutically, which can be done after the improvement of the environment and the treatment of possible parasitic infections have taken place. Viral infections usually cannot be treated by therapeutics, but in some viral diseases changing the water temperature or the salinity of the water, for example, can lead to reduced mortality in a fish population.

This shows that prophylactic measures in the form of good management practice, maintenance of optimal environmental conditions, optimal feeding of the fish and the avoidance of stressful situations are most significant to avoid clinical outbreaks of fish diseases caused by facultative pathogens. Also in many cases optimisation of the husbandry conditions leads to an improvement of the situation. Water changes, lowering the stocking densities in the keeping units, stopping feeding in cases of low oxygen saturation in the water and increasing the oxygen level are often very helpful to reduce mortalities in fish populations. In cases of infections with protozoan parasites, such as *Ichthyophthirius multifiliis*, increasing the water current and water exchange rates can prevent the theronts of the parasite attaching to the fish's skin or gills. Improvement of the hygienic conditions, such as removing waste from the bottom of the tanks, installing a sedimentation chamber and, in some cases, also the use of UV-C radiation or the application of ozone to the water in a chamber separated from the fish, can reduce the load of heterotrophic bacteria in the systems and thereby reduce the risk of bacterial infections in the fish.

If infections with parasites or bacteria have already progressed too far, management procedures are sometimes not successful and therapeutic agents have to be used. In general, fish can be treated with therapeutics orally, through the water or by injection, for example in the abdominal cavity (Fig. 5.1A) and the muscle (Fig. 5.1B). Fish for human consumption are usually treated by therapeutics that are given with the food, but several legal requirements must be met before these treated fish are put into the human food chain. The dosage of such treatments is calculated per body weight of the fish. Commercially available medicated foodstuffs can be given, or therapeutics can be mixed in the food and coated with oil. Medicated feed is normally given for a certain period, mostly 7–10 days. Feeding fish with medicated food causes less stress for the animals and less therapeutics are needed compared with bath treatments. However, severely diseased fish that are not feeding create a problem, as they are often not treated. Treatment of the water is difficult in ponds that are in direct connection to natural waters. The therapeutics can contaminate the natural environment and, as the water is exchanged regularly, the necessary dosage cannot be maintained over a sufficient period. Antibiotic treatments would also disturb the physiological microbiome in the systems in a way that might lead to severe problems with chemical water parameters. As such, the application of antibiotics to waterbodies in many countries is controlled and limited by government agencies and legislation. Short-term and long-term treatments in water baths can be used in separated tanks if there are parasitic infections. For short-term treatments higher amounts of the therapeutic agent are used and

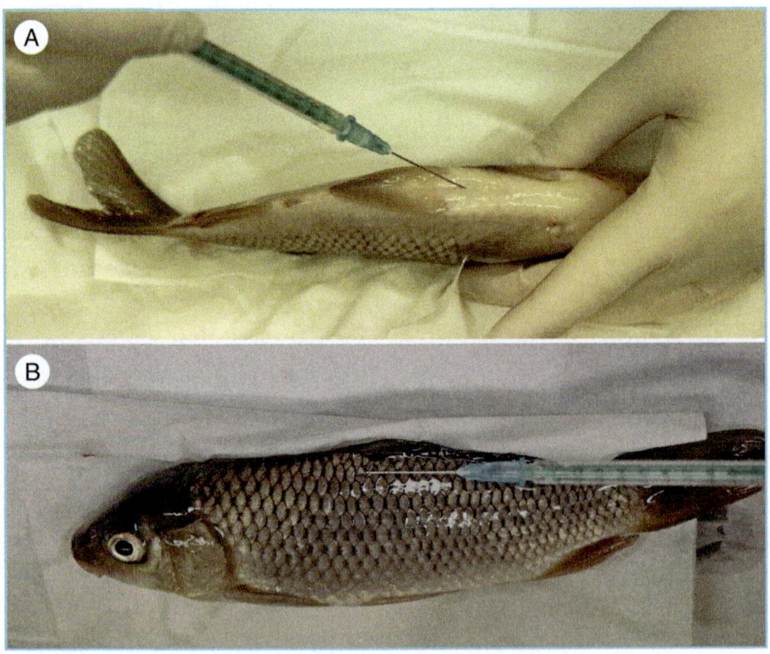

Figure 5.1 Injection of therapeutics. (A) Intraperitoneal injection technique. (B) Intramuscular injection technique.

the procedure is more stressful for the treated fish but the effect on parasites is faster. Long-term treatments in water baths are gentler and, especially, already weakened fish might better tolerate these treatments. In general, the use of bath treatments also means fish that are not feeding can be treated. Injections are usually used only in ornamental pet fish where the individual animal has an emotional value for the owner. In this case the dosage of the therapeutic can be exactly calculated by weighing the fish before injection. The uptake of the active substance depends on the injection site (muscle or body cavity (Fig. 5.1)) and the substance used. In the case of skin lesions damaged scales can be removed and the wound cleaned and cured by the local application of antiseptic paste (Fig. 5.2).

Treatments with antibiotics should be carefully considered, planned, and only used if other management strategies are not expedient. Bacteria isolated from fish or the aquatic environment are very often resistant against several antibiotic substances. If an antibiotic treatment is required, the isolation and identification of the responsible bacterial species as well as an antibiogram of all detected bacterial isolates should be made. Treatments without testing the isolates for resistances might not only lead to the lack of success, but also result in the promotion of bacterial resistance in the aquatic environment. Antibiotic treatments should, therefore, only be performed with a sufficient and appropriate dosage, and over a sufficient period.

Different therapeutic agents are registered for use in fish in different countries. For an overview on the approved therapeutics it is helpful to enquire about the situation in the respective country, and the possibilities of importing and using therapeutics from other countries that are approved for fish treatments. Future control measures may utilise bacteriophages and novel antimicrobials such as antimicrobial peptides. The use of these measures may circumvent the issue of antibiotic resistance which is developing in several fish pathogenic bacteria. Bacteriophages act on the bacteria in a species specific manner, and therefore cocktails may be needed. Antimicrobial peptides can be isolated from biological organisms or designed in silico and produced synthetically. These short peptides were shown to have strong antimicrobial acitivy and have a potential

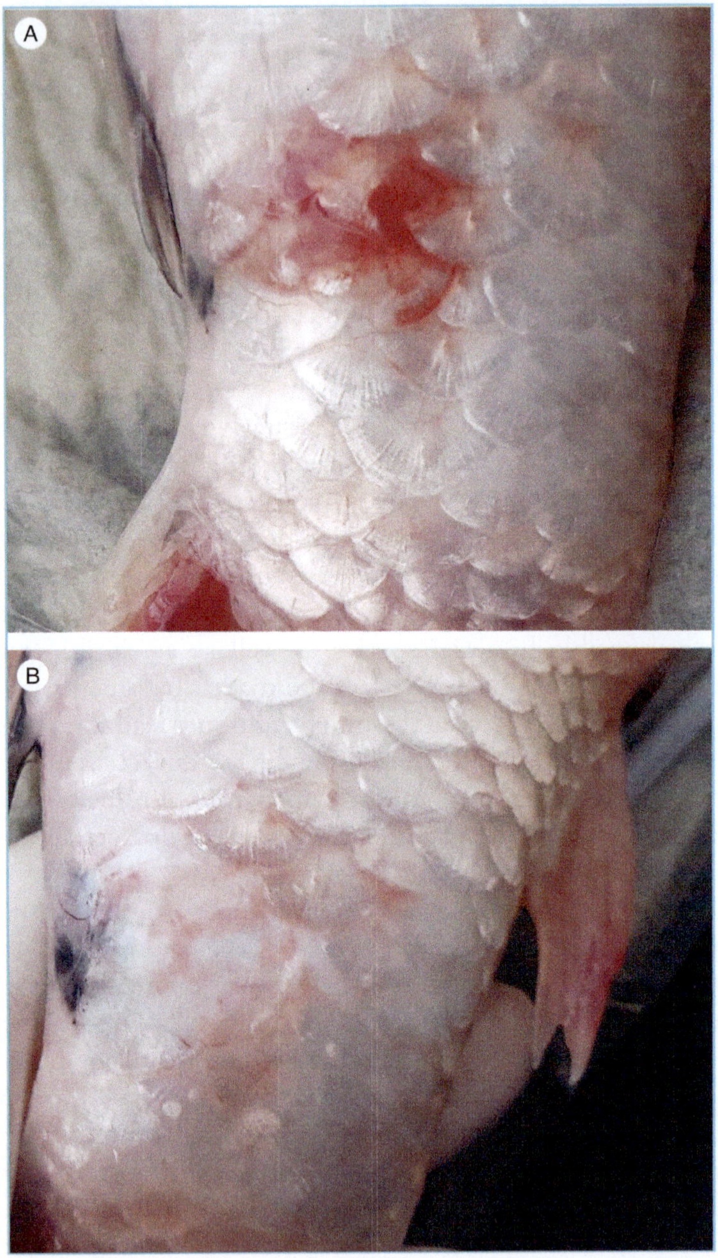

Figure 5.2 Effect of treatment of a skin lesion. (A) Untreated skin lesion of a koi with inflamed scale pockets. (B) The same skin region with treated skin lesion and damaged scales removed.

to replace antibiotics. However at present bacteriophage therapy or the application of antimicrobial peptides is still at the laboratory stage and commercial products are not available.

Vaccinations are also possible in fish. In most countries only a few vaccines are approved for fish and most of those are vaccines to prevent bacterial infections, but there are vaccines against some viral infections as well. Fish can be vaccinated through the water (Fig. 5.3), orally or by injections (Fig. 5.4). Bath vaccination is usually used in fry. The vaccine, regularly consisting of formalin-inactivated pathogens, is given in a defined water volume and the fish are put into the water–vaccine mixture for a certain period. At the time of vaccination, the fish must be immunocompetent. Therefore, fry can be vaccinated with a body weight of at least 2–5 g. As immunity in fish is usually not very long-lasting, a booster vaccination may be necessary. This is usually given orally with the feed, but in countries with large fish production industries, vaccination by injection is also common. If there is no commercially approved vaccine available against a bacterial pathogen that is regularly causing disease outbreaks in a specific farm, a farm-specific vaccine can be produced. To facilitate this the infectious agent that is causing the clinical signs must be isolated and bath vaccines, as well as oral vaccines, can be made with these specific isolates. The use of either commercially available or farm-specific vaccines is recommended in

Figure 5.3 Bath vaccination of fish.

Figure 5.4 Vaccination of fish by intra-peritoneal injection. Courtesy of Grit Bräuer, Sächsische Tierseuchenkasse, Dresden, Germany.

cases where otherwise regular antibiotic treatments would have to be performed.

5.2 Management Practices

Perhaps the most cost-effective and most efficient way of reducing the impact of disease on a fish population is to adopt management practices that support the concept that 'prevention is better than cure'. This applies not only to the hobbyists who have a few ornamental carp species, but also to large production units with thousands of carp. Usually, it is the fry stage that is most susceptible to infection, but disease can be introduced into a system through a range of routes where some simply adopted principles have not been adopted or have been circumvented to save either money or time. There are some highly recommended management strategies that should be adopted to prevent infection being introduced, or to reduce the impact of disease on a fish population. These include, for example, the following.

1. Always endeavour to obtain introductions (eggs, fry or adults) from recognised disease-free stock, and always obtain validation of such from an appropriate authority.
2. Where appropriate, insist that fish are treated prior to any movement into new waters.
3. Always ensure that footwear, nets, tanks etc. have been disinfected prior to being used.
4. If available, use a recognised fish transporting company that has a good track record in fish health management practices.
5. Withdraw feeding of the fish to be transported 1–2 days prior to transport in order to slow down metabolism and to avoid high ammonia concentrations in water during transport.
6. Try to ensure that transport water is clear of any sediment or invertebrates that may be acting as disease intermediate hosts.
7. Ensure that the transport water is well oxygenated and that the time to transport the fish is as short as possible. Avoid rapid changes in water temperature, or exposure of fish to the sun or to cold during transport. When necessary, use temperature insulation.
8. It is highly recommended that fish are quarantined in a separate facility either prior to movement or after they arrive at the new fish site. They should not be put directly into established and existing fish stocks.
9. Quarantine sites should be separated from existing fish units, have separate water flows and have separate routine equipment for fish maintenance or manipulation.
10. If dead, moribund or stressed fish are noted, consult an appropriate veterinary authority as soon as possible. Quick action may reduce fish loss and the economic impact.
11. Implement strict biosecurity practices throughout, such as using separate nets etc. in ponds/tanks from different 'epidemiological units' and insisting that all staff use appropriate disinfection protocols.
12. Restrict visitors' access to facilities. If visitors have access, their footwear and, for instance in fisheries, equipment such as landing nets, fishing rods etc. should be disinfected before entering.

5.2.1 Disinfection measures

After use, all objects should be cleaned thoroughly to remove dirt, fish faeces, scales or blood and then dried in air or by exposing them to the sun. To prevent the spread of viruses, objects should be disinfected. For this, substances should be used that have no toxic effects on users and fish. Peroxide compounds in combination with organic acids are recommended. When selecting preparations, observe the health and safety regulations in place and the recommendations of the appropriate bodies for the use of disinfectants.

5.3 Selected References for Further Reading

Bhat, R.A.H., Khangembam, V.C., Pant, V., Tandel, R.S., Pandey, P.K. and Thakuria, D. (2024) Antibacterial activity of a short de novo designed peptide against fish bacterial pathogens. *Amino Acids* 56, 28.

Noga, E.J. (2010) *Fish Disease*, 2nd edn. Wiley-Blackwell, Ames, Iowa.

Stoskopf, M.K. (1993) *Fish Medicine*. Saunders, Philadelphia.

Wildgoose, W.H. (2001) *BSAVA Manual of Ornamental Fish*, 2nd edn. British Small Animal Veterinary Association, Quedgeley, Gloucester, UK.

Xia,J., Ge, C. and Yao, H. (2024) Antimicrobial peptides: An alternative to antibiotic for mitigating the risks of Antibiotic resistance in aquaculture, *Environmental Research*, 251, 118619.

Yang, L., Yang, Q., Hu, R.-G, Cong, W., Li, S. and Kang,Y.-H. (2024) The evaluation of bacteriophage therapy in aquaculture: A systematic review and meta-analysis, *Aquaculture*, 588, 740L925,

Chapter 6

Viruses

6.1 Introduction

Viruses are non-living (abiotic) infectious particles; they are sophisticated forms of parasites, which utilise host metabolism and ultimately rely on host cells for replication. They often coevolve with their host, adapting to exploit effectively their host cells for replication without causing severe clinical signs. Virus replication thus depends on hijacking the host cell metabolism for viral particle production, although fish, like other host organisms, are able to fight this process. These pathogen–host interactions often result in pathological changes leading to the development of disease. For example, a replicating virus needs to escape from a cell to infect a new cell of the host, a process that often leads to the death of the cells. If these cells play important roles in fish physiology and are vital for survival of the fish for example, epithelial cells of gills, this damage can lead to life-threatening pathology. In addition, the host's immune responses can compound these pathological changes, resulting in a strong inflammatory response or the so-called cytokine storm.

Viruses are the fastest-spreading forms of pathogens and thus have the highest probability of leading to the emergence of new unknown agents. Novel viruses that have jumped from one host species to the naïve population of new host species are the most dangerous for the host animals. These viruses can cause severe clinical signs, often leading to mass mortality, as the virus is not as yet adapted to the host. Generally, viral infections in fish are responsible for about 20–25% of total infectious diseases in aquaculture. This puts them in second place after bacteria (50%) and ahead of parasites and fungi. In some cases, like the emergence of *Cyvirus cyprinidallo*

 Carps: A Practical Approach to Disease Control and Health Management (eds David Hoole et al.)
DOI: 10.1079/9781836990024.0006

2 (Cyprinid herpesvirus 2) and *Cyvirus cyprinidallo 3* (Cyprinid herpesvirus 3), viruses cause extreme mortalities, which can have a very drastic effect on fish populations. In cyprinids, there is an additional force driving the spread of viral disease. The global trade in ornamental cyprinids like koi, a colourful variety of common carp (*Cyprinus carpio*), and goldfish (*Carassius auratus*), a close species to crucian carp (*Carassius carassius*) and Prussian carp (*Carassius gibelio*), leads to these valuable species being shipped by air transport all over the world. This global trade allows the viruses to spread very effectively in the absence or insufficiency of biosecurity and diagnostic measures.

Viruses are very diverse and abundant biological entities, and fish can be carriers of multiple infections of divergent viruses. The viriome (collection of all viruses) inhabiting a single fish host or a fish population can consist of both virulent and non-virulent viruses. The virulent viruses appear to belong to defined (although somewhat limited) numbers of families that cause diseases in animals. This section will focus on the most virulent viruses of cyprinids however, examples of viruses only loosely associated with diseases will be also presented.

6.2 Diseases of Confirmed Viral Aetiology

6.2.1 Cyvirus cyprinidallo1 – Cyprinid herpesvirus 1 – Carp pox

Introduction

Existence of a disease causing pox-like changes on the skin of fish was recorded as early as in the 16th century. The most common form in carp was called carp pox and is now known to be caused by infection with *Cyvirus cyprinidallo1* – Cyprinid herpesvirus 1 (CyHV-1). The disease is characterised by skin changes, which are recorded at relatively low water temperatures and which reduce the visual attractiveness of the fish. The disease is of lower concern and can often be self-limiting, but in some cases can lead to severe consequences resulting in skin tumours, which ultimately could cause the death of the fish. There seem to be several herpesviruses that cause similar changes in the skin epithelium and have coevolved with their hosts, although in common carp the disease has clearly been linked with CyHV-1 infection. This virus is also capable of causing acute disease of common carp fry, which has been associated with tumorigenesis in persistently infected fish and high mortalities.

Disease description

CyHV-1 causes proliferative skin disease (papilloma-like growths) in common carp, which is generally considered a visual defect and was initially considered as harmless. White–creamy-pink growths occur which increase in size over time on the skin and fins (Fig. 6.1). Histologically these benign growths show extensive epidermal hyperplasia containing mainly keratinocytes and collagen fibres. The cells in the growths often display margination of the chromatin and pale eosinophilic cytoplasm, which are characteristic of herpesvirus infections. The changes are not permanent and are reversed with increase of water temperature. Recent studies

Figure 6.1 White-creamy-pink growths on carp skin caused by an infection with *Cyvirus cyprinidallo1*. Courtesy of the Environment Agency, National Fisheries Laboratory, UK.

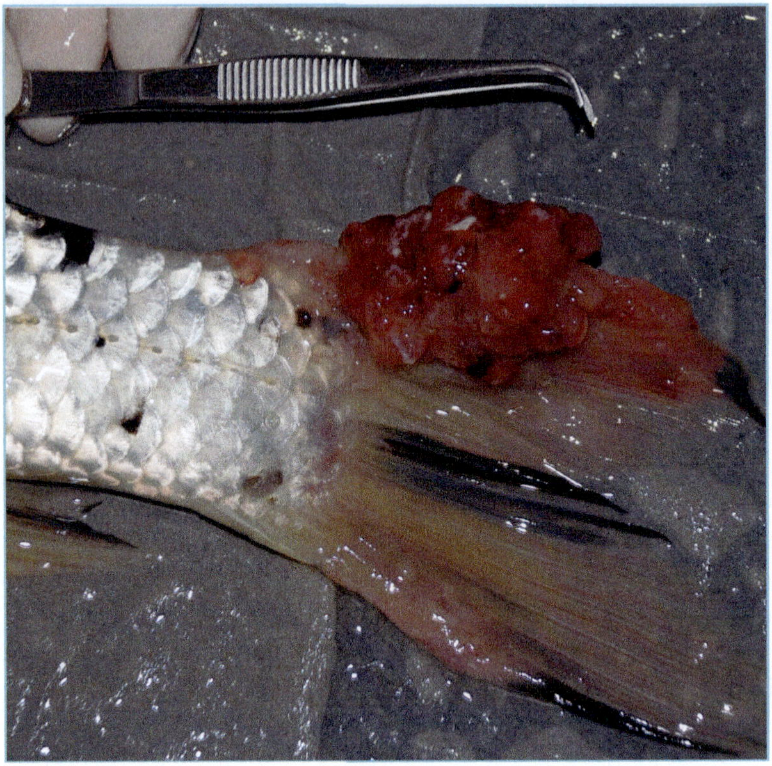

Figure 6.2 Skin tumour of koi caused by an infection with *Cyvirus cyprinidallo1*.

have shown that the virus could be involved in the generation of skin tumours (Fig. 6.2), as the CyHV-1 DNA and viral RNA expression was found to be present in skin tumours (cutaneous squamous cell carcinomas) of koi (Fig. 6.3) The mechanism of tumorigenesis aided by CyHV-1 is not known, but the genome of the virus encodes several paralogues of the known oncogene JunB. Involvement of CyHV-1 JunB, JunC and JunD genes requires further study, as they could be putatively linked to generation of some skin tumours in koi.

The other more severe disease of common carp fry has been recorded during CyHV-1 experimental infections of juvenile common carp. This acute form of disease leads to mortality of over 85% in carp fry. Infections in older fish result in lower mortality

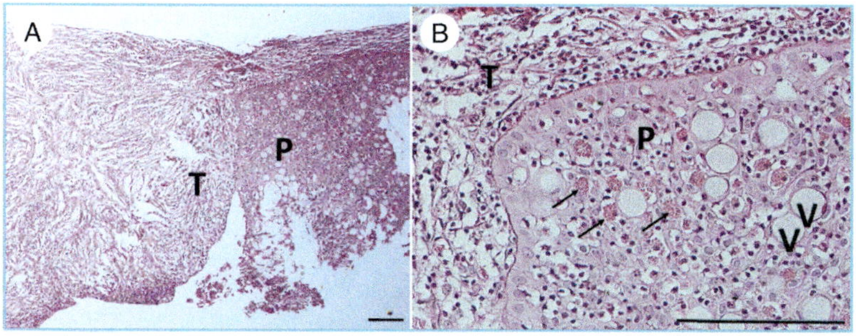

Figure 6.3 Histology of skin growths caused by infection with *Cyvirus cyprinidallo1*. (A) Low-power magnification of tumour (T) and papilloma-like growth (P). (B) High magnification showing virus-infected cells (V) and infiltrating granulocytes (arrow). Indicative scale bars 50 μm. Courtesy of Pedro H.O. Viadanna.

rates, with all fish surviving the infection at the age of 8 weeks. The acute CyHV-1 infection in fry leads to systemic disease characterised by lethargic behaviour and darkening skin, with occasional haemorrhages on the operculum and abdomen. Clinical signs of increased water retention, like distended abdomen and exophthalmia, can also occur. During acute infection of fry, the virus causes severe necrotic hepatitis and extensive tubular nephritis.

Characteristically for all herpesviruses, convalescent fish are most likely lifelong carriers of the latent infection, in which the virus is present in the cells but is hidden from the immune system and does not replicate. The infection in its lytic form can be reactivated and can produce new growths in permissive temperatures. Therefore, the disease is a reoccurring event in the infected population and cannot be eradicated.

The herpesviruses usually have quite a narrow host range, although carp pox-like disease has been recorded in multiple cyprinid species; for example, CyHV-1 has been confirmed in common carp and barbel (*Barbus barbus*). The proliferative skin changes characteristic for carp pox in roach (*Rutilus rutilus*) and asp (*Leuciscus aspius*) have also been associated with a novel cyprinid herpesvirus (CyHV-5). Furthermore, the experimental infection of grass carp and crucian carp fry with CyHV-1 has not led to an outbreak of the disease and increased mortality.

Pathogen

CyHV-1, a member of *Cyvirus* genus in *Alloherpesviridae* family, possesses an enveloped 120 nm virion encapsulating the dsDNA genome. The genome has 291 kb and encodes for 137 proteins.

Detection

Despite the lack of severity of the disease in adult common carp, multiple methods have been developed to detect the virus. PCR-based methods are available for detecting the virus DNA and although several protocols are available, the quantitative double-labelled probe-based PCR should be treated as a gold standard. Electron microscopy can be employed to visualise the viral particles in affected cells, and immunohistochemistry can be performed in order to label the infected cells with virus-specific antibodies. Several cell lines have also been used to isolate the virus. For example, koi fin 1 (KF-1), koi fin cells (KFC), epithelioma papulosum cyprini (EPC) and fathead minnow cells (FHM) are all susceptible to CyHV-1 infection, which causes cytopathic effects (CPE) ranging from morphological changes in cells (cell rounding with visible forming of vacuoles) to focal lysis of the cells, which can be observed by day 7 post-infection.

Prevention and treatment

Due to the lack of widespread testing, there are no preventive measures that can be adopted to limit the impact of the virus. The virus should be considered endemic to all places where common carp are residing. The disease is normally self-limiting in older fish, and the clinical changes do not cause any significant harm to the fish besides the visual detriment. Only extensive growths over the gills could directly threaten the life of the fish. The increase of water temperature (> 20–25°C) causes the pathological changes to disappear rapidly. Extensive pox-like changes and skin tumours caused by the virus can be surgically removed by a veterinary professional. This does not guarantee a cure, as the herpesviruses of fish possess the ability for persistent and latent lifelong infection. The management of the tumours depends on the cancer-free surgical margins and aggressiveness of the cancer type. No vaccine or virus-resistant strains are available.

FACTS SHEET

Disease: Carp pox
Pathogen: *Cyvirus cyprinidallo1* – Cyprinid herpesvirus 1 (CyHV-1)
Susceptible hosts: Common carp, barbel
Clinical signs: Young fish: distended abdomen, exophthalmia; older fish: papilloma-like growths, skin tumours.
Mortality: 85% in 2-week-old fry, low or none in fish older than 8 weeks
Temperature: 8–18°C
Prevention and treatment: Disease self-limiting – increase of the temperature to > 20°C could limit the virus replication and cause the reversal of skin changes. Surgical treatment of the skin changes or tumours can be performed

6.2.2 Cyvirus cyprinidallo3 – Cyprinid herpesvirus 3 – Koi herpesvirus – Koi herpesvirus disease

Introduction

This deadly disease of the common carp was first recorded in the Middle East, Europe and North America in the late 1990s. Infected common carp suffered gill and skin necrosis leading to sudden and uncontrolled massive mortalities. Koi herpesvirus disease (KHVD), which was named after the initial name of the virus (koi herpesvirus – KHV), is now recognised as one of the most severe threats to common carp aquaculture. Since its detection, the virus and the disease it causes have been recorded in common carp production units from five continents. The only virus-free continent is Australia, where common carp is considered an invasive species.

Disease description

Koi herpesvirus disease is caused by the alloherpesvirus *Cyvirus cyprinidallo3* – Cyprinid herpesvirus 3 (CyHV-3). CyHV-3 causes systemic infection resulting in necrosis of gills and kidney, leading to the loss of ions. Infected fish are lethargic, gasp for air and show changes of skin epithelium, such as proliferation and skin flaking (Fig. 6.4A). There is also a reduction of mucous layer leading to the so-called 'sandpaper skin', which can be rough to the touch. Pathological changes also develop in the gills, with initial changes of colouration and infected fish also display enophthalmia. The progression of external clinical signs can be drastic, with the sloughing of the skin epithelium and massive necrosis of skin (Fig. 6.4B) and gills (Fig. 6.5).

Histologically, skin demonstrates deterioration of epithelium, a lack of mucus and empty goblet cells (Fig. 6.6). In gills the virus causes occlusion of intralamellar spaces by hypertrophic cells and the clinical signs progress to necrosis (Fig. 6.7). The intestinal lamina propria is often also necrotic or inflamed. Internal organs are also affected with focal necrosis occurring in the kidney, spleen and liver.

Fish of all ages are susceptible to the disease, with extremely high mortalities (80–100%) affecting all age groups. Death usually occurs 7–14 days post-infection and mortality is thought to occur from multi-organ failure. For example, dysfunction in the gills and kidneys results in a severe osmotic imbalance and alteration of sodium levels. These levels are lethal for fish when they persist for longer time. The survivors are most likely lifelong carriers of the latent infection, which is common for all herpesviruses. Infection in its lytic (symptomatic, deadly) form can be reactivated by, for example, fish manipulation, transport or any other type of stress. Latent carriers can thus pose a severe threat to naïve populations of common carp as the reactivated virus can be shed, causing infection of other fish. Several sites harbour the latent phase of infection, with brain and B leucocytes being the most studied. The existence of the latent phase of the infection, primarily in the internal organs of fish, has implications for diagnostics, as sampling of exterior organs is not always successful for the detection of persistent infection. Therefore, asymptomatic virus carriers can be difficult to detect *intra vitam*.

The presence of CyHV-3 infection is very often associated with the presence of secondary bacterial and parasitic infections. This is related to the disruption of mucosal layers in the skin, gills and intestine, which very often leads to bacterial co-infection, complicating

Figure 6.4 Pathological changes induced by infection with *Cyvirus cyprinidallo3*. (A) Note presence of white flaking skin on the surface of the fish. (B) Erosion of skin and fins caused by CyHV-3(V).

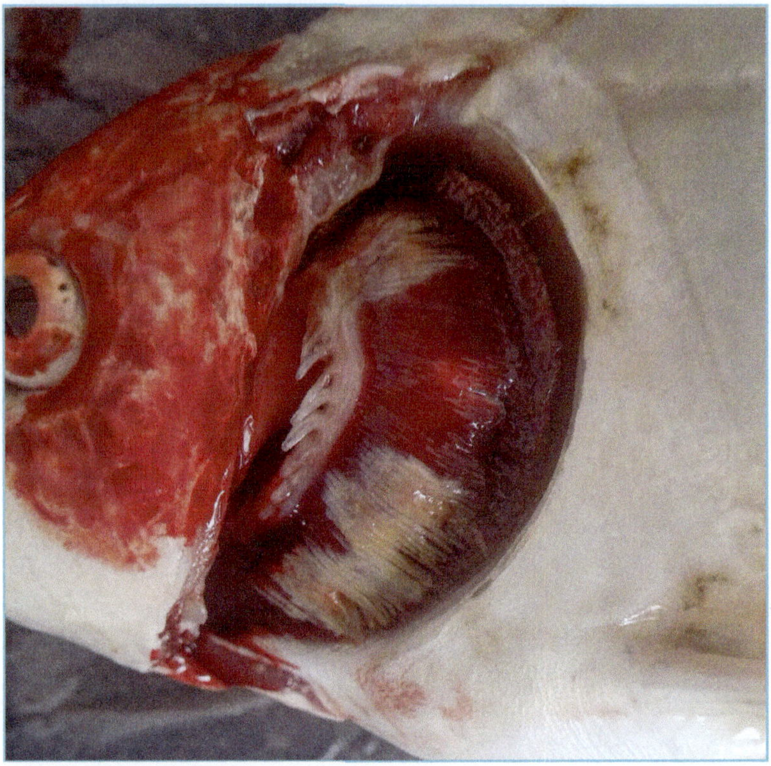

Figure 6.5 Koi carp with operculum removed, showing pathological changes induced by infection with *Cyvirus cyprinidallo3* in gills leading to severe necrosis.

the diagnosis and possible treatment options. Secondary superinfections with *Aeromonas* spp. or *Flavobacteria* spp. often increase the speed and severity of mortality. Several ectoparasites, such as *Ichthyobodo necator*, *Ichthyophthirius multifiliis*, *Dactylogyrus* spp. and *Gyrodactylus* spp., can also compound the general health of the fish.

Pathogen

Cyprinid herpesvirus 3 (CyHV-3), an alloherpesvirus, is a member of the genus *Cyprinivirus* in the family *Alloherpesviridae*. The virus comprises an enveloped 80–120 nm virion encapsulating the

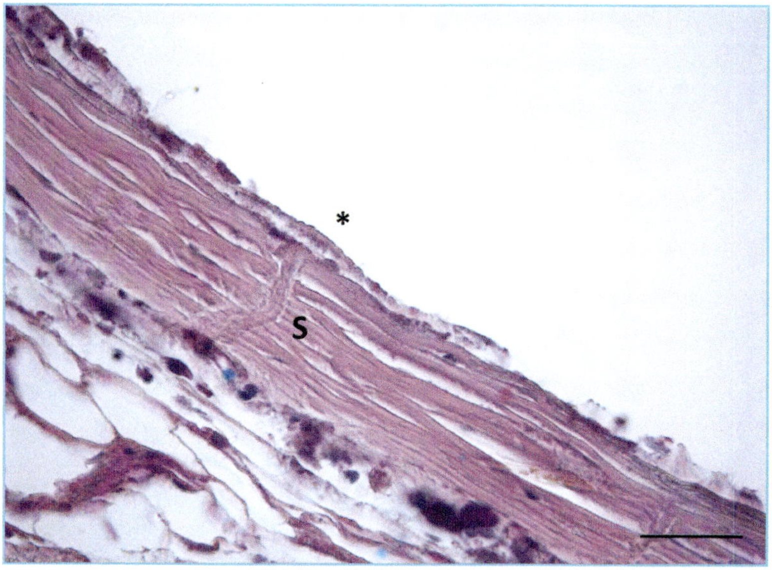

Figure 6.6 Histopathological changes in the skin induced by infection with *Cyvirus cyprinidallo3*. Note the absence of epithelium (*) which should be adjacent to the stratum compactum (S). Indicative scale bar 30 μm.

dsDNA genome (Fig. 6.8). The genome has 295 kb with 155/156 open reading frames (ORFs) encoding for possible proteins.

Detection

Initially, detection was made by the isolation of the virus in fish cell lines, common carp brain (CCB) and KF-1 cells, and electron microscopy. Since then, multiple cell lines have been developed and can be used for the detection of acute infections. Cell lines, however, have lower detection thresholds than other molecular methods, and virus-induced vacuolisation and focal lysis can be observed by days 3–7 post-infection. PCR-based methods are available for detecting the virus DNA and viral RNA in the lytic and latent phases of the infection. Multiple PCR protocols are available with classical (end-point) PCRs, semi-nested PCRs, nested PCRs and quantitative PCRs being continually developed and improved. Quantitative double-labelled probe-based PCR is treated as a gold standard.

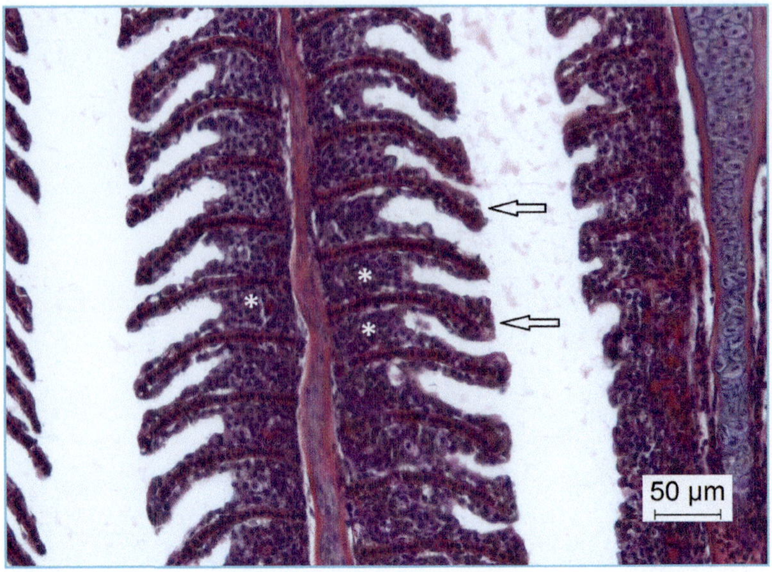

Figure 6.7 Histopathological changes in the gill showing occlusion of intralamellar spaces (*) and thickening of respiratory epithelium in the secondary lamella (arrow). Indicative scale bar 50 μm.

Electron microscopy can be employed to visualise the viral particles and several monoclonal antibodies have been developed for detection of the virus in immunohistochemistry. ELISA, loop-mediated isothermal amplification (LAMP) assay, lateral flow devices based on PCR, recombinase polymerase amplification (RPA) or antibodies against the virus can be used for rapid detection of the virus. Also gold nanoparticle-based hybridization assay provides quick detection even at the pond side.

Prevention and treatment

CyHV-3 is highly prevalent, can lead rapidly to mortality and can be spread by asymptomatic, latently infected carriers. Very robust testing and strict biosecurity measures have to be employed, therefore, to protect virus-free populations.

Multiple vaccines based for example on inactivated virus, DNA vectors and live attenuated virus have been developed and are

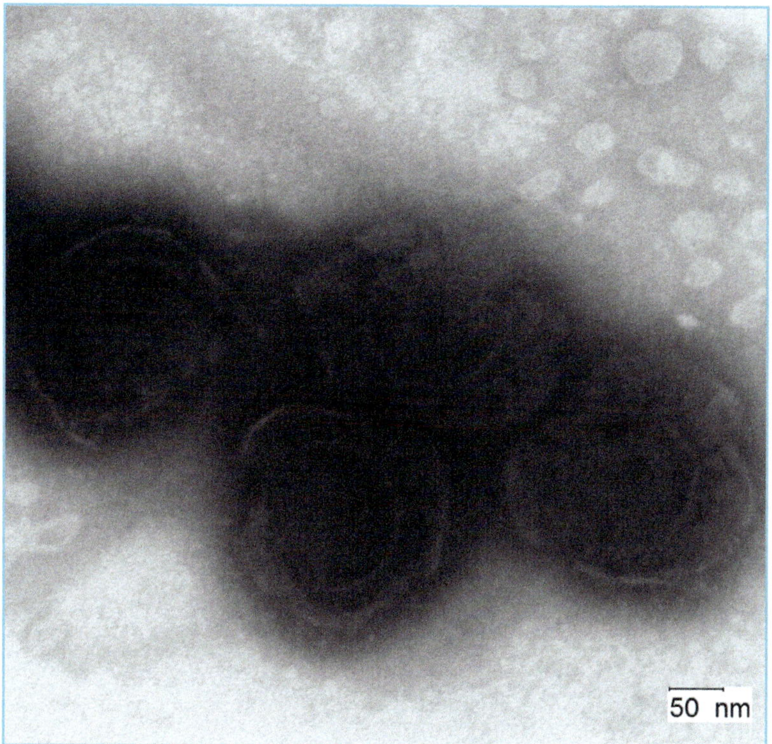

Figure 6.8 Transmission electron micrograph of negatively stained purified *Cyvirus cyprinidallo3*. Scale bar 50 nm.

available in selected markets. Especially, genetically modified live attenuated virus-based vaccines are proving very promising, providing high protection rates and the possibility of self-spreading in the population, but legal limitations will curtail their use in some parts of the world.

Disease-resistant common carp strains such as Amur wild carp, the Ropsha strain and newly developed strains are available and could be introduced into areas that are contaminated with the virus and cannot be returned to virus-free status. Unfortunately, these disease-resistant strains do not always have the best phenotype and growth rate, and therefore may not be full replacements for local breeds. There is, however, noticeable progress in the development of marker-assisted selection (MAS) for resistance to KHVD, which may lead to the development of better strains to limit the losses in common carp aquaculture.

If the infection is caught early enough, before severe clinical signs develop, increasing the water temperature to 30°C could limit the virus replication, but the fish will become latently infected. These survivors of natural infection could be resistant to reinfection for some time. They are most likely lifelong carriers of latent (hidden) infection when the virus genetic material is inside the cell nucleus and therefore, the virus is not apparent to the immune system of the fish.

Several drugs, including acyclovir, used for mammalian herpesviruses have been tested and confirmed to limit virus replication *in vitro*. Also, new-generation treatments such as short interfering RNA show promise but have been tested only *in vitro*. The removal or modification of cholesterol was also proven to be effective in blocking CyHV-3 entry. However, the use of any drugs or new-generation treatments *in vivo* might be difficult for implementation in aquaculture, due to high costs and the specialist labour required.

As indicated above, infection with CyHV-3 can often be associated with secondary superinfections of bacteria, such as *Aeromonas* spp. and *Flavobacteria* spp., and/or ectoparasites like *Ichthyobodo necator*, *Ichthyophthirius multifiliis*, *Dactylogyrus* spp. and *Gyrodactylus* spp. The treatment of these secondary infections can, in some cases, increase the survival chances of the fish by lowering the general pathology related to the secondary pathogen burden.

FACTS SHEET

Disease: Koi herpesvirus disease (KHVD)
Pathogen: *Cyvirus cyprinidallo3* – Cyprinid herpesvirus 3 (CyHV-3)/Koi herpesvirus (KHV)
Susceptible hosts: Common carp
Clinical signs: Lethargy, gasping for air, skin epithelia sloughing, necrosis of gill, kidney and liver
Mortality: Up to 100%
Temperature: 16–28°C
Prevention and treatment: Biosecurity and testing for the presence of the virus before purchase, vaccination, utilisation of resistant strains of fish. Treatments have not been widely tested, although increase of the temperature to 30°C could limit virus replication

6.2.3 Carp oedema virus – Koi sleepy disease (KSD)

Introduction

This disease, which is characterised by a characteristic sleepy behaviour of the common carp, was described in the 1970s in Japan, where it was named koi sleepy disease because of its association with the koi trade. The disease later emerged in North America and Europe, and thanks to the development of diagnostic methods the disease can be traced to infection cases where higher mortalities were noted. Koi sleepy disease is fast becoming the second most dangerous viral disease of common carp. Carp oedema virus (CEV), the causative organism of koi sleepy disease, is considered to be a poxvirus and a member of *Chordopoxvirinae* subfamily in the family *Poxviridae*. Together with the other member of the *Chordopoxvirinae* subfamily, salmon gill poxvirus (SGPV), it causes a branchial disease with the infection predominantly limited to gills, which leads to the rapid onset of mortality.

Disease description

Common carp affected by KSD are initially lethargic and, with the progression of the disease, fish lie at the bottom of the tank, stop responding to external stimuli and eventually enter into a coma-like state (Fig. 6.9). The onset of mortality follows. The virus infection can cause massive mortality within a week from exposure, but it is dependent on the water temperature. The disease, caused by the CEV from the genogroup I in temperate climate, is common in spring and autumn, and is generally more prominent at lower temperatures (< 18°C) than KHVD. In koi, however, the disease caused by CEV genogroup II is less temperature restricted and has been recorded in a wide range of temperatures up to 25°C.

Gills are the primary target for the virus; therefore, most of the clinical signs and pathological changes are related to this organ. Infected gills appear pale and swollen with increased level of mucoid excretion (Fig. 6.10). Histologically, proliferation of intralamellar cellular mass with progressing inclusion of intralamellar spaces by

Figure 6.9 Behavioural changes in common carp infected with carp oedema virus. Note two individuals in the 'sleepy' position on their lateral side.

Figure 6.10 Common carp with operculum removed, showing pale and swollen gills with excessive amount of mucoid excretion caused by CEV infection.

hyperplasticity of epithelial cells occurs. The gill epithelium also shows oedema and lifting of the cells, some of which have foamy contents. Infiltration of granulocytes and presence of apoptotic cells also occur in the gills of CEV-infected fish (Fig. 6.11). Other organs may also be affected, primarily due to the loss of hydro-mineral balance or ammonia intoxication.

As carp are capable of tolerating low oxygen levels, the fish can partially compensate for KSD infection by limiting its activity and growth, and shifting the metabolism to more oxygen-independent pathways. Fish are less capable of coping with accumulating ammonia, which cannot be excreted by damaged gills. Ammonia, which is toxic for a large number of cells and particularly impairs the function of nervous system, may be the cause of the neurological clinical signs like sleepiness, the inability to hold balance and the onset of a coma-like state. The accumulation of ammonia can also cause brain swelling, which occurs in some diseased fish.

The gills are uniquely involved in osmoregulation by actively taking up sodium ions from the water. This process is most likely impaired during CEV infection, therefore severe hyponatraemia with levels as low as 70 mmol/l can occur. These sodium levels are highly lethal if the sodium decline is rapid, as in the case of KSD when sodium levels start to drop after 3 days post-infection. Low sodium levels are also linked with brain swelling and can cause neurological clinical signs. The link to mortality has been shown in the successful treatment of fish with very advanced clinical signs by bathing in salt (0.5% NaCl) solution.

Although fish infected by the poxviruses can survive, most are likely to be lifelong carriers, and the virus is capable of persisting in fish for at least several (4–5) months after salt treatment. Fish should always be tested before purchase.

Infection with CEV leads to immunosuppression within the gill tissue as well as the whole body and thus causes damage to the gill barrier, leading to changes in bacterial flora and the occurrence of secondary infections. Pathology associated with KSD can be exacerbated by the occurrence of secondary infections such as *Flavobacterium branchiophilum* and *Aeromonas* spp., which can often complicate diagnosis and makes the possible treatment options

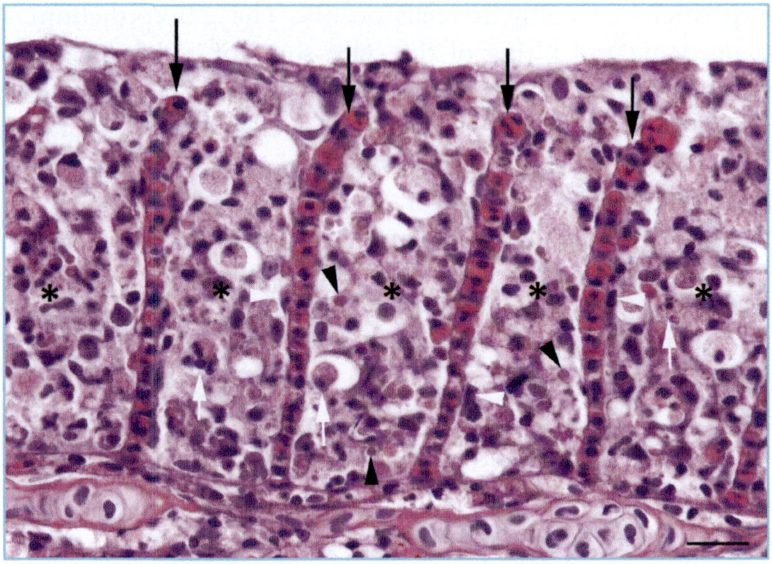

Figure 6.11 Histopathological changes in common carp gills infected with CEV. Note the presence of clubbing and fusion of gill secondary lamellae (black arrows) with complete occlusion of the interlamellar spaces (*) due to accumulation of cellular debris (white arrows); hypertrophy of epithelial cells (white arrowheads); cytoplasmatic eosinophilic inclusions (black arrowheads). Scale bar 40 µm. Source: Adamek *et al.* (2017) *Veterinary Research* 48, 12 (doi 10.1186/s13567-017-0416-7).

more complex. Several ectoparasites, such as *Ichthyobodo necator*, *Ichthyophthirius multifiliis*, *Dactylogyrus* spp. and *Gyrodactylus* spp., can also be associated with KSD. The treatment of the secondary infections can be beneficial, as it limits the secondary pathogen burden.

Pathogen

Carp oedema virus is not fully classified, but it is most likely a poxvirus. CEV is therefore possibly a member of *Chordopoxvirinae* subfamily in the family *Poxviridae*. The virion (Fig. 6.12), measuring 260–413 nm, possesses an enveloped, marble-shaped, ovoid pleomorphic particle encapsulating the dsDNA genome. At least two genetically divergent genogroups are known: genogroup I

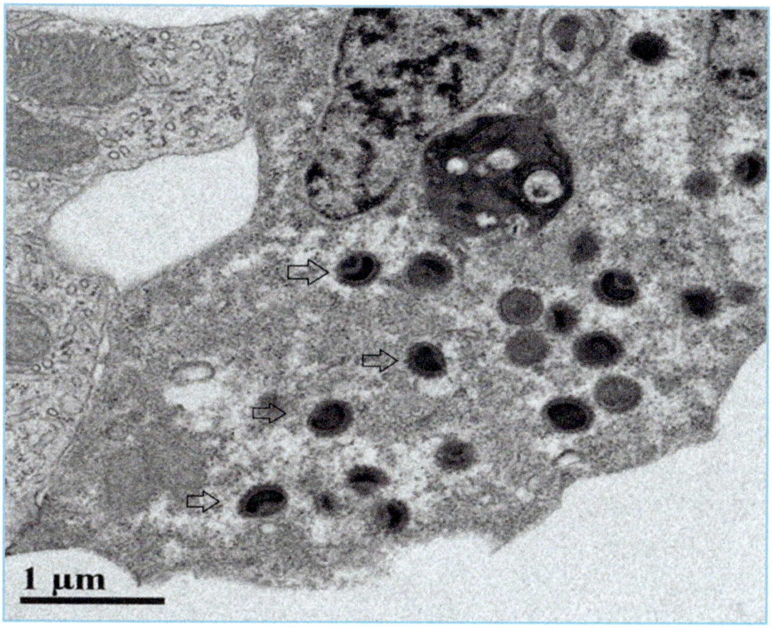

Figure 6.12 Transmission electron micrograph of CEV inside the gill epithelium of koi carp. Note large intracytoplasmic spheroid CEV virions (arrows). Scale bar 1 µm. Source: Jung-Schroers *et al.* (2015) *BMC Veterinary Research* 11, 114 (doi:10.1186/s12917-015-0424-7).

(European genotype), which is present in common carp in Europe and North America, and genogroup II (Asian genotype), present in carp in Asia and in koi all over the world. Currently the data for the genome of genogroup II are available and the size is estimated to be 456,821 bp encoding for 392 potential proteins.

Detection

The biggest limitation for the initial detection of the virus is the fact that the virus cannot be replicated in any of the most commonly used cell cultures, for example CCB, FHM, EPC, KF-1. Although there have been several attempts to develop cell lines that support replication of this virus, none have been successful. Diagnosis is based on PCR methods for detection of virus DNA. Several protocols have been developed, although attention should be paid to which genogroup requires detection. Quantitative double-labelled probe-based PCR (qPCR) and droplet digital PCR (ddPCR) allow the detection of all known genogroups. Loop-mediated isothermal amplification (LAMP) assay can also be used for detection of viral DNA and additional methods are being developed to enable detection of the virus at the pond side.

Prevention and treatment

Carp oedema virus is very widespread in common carp culture systems and koi populations, making eradication of the virus challenging. Rapid and robust testing can be applied for limiting the spread of the virus to potentially naïve populations. Due to the fact that poxviruses replicate in the cytoplasm of the cells, they are difficult to diagnose in latency, as in the alloherpesviruses (CyHV-1, CyHV-2 and CyHV-3). Still, there are indications of a persistent presence of the virus in fish populations with subclinical (asymptomatic) virus carriers, which might be difficult to diagnose with some methods.

There are indications that the convalescent fish are resistant to reinfection, giving a basis for the development of a vaccine, but a vaccine is not currently available. Work on vaccine development is impaired by the lack of susceptible cell lines and insufficient research funding. With new biotechnological developments, it is possible that vaccines based on DNA or mRNA will become available in some markets. However, with the large genome, the virus can have many potential immunogenic proteins, which form a target for vaccine development. If virus-susceptible cells are established in the future, then development of simple, inactivated vaccines will be possible. In addition, susceptible cells could be used for development of recombinant live attenuated vaccines.

Common carp strains such as Amur wild carp (*Cyprinus rubrofuscus*) or the Ropsha hybrids derived from it appear to be much more resistant to the infection and do not develop clinical signs of KSD. Utilisation of these fish strains can potentially limit the losses, but due to their less attractive phenotype and speed of growth rates they may not replace local breeds. Very little development has been undertaken on marker-assisted selection (MAS) for resistance to KSD, but the efforts for developing koi herpesvirus disease-resistant strains could be helpful for combating the impact of KSD.

Although antiviral drugs are currently not available for fish poxviruses, other treatments have been utilised. Salt (0.5% NaCl) baths are very effective in treating the clinical signs of the disease, but do not stop virus infection. Salt baths balance the loss of ions (e.g. sodium) and allow ammonia secretion, thereby removing the causes of mortality. Additionally, increasing the water temperature to ≥ 29°C can limit virus replication. This is the basis for so-called 'natural immunisation' used by some koi traders, which is based on exposure of the fish to the virus by cohabitation with infected fish and then transferring the exposed (freshly infected) fish to warm water (≥ 29°C) containing 0.5% NaCl. This procedure must be used with caution and is strongly discouraged if fish are going to be sold to other farms or private customers keeping koi. The broad use of this procedure by koi breeders and dealers has most likely led to spreading of the virus by persistent virus carriers, which will be created by this procedure and which has resulted in a high prevalence of the virus around the world.

In cases of severe co-infections, treatment of the secondary pathogen (parasites or bacteria) can be beneficial as it limits the secondary pathogen burden.

FACTS SHEET

Disease: Koi sleepy disease (KSD)
Pathogen: Carp oedema virus (CEV)
Susceptible hosts: Common carp
Clinical signs: Sleepiness, fish lying at the bottom of the tanks, gill swelling, brain swelling
Mortality: Diverse mortality up to 80–100%
Temperature: 8–25°C
Prevention and treatment: Biosecurity and testing for the presence of the virus before purchase, utilisation of resistant strains of fish. Treatments are not widely tested. An increase of the temperature to 29°C could limit virus replication and the addition of salt (NaCl) up to 0.5% will balance the loss of ions and allows ammonia secretion

6.2.4 Spring viraemia of carp virus – Spring viraemia of carp

Introduction

This disease, which is characterised by pale gills, signs of ascites and haemorrhages of the skin, was first described in the 1960s, when mortalities of common carp occurred in the first year of production in spring after the first wintering. Due to this seasonality, the disease was named spring viraemia of carp. A rhabdovirus called spring viraemia of carp virus (SVCV) was identified as a causative agent in the early 1970s in Europe but since then the pathogen has spread all over the world. The disease was a serious problem in common carp culture in the 20th century, but in continental Europe, where it initially emerged, the disease is now seen rarely and does not cause significant problems.

Disease description

As with many other rhabdoviruses in fish, SVCV has a rather broad host range, being able to infect multiple cyprinids and non-cyprinid species. The initial detection of the virus was in common carp, but grass carp (*Ctenopharyngodon idella*), Prussian carp, crucian carp, silver carp (*Hypophthalmichthys molitrix*), bighead carp (*Hypophthalmichthys nobilis*) and several other cyprinids have also been confirmed to be susceptible to the disease. SVCV induces a systemic infection, with the virus replicating in many vital organs of the fish. It has been detected in skin, gill, trunk and head kidney, heart, liver, spleen and intestine of infected crucian and common carp. It seems that the virus is endothelialotrophic, especially towards the endothelium of blood vessels, where it replicates, causing vasculitis (inflammation of the vessel wall). This leads to rupturing and haemorrhaging of the vessels. In addition to external and internal bleedings (Figs. 6.13 and 6.14), the virus causes necrotic changes in hepatocytes in the liver and tubular cells in the kidney. The pathology is also induced in other tissues, such as the heart, where all forms of inflammatory changes have been recorded with pericarditis, endocarditis and myocarditis progression, which can lead to focal necrosis of the heart muscle. The intestine and

Figure 6.13 Pathological changes induced by infection with SVCV. Note signs of ascites and haemorrhages of skin.

swim bladder are also affected, with haemorrhagic changes leading to changes in the gut and ulceration of the swim bladder.

The clinical signs of infection occur in low temperatures and progression of the disease tends to be time dependent. The first clinical signs occur in the second week post-infection and mortalities appear in the third and fourth weeks post-infection. The virus can persist in the fish for a long time (2–3 months), producing virus carriers, which could shed the virus to naïve fish. The virus is shed via mucus, urine and faeces.

Due to the virus characteristics and low water temperature when the particle is shed, SVCV is known to be relatively stable in water and pond ground, and can be infective for 1–1.5 months.

The temperature and seasonality dependence are well established for this disease. The outbreaks occur in spring and autumn in temperatures of 5–17°C. In the 1960s to 1980s in Europe, mortality of up to 70% was recorded in the field, but since then the virus seems to be less virulent and causes sporadic low-rate mortalities. Differences in

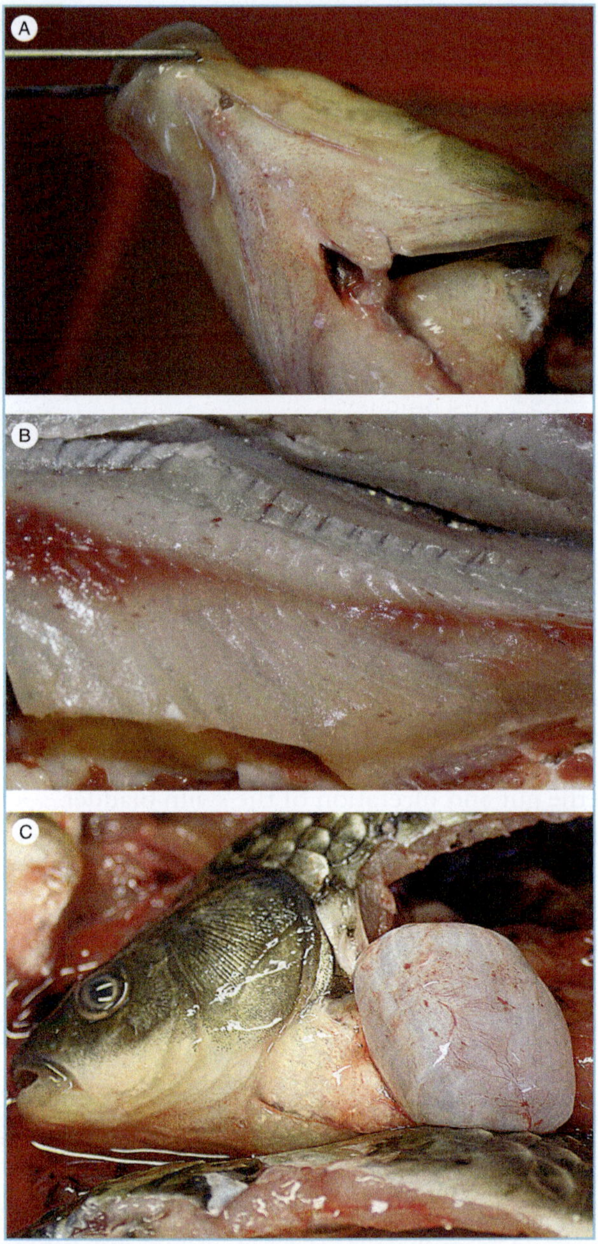

Figure 6.14 Organs possessing haemorrhages due to infection with SVCV. (A) Skin on lower jaw of infected carp. (B) Muscle with isolated petechiae. (C) Signs of haemorrhages on the surface of the swim bladder. Image (B) courtesy of the Environment Agency, National Fisheries Laboratory, UK.

virulence between isolates are present and experimental infections with most of these isolates generally point to a lower mortality of about 20%.

Pathogen

Spring viraemia of carp virus is a rhabdovirus. SVCV is a member of the family *Rhabdoviridae*, genus *Sprivivirus*, species *Carp sprivivirus*. It has 11 kb negative-sense, single-stranded RNA genome encoding five structural proteins. The bullet-shaped morphology of SVCV resembles a classical shape of vertebrate rhabdoviruses. The virus measures 80–180 nm in length and 60–90 nm in diameter.

Detection

SVCV replicates very well in multiple cell cultures. EPC and FHM were initially used for virus detection and replication, although now dozens of cell lines have been shown to be susceptible for virus replication. The virus rapidly lyses the cells, leading to the total destruction of the cell monolayer within 3–7 days. Confirmation of the virus from tissue can be undertaken using immunocytochemistry virus-specific antibodies (immunofluorescent antibody technique – IFAT), reverse transcriptase (quantitative) RT-(q)PCR, and enzyme-linked immunosorbent assay (ELISA). For rapid laboratory-based detections, RT-(q) PCR can be performed from the samples collected from the fish. Reverse transcription loop-mediated isothermal amplification (RT-LAMP) has also been developed and could be used in the field. The isothermal temperature requirements, high specificity and sensitivity, and short incubation time of the RT-LAMP assay make it an excellent choice as a field diagnostic test for SVCV.

Prevention and treatment

Although a new generation of treatments have been considered, such as short interfering RNA, these have only been tested experimentally *in vitro* and *in vivo* and will prove difficult to implement in aquaculture situations, due to the cost and technical difficulties as they have to be injected.

Despite the fact that rhabdoviruses generally generate strong immune responses, which lead to high titres of virus-neutralising antibodies that protect against re-infection, the response in low temperatures in common carp is relatively slow and the antibodies develop over a long period of time. Vaccination against SVCV has been shown to be quite effective using DNA vaccines, but currently a commercial vaccine is not available. Most likely, this is because the virus is no longer considered a priority and application of effective vaccination protocol requires intramuscular injection of the DNA-based vaccine.

Ropsha, a strain of common carp, seems to be much more resistant to infection with SVCV and does not develop clinical signs of the disease. This shows that breeding work for resistance is possible and Ropsha-based crosses can be implemented in farms under SVCV pressure. The relative lack of massive outbreaks of the disease in Europe could be related to adaptation of fish to the virus.

FACTS SHEET

Disease: Spring viraemia of carp
Pathogen: Spring viraemia of carp virus (SVCV)
Susceptible hosts: Common carp, grass carp (*Ctenopharyngodon idella*), *Carassius* spp., as well as several other cyprinids and non-cyprinids
Clinical signs: External and internal bleedings, necrosis of hepatocytes in liver and tubular cells in the kidney
Mortality: Up to 20%
Temperature: 8–17°C
Prevention and treatment: Biosecurity and testing for the presence of the virus before purchase. Vaccination is possible but currently not commercially available. Utilisation of resistant strains of fish. Clinical signs subside with increase of water temperature; therefore, this could be used for limiting the virus replication

6.2.5 Cyprinid herpesvirus 2 – Herpesviral haematopoietic necrosis

Introduction

Herpesviral haematopoietic necrosis (HVHN) is the main disease of *Carassius* spp., including crucian carp and Prussian carp. The *Cyvirus cyprinidallo2* – cyprinid herpesvirus 2 (CyHV-2), which emerged in ornamental goldfish (*Carassius auratus*) in the first half of the 1990s, has an impact very similar to CyHV-3, where mortalities reaching 100% occur in infected fish populations. However, in contrast to the CyHV-3, HVHN causes mortality without sometimes inducing external clinical signs, which are mainly necrosis of the haematopoietic tissues of the head and trunk kidney and spleen. The susceptibility of ornamental goldfish has helped to spread the virus globally to all continents where *Carassius* spp. are present. Due to the high mortality caused in cultured *Carassius* spp., the disease is recognised as one of the biggest threats to cyprinid aquaculture, especially in China, which is the biggest producer of crucian carp and Prussian carp. HVHN is currently recognised as a disease with huge economic and social impacts on the food-fish aquaculture industry.

Disease description

During the first outbreaks of herpesviral haematopoietic necrosis (HVHN), fish externally show only minor clinical signs of apathy and staying at the bottom of the tank. Later, cases with external disease presentation such as skin and gill lesions and necrosis occur. During necropsy, ascites, swollen and pale kidney and increased size of spleen are noted. The occurrence of outbreaks is temperature dependent, with a water temperature range of 15–25°C being considered permissive for the disease.

Histopathology consists primarily of necrosis of the haematopoietic tissues of the anterior and posterior kidney and the splenic pulp. Other lesions include focal necrosis and inflammation of the lamina propria and submucosa of the intestine, and occasionally pancreatic necrosis. In certain cases, gills show pathogenic changes with occlusion, clubbing and fusion of secondary gill lamellae, sometimes leading to gill necrosis with focal patches of necrotic tissue in gill lamellae. The virus can induce high mortality in juvenile fish, but all age classes of *Carassius* spp. are susceptible (Fig. 6.15).

Characteristic for all herpesviruses is the ability to induce latency. CyHV-2 is also capable of inducing latent infections, which aids virus spread. In experimental situations, latency of the virus can be

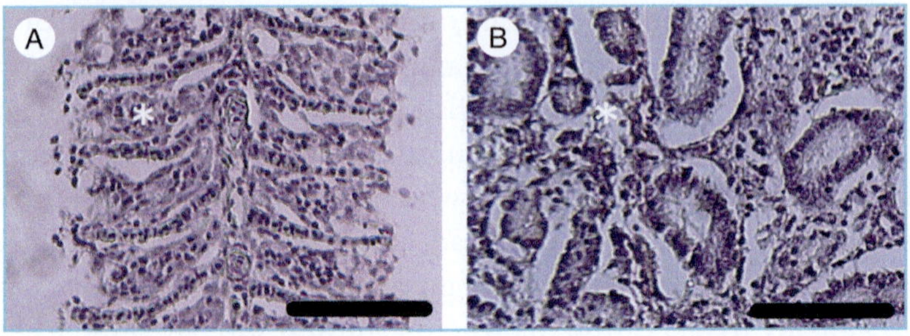

Figure 6.15 Pathological changes (*) induced by *Cyvirus cyprinidallo2* in gills. (A) Occlusion of secondary lamella. (B) Kidney with necrosis of nephric tubules. Indicative scale bar 100 µm.

reactivated to a lytic state by temperature stress, but other stressors such as transport, reproduction and lower water quality may also putatively trigger outbreaks.

Co-infections with facultative pathogenic bacteria such as aeromonads are common. The existence of additional pathogens could modify the clinical presentation of the disease, for example the presence of ascites and increased pathology in gills.

Pathogen

Cyvirus cyprinidallo2, a member of the *Cyprinivirus* genus in the family *Alloherpesviridae*, comprises an enveloped 170–220 nm virion encapsulating the dsDNA genome. The genome has 290 kb with 150 ORFs encoding for possible proteins.

Detection

Due to the importance of the disease, a broad arsenal of diagnostic methods has been developed over the years. PCR-based methods are available for detecting the virus DNA and viral RNA in the lytic phase of the infection. Multiple PCR protocols are available, including classical (end-point) PCRs, semi-nested PCRs, nested PCRs and quantitative PCRs. These PCR-based methods are being continually developed and improved to include digital droplet PCR. Quantitative double-labelled probe-based PCR is treated as a gold standard.

Initially cell culturing of CyHV-2 was more difficult than other herpesviruses of fish. However, over the years multiple cells lines have been developed, for example goldfish fin (GFF), gibel carp brain (GiCB), gibel caudal fin (GiCF), fantail goldfish fin (FtGF), fantail goldfish gill (FtGG), fantail goldfish liver (FtGL) and fantail goldfish brain (FtGB), and have proven to be susceptible to CyHV-2 infection. Furthermore, with increasing accessibility these cell lines can be used for the detection of acute infections. Cell lines, however, tend to have a lower detection threshold than other molecular methods. The virus causes vacuolisation and focal lysis of the cells up to total destruction after days 3–7 post-infection.

Electron microscopy can be employed to visualise the viral particles. Several monoclonal antibodies have been developed for the detection of the virus in immunohistochemistry. ELISA, loop-mediated isothermal amplification (LAMP) assay, lateral flow devices based on PCR, recombinase polymerase amplification (RPA) may be used. Gold nanoparticle-based hybridization assay provides a quick detection even at the pond side.

Prevention and treatment

Due to the widespread nature of the virus, its ability to lead rapidly to mortality and the presence of asymptomatic (latently infected) carriers, widespread testing and strong biosecurity should be applied for introductions, especially into virus-free areas.

Similar to other serious herpesviruses of cyprinids, vaccines against CyHV-2 based on inactivated (killed) virus, DNA vectors, recombinant bacteria vectors, live attenuated virus and other types are being developed and are already available in selected markets. Most of the vaccines provide significant increase in survival and could be used to reduce the impact of the pathogen.

Experimental infections have also shown different susceptibility levels in different varieties of goldfish, and there is the possibility that genetic resistance can occur in some carp strains. These strains, when identified, could be used for selection and crossing to obtain higher survival of fish in areas where the virus is endemic.

Mortalities during an outbreak can be limited by moving the fish from a water temperature that supports the infection to either a higher (e.g. 32–35°C) or lower (e.g. 13–15°C) non-permissive temperature. Lower temperatures will most likely slow down the speed at which the virus is controlled and cleared by the fish, and lead to the very sluggish development of long-lasting immunity. Therefore, temperature shift should be used with care as it could lead to the development of latently infected individuals.

In cases of severe co-infections, treatments that reduce the secondary pathogen burden can be useful, but in cases of very virulent CyHV-2 they might not improve survival rates.

FACTS SHEET

Disease: Herpesviral haematopoietic necrosis (HVHN)
Pathogen: *Cyvirus cyprinidallo2* – Cyprinid herpesvirus 2 (CyHV-2)
Susceptible hosts: Crucian carp, Prussian carp
Clinical signs: Lethargy, occasional changes on the gills, rapid and high mortality
Mortality: Up to 100%
Temperature: 15–25°C
Prevention and treatment: Biosecurity and testing for the presence of the virus before purchase, vaccination, utilisation of resistant strains of fish. Treatment remains unknown

6.2.6 *Carassius auratus herpesvirus (CaHV) – Carassius auratus herpesvirus disease (CaHVD)*

Introduction

Carassius auratus herpesvirus (CaHV) is responsible for a disease of crucian carp that is characterised by acute gill haemorrhages and high mortality. The virus is very closely related to Cyprinid herpesvirus 2 causing herpesviral haematopoietic necrosis (HVHN) and the separation of these two viruses and diseases is challenging.

Disease description

Details of the pathological changes caused by the virus are limited, although involvement of gill pathology has been noted in crucian carp. The disease has been reproduced in experimental infections and there are suggestions that mortalities associated with HVHN may in fact be caused by infection with CaHV. In contrast to HVHN, CaHVD has been linked to a haemorrhagic disease displaying bleeding in gill filaments, skin and swim bladder, with blood congestion in the jaw and intraperitoneal cavity. Liver, spleen and kidneys show signs of swelling and white nodules. Histologically epithelial cells of the gill lamellae become detached and the coagulative necrosis leads to a narrowing of the tubular lumen.

The second main difference allowing differentiation between CaHV and CyHV-2 is existence of multiple insertions, deletions and the absence of a terminal direct repeat in the CaHV genome when compared with CyHV-2. The lack of terminal direct repeat means that the genome is smaller (275 kb), but still contains the same number of 150 predicted ORFs. The genetic differences could reflect divergence in virulence and host adaptation of CaHV and CyHV-2. However, confirmation of this requires further studies.

Pathogen

Carassius auratus herpesvirus (CaHV) could be considered as a variant of cyprinid herpesvirus 2 and therefore it is a member of genus *Cyprinivirus* in the family *Alloherpesviridae*. CaHV is an enveloped 170–200 nm virion encapsulating the dsDNA genome. The genome has 275 kb with 150 ORFs encoding for possible proteins.

Detection

Reliable cell cultures for CaHV are currently not available, although initial isolation has been performed using *C. auratus* embryo cell line (CGB). The virus detection can be performed by PCR-based methods which can be adopted from CyHV-2. Therefore, multiple PCR protocols are potentially available with classical (end-point) PCRs, semi-nested PCRs, nested PCRs, quantitative PCRs. However, their applicability should be confirmed *in silico* before use.

Prevention and treatment

Different gynogenetic clones of gibel carp (*Carassius gibelio*) have shown distinct resistance and increased survival levels upon experimental challenge with CaHV. This suggests that breeding of resistant fish species may be feasible. Work on vaccines developed for CyHV-2 could be extended to work for CaHV.

FACTS SHEET

Disease: *Carassius auratus* herpesvirus disease (CaHVD)
Pathogen: *Carassius auratus* herpesvirus (CaHV)
Susceptible hosts: *Carassius auratus, Carassius gibelio*
Clinical signs: Acute gill haemorrhages
Mortality: 80–100%
Temperature: 22–24°C
Prevention and treatment: Biosecurity and testing for the presence of the virus before purchase. Vaccination is possible but currently not commercially available. Utilisation of resistant strains of fish. Treatment is unknown.

6.2.7 Grass carp reovirus (GCRV) – Grass carp haemorrhagic disease (GCHD)

Introduction

Grass carp reovirus (GCRV), also known as grass carp haemorrhage virus (GCHV), is the main viral pathogen causing disease of grass carp (*Ctenopharyngodon idella*). The first report of the haemorrhagic disease in grass carp occurred in the 1950s. The viral ethology was confirmed in the 1970s, making grass carp haemorrhagic disease (GCHD) the first viral disease affecting fish in mainland China. To this day China leads research into GCRV, making this virus one of the most-studied viral pathogens in fish. The research demand is driven by substantial economic losses caused by the virus, estimated to be at least one billion RMB per year in the 2000s. In traditional Chinese aquaculture the disease is called 'red muscles', 'red intestine', or 'red fin and gill cover', which is a reflection of the different presentation of the disease in certain cases. Also 'red spot disease' (RSD) is a synonym of GCHD originating from the description of clinical signs shown by infected grass carp in Vietnam. What is interesting is that GCRV is a good example of a pathogen that has not spread globally as effectively as CyHV-2 or CyHV-3, which also infect ornamental varieties of culture fish. Outbreaks are currently limited to Asia, although the virus was most likely also introduced to North America where it is called golden shiner virus (GSV). However, the virus could be on other continents, as the development of GCHD may be limited by the water temperature, which seldom reaches the levels permissive for the disease development (> 24°C).

Disease description

Grass carp haemorrhagic disease (GCHD) can be clarified, as the name suggests, as contagious haemorrhagic disease with high mortalities in grass carp. The virulence of GCRV was primarily reflected in the destruction of the blood circulatory system leading to unprecedented levels of bleeding (Fig. 6.16). This originates due to the disruption of the endothelium of blood vessels leading to multiorgan spot haemorrhages, which may be related to a reduction in blood coagulation. This leads to lower counts of red and white blood cells. The fish becomes lethargic with increasing unresponsiveness leading to death. Significant darkening of body colour, occasional exophthalmia, extensive haemorrhages at the base of fins, gill covers and at the bottom and inside of the mouth are clearly visible in GCHD-affected fish. The gills could be either haemorrhagic or pale, depending on the stage of the disease. A significant reddening of the vent is also common. Internal organs show signs of haemorrhagic disease with bleedings and haemolysis throughout the musculature, leading to a red colour of the muscles. The intestine also becomes much redder in affected fish, which is a sign of enteritis. In the peritoneum a massive abdominal haemolysis and haemorrhage in organs such as liver, spleen, kidney and intestine occur. The spleen tends to be the most affected internal organ, showing severe necrotic lesions, and vacuolization with lymphocyte and macrophage infiltration. Degeneration and necrosis of hepatocytes, dilated spaces between hepatic sinuses, and haemorrhages from disrupted blood vessels occur in the liver. Muscles show signs of

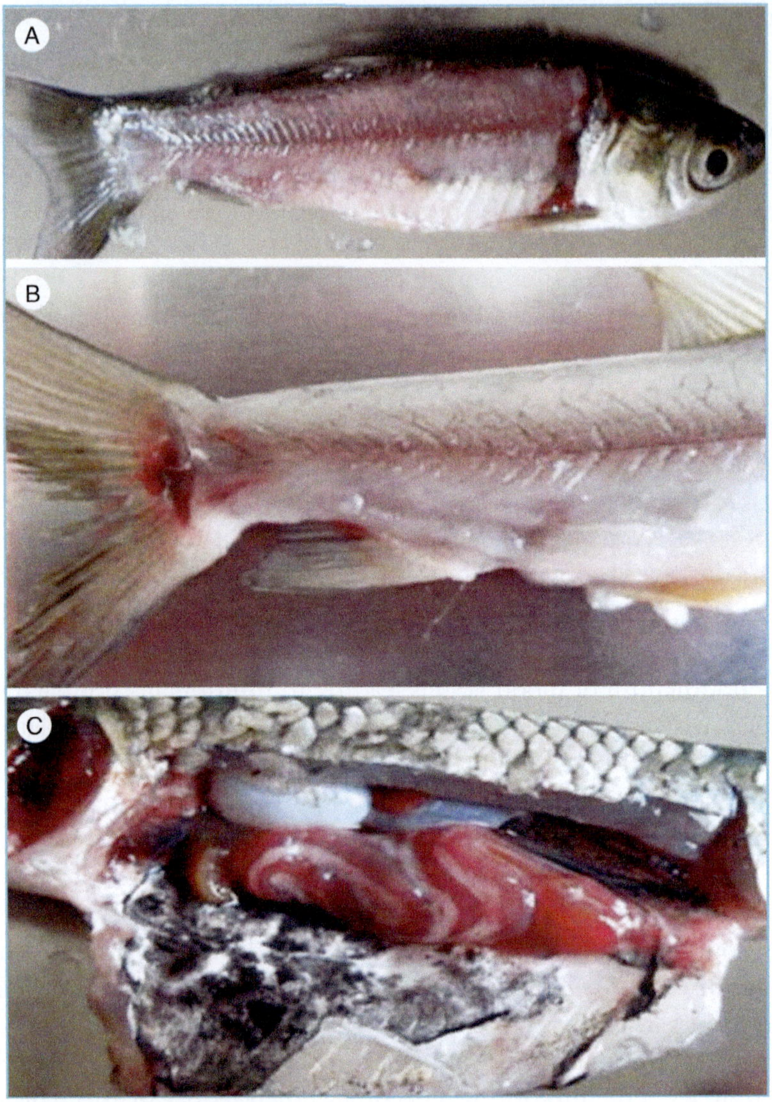

Figure 6.16 Pathological changes induced by grass carp reovirus. (A) Extreme bleeding into the muscle induced by infection with GCRV. (B) Bleeding at the base of the tail fin. (C) Haemorrhage of the intestine. Figure modified from material published under Creative Commons Attribution 4.0 International License (https://creativecommons.org/licenses/by/4.0/) by Liang, H.R., Li, Y.G., Zeng, W.W., Wang, Y.Y., Wang, Q. and Wu, S.G. (2014) Pathogenic and tissue distribution of grass carp reovirus after intraperitoneal administrations. *Virology Journal* 11, 178. DOI: 110.1186/1743–1422X-1111-1178.

cell lysis, which leads to increasing intracellular spaces. In the gills, the primary lamellae are swollen and deformed, and show signs of necrosis and blood congestion. Secondary lamellae are either occluded by intralammellar cell debris or completely destroyed by necrotic changes. In the kidney, necrosis of renal tubular epithelial cells is present. The intestinal mucosa can also be affected, with atrophied and ruptured villi with the accumulation of congested blood at their tips.

The susceptibility of fish is highly dependent on age and water temperature. Juvenile fish less than 1 year old are more susceptible, hence high mortalities (90–95%) occur in fish in the first year of production if there is a GCHD outbreak. Fish that are 3 years old tend to be more resistant and mortality rates are less severe. The virus seems to be thermophilic and viral replication is most effective in relatively high temperatures, with the optimum for viral enzymes around 28°C; disease occurs primarily at temperatures of 24–30°C. Death occurs 7–14 days post-infection.

The virus is less important in other cyprinids, such as black carp (*Mylopharyngodon piceus*) and the topmouth gudgeon (*Pseudorasbora parva*). However, the latter fish species can be a very effective vector of the virus, leading to uncontrolled spread of the disease between ponds or farms.

Pathogen

Grass carp reovirus (GCRV) belongs to the genus *Aquareovirus* in the family *Reoviridae*. It has a non-enveloped icosahedral spherical virion 60–80 nm in diameter and a double-layered capsid containing a genome composed of 11 segments of double-stranded (ds) RNA.

GCRV isolates in China can be classified into three genotypes, GCRV-873 (GCRV-I), GCRV-HZ08 (GCRV-II) and HGCRV (GCRV-III), which differ in virulence level, with GCRV-II causing severe disease and mortalities (Fig. 6.17).

Detection

Classical electron microscopical observation can be utilised for detection of the viral particles. Monoclonal antibodies to the virus

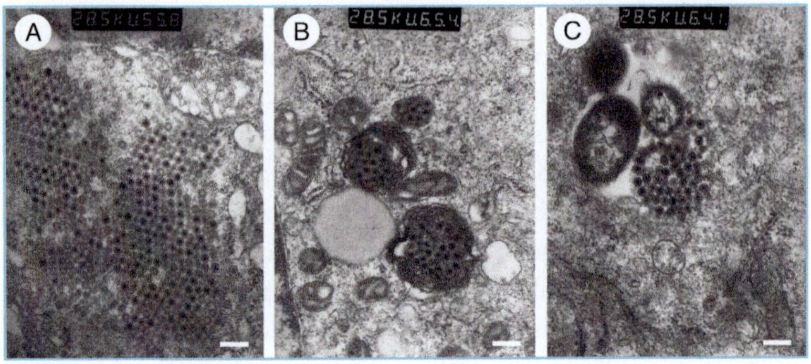

Figure 6.17 Ultra-thin sections from grass carp swim bladder (GSB) cells infected with GCRV. (A) GCRV JX-0901 (Genotype I). (B) GCRV HZ08 (Genotype II). (C) GCRV H (Genotype III). Indicative scale bars 200 nm. Courtesy of Prof Qing Wang, Pearl River Fisheries Research Institute, China.

are available allowing confirmation of the presence of the virus in cells or tissues by immunocytochemistry and immunohistochemistry. In addition, the presence of the virus can be confirmed by *in situ* hybridisation or *in situ* PCR. The availability of grass carp IgM also allows the use of ELISA for detection of the virus in grass carp.

The virus can also be isolated using *C. idella* kidney (CIK) cells and grass carp swim bladder (GSB) cells, but not all genotypes produce a clearly visible cytopathic effect in these cell lines. For example, some isolates of the most virulent genotype, GCRV-II, do not induce lysis of the cells and therefore do not produce the CPE effect.

Molecular methods tend to be the most favoured for the detection of the virus, and RT-PCR techniques and quantitative RT-PCR enable quick and reliable detection of GCRV. The quantitative methods have the additional advantage that they could indicate if GCRV is the primary pathogen causing the mortality. The even more precise droplet digital PCR that is being developed allows the absolute quantification of the viral particles in the fish tissue or the environment, even when the viral particles are present in very low numbers.

Recently, reverse transcription loop-mediated isothermal amplification (RT-LAMP) has been developed, allowing rapid and easy

GCRV diagnosis without using complicated instrumentation. A rapid detection of the virus at the side of the pond is also possible with diagnosis lateral flow devices.

Prevention and treatment

Strict biosecurity measures based on epidemiological surveillance and diagnostic testing have to be employed to protect virus-free fish populations.

Outbreaks can also be prevented or mortalities can be limited by vaccination. Inactivated virus vaccines were developed very early after the discovery of GCRV, facilitated by the fact that the virus is capable of replicating in the CIK cells, allowing production of a high number of virus particles, which after inactivation can be used for immunisation of the fish. Cell cultures also enabled the development of live attenuated virus vaccine, which was first used for aquatic animals in China with a State Medicine Manufacturing Approval Number. Nowadays the vaccine for GCRV-892 strain is widely applied in China.

The next-generation vaccines, such as recombinant subunit (viral protein-based) vaccine or DNA-vector recombinant baculovirus vaccine, are still in an experimental stage.

However, laboratory tests show that they have the potential to provide relatively high protection against the virus. The main disadvantage is that these vaccines have to be injected, as their effectiveness in bath application is rather limited. This increases the labour intensity required for administration of the vaccines. One of the solutions would be oral administration of the vaccine. The experimental vaccine based on orally applied *Bacillus subtilis* spores has already been tested, but it provided only limited protection. Nonetheless, with biotechnological progress more effective vaccines that are easy to apply may be developed in the future, providing fish with resistance to development of the disease caused by GCRV.

The other solution could be the use of disease-resistant fish. The existence of genetically driven resistance/susceptibility to GCRV has been confirmed, and SNPs were identified mainly in genes responsible for successful response to viral infection. These genetic markers could be a basis for development of strains resistant to the infection.

Another supplementation of vaccination or using disease-resistant fish could be the application of antiviral drugs. Epigallocatechin-3-gallate (EGCG) from natural sources has been tested for its ability to block the entry of the virus into cells *in vitro*. In addition, ammonium chloride, dynasore, pistop2, chlorpromazine and rottlerin were shown to block virus entry. A new generation of treatments that impede viral RNA replication is being considered. Both short interfering RNA (siRNA) and small hairpin RNA have been successfully tested *in vitro*. The application of such techniques to aquaculture may be difficult to implement due to cost and limited effectiveness, as they have to be injected.

FACTS SHEET

Disease: Grass carp haemorrhagic disease (GCHD)
Pathogen: Grass carp reovirus (GCRV)
Susceptible hosts: Grass carp (*Ctenopharyngodon idella*)
Clinical signs: Haemorrhage in muscle, skin, intestine and gill
Mortality: 95% Temperature: 24–30°C
***Prevention and treatment*:** Biosecurity and testing for the presence of the virus before purchase, vaccination, utilisation of resistant strains of fish. Treatment is through several antiviral drugs

6.2.8 Grass carp rhabdovirus (GrCRV) – Grass carp rhabdovirus disease (GrCRVD)

In addition to the grass carp reovirus, another rhabdovirus was detected in grass carp in Europe in the 1970s. The virus was isolated from a moribund 3-year-old grass carp. The diseased fish showed ventral haemorrhagic inflammation, bleedings in the scale bases, necrotic fins, inflammation and liquid in the abdomen, swollen spleen, pale liver with swollen capillary vessels, opaque internal wall of the swim bladder with petechial haemorrhages and enlarged blood vessels. The temperature range of this new disease was confirmed *in vitro* to be between 16°C and 23°C. In fish injected with this new viral isolate and maintained at 23°C, 100% mortality occurred. The virus was clearly different from GCRV, but possessed a typical rhabdovirus morphology with a bullet-shaped virion measuring 70 nm in width and 120 nm in length. This GrCRV was inactivated by lipid solvent, confirming the existence of an envelope, which is missing in GRCV. It was also serologically significantly different from other cyprinid rhabdoviruses, for example SVCV, as it was not neutralised with the sera SVCV-infected common carp. Genome sequencing showed that this new rhabdovirus has a very close genetic relationship with tench rhabdovirus (TenRV) and pike fry rhabdovirus (PFRV). These viruses should perhaps be considered to represent a separate lineage within a single species called 'Pike fry rhabdovirus'. This is substantiated by the fact that one of the PFRV isolates (PFR-80560) was virulent for grass carp, causing 18% mortality in fry at 17°C. The induced disease was characterised by abnormal behaviour, exophthalmia and haemorrhage in the fins and muscles, as well as a swollen abdomen. The virus genome is nearly fully sequenced, therefore there is an opportunity to develop RT-PCR-based methods for monitoring the prevalence of this virus. Although present in grass carp for the past 50 years, the virus has not emerged to be a serious pathogen for this fish species. This is most likely related to the fact that it is not present in China, where most of grass carp production takes place.

FACTS SHEET

Disease: Grass carp rhabdovirus disease
Pathogen: Grass carp rhabdovirus or Pike fry rhabdovirus – *Sprivivirus esox*
Susceptible hosts: Grass carp and additional non-cyprinids
Clinical signs: Bleedings in the scale bases, necrotic fins, ascites
Mortality: 18–100%
Temperature: 17–23°C
Prevention and treatment: Biosecurity and testing for the presence of the virus. Treatment is unknown

6.2.9 *Koi ranavirus (KIRV)*

Ranavirus was detected during a mortality event in ornamental common carp in South India. Infected fish showed clinical signs

Figure 6.18 Koi carp infected with KIRV showing emaciated fish gasping for air at the water surface. Courtesy of Kollanoor Riji John, Kerala University of Fisheries and Ocean Studies, Panangad, Kochi, India.

such as skin darkening, loss of scales, uncoordinated erratic swimming and lying at the bottom of the tank, followed by death. The ranavirus was isolated from the diseased fish using snakehead kidney (SNKD2a) cells and genetic analysis revealed a major capsid protein gene, which showed an identity of 99.9% with that of largemouth bass virus. The experimental infection produced some puzzling results: whilst the spleen and kidney homogenates caused 100% mortality, homogenates of the brain and gill, as well as KIRV replicated in cell cultures, did not cause any clinical signs. Follow-up studies showed that the virus was highly prevalent in several fish species and the cell culture-driven virus did not induce any mortality. Therefore, it is questionable if KIRV was the primary pathogen during the initial infection. Nonetheless, this pathogen should be considered when investigations of epizootic ulcerative syndrome (EUS)-like clinical signs are presented by the fish (Figs. 6.18 and 6.19).

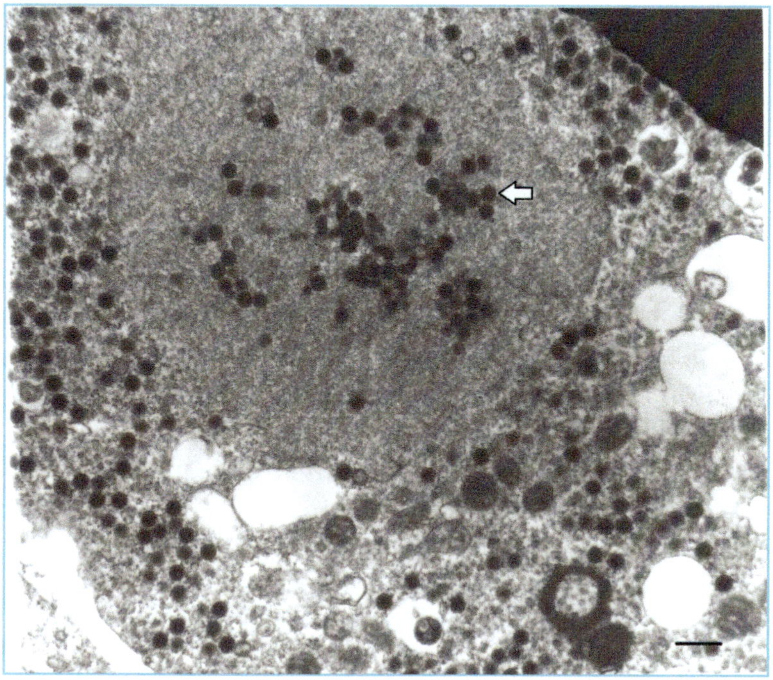

Figure 6.19 Ultra-thin sections of koi carp tissue infected with KIRV (arrow). Indicative scale bar 300 nm. Courtesy of Kollanoor Riji John, Kerala University of Fisheries and Ocean Studies, Panangad, Kochi, India.

FACTS SHEET

Disease: Unknown
Pathogen: Koi ranavirus (KIRV)
Susceptible hosts: Common carp, olive barb (*Systomus sarana*), Pengba/Belengee (*Osteobrama belangeri*)
Clinical signs: Skin darkening, loss of scales, uncoordinated erratic swimming (questionable)
Mortality: 0–100% (questionable)
Temperature: Unknown
Prevention and treatment: Unknown

6.2.10 Lymphocystis disease virus (LCDV)

Cyprinids seem to be also susceptible to lymphocystis disease (LCD). In particular, adult grass carp (*Ctenopharyngodon idella*) and rohu (*Labeo rohita*) develop the disease. Typical clinical signs that have been observed include benign, wart-like tumours on the fins and skin, clusters of shiny bluish nodules on the body surface, and raspberry-like clusters of cells, which are caused by infection with lymphocystis disease virus (LCDV). In the blood morphology, lymphocytes and erythrocytes containing single or multiple intracytoplasmic inclusion bodies have been observed in grass carp. Histopathologically, LCDV-infected grass carp show extensive hypertrophy and lamellar fusion of gills, liver necrosis, myocardial inflammation, fused intestinal villi and glomerular degeneration of the kidney. In electron microscopy, cells with intracytoplasmic inclusion bodies contain large numbers of LCDV virions, which are 200–280 nm in diameter with icosahedral morphology possessing an electron-dense core. The Koch's postulates have not been yet fulfilled for LCDV in cyprinids, as experimental infection using the 0.45 µm filtered LCDV-positive tissue suspension did not reproduce the disease's clinical signs in grass carp fingerlings.

FACTS SHEET

Disease: Lymphocystis disease (LCD)
Pathogen: Lymphocystis disease virus (LCDV)
Susceptible hosts: Grass carp, rohu, probably other cyprinids
Clinical signs: Wart-like tumours, raspberry-like clusters of cells
Mortality: Most likely no mortality
Temperature: Unknown
Prevention and treatment: *Unknown*

6.3 Viruses of Unconfirmed Virulence Belonging to the Cyprinid Fish Viriome

The viruses listed below have one common characteristic: although they replicate in multiple cell lines, they do not have the ability to induce a disease or mortality *in vivo* even though they can be easily re-isolated. Such viruses can be a problem in that they can mislead diagnostic investigations.

6.3.1 Common carp paramyxovirus (CCPV)

Common carp paramyxovirus (CCPV) is not a well-characterised virus, although it has been isolated repeatedly from the gills of koi and common carp (Fig. 6.20.). The virus generates a very distinctive cytopathic effect in common carp cells, forming large syncytia. Recent epidemiological studies from Germany and Poland indicate that the virus could be an additional pathogen during koi sleepy disease (KSD), primarily caused by carp oedema virus (CEV). As CEV is currently not able to be cultured in available cell lines, the paramyxovirus could have been wrongly associated with gill changes by successful re-isolation of this virus during KSD. This is supported by the fact that the link between CCPV and the disease could not be reproduced. Experimental infections cannot replicate

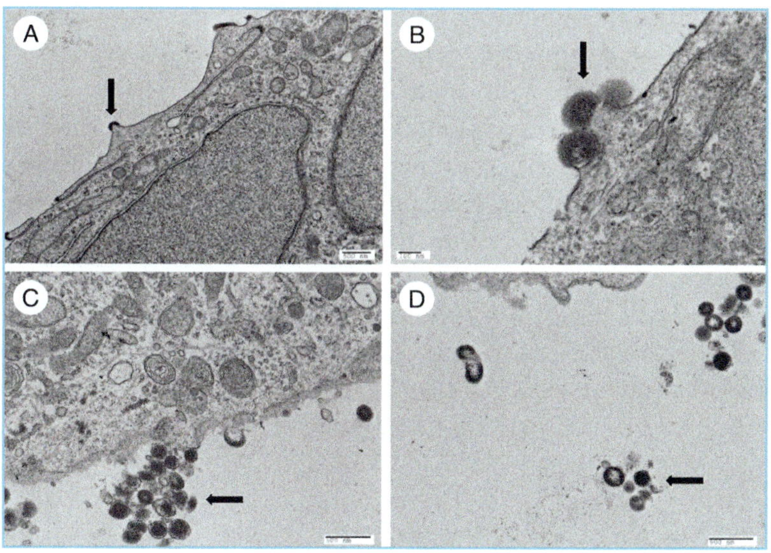

Figure 6.20 Electron microscopy of release of common carp paramyxovirus from infected cells. (A) Formation of virion on cell surface (arrow). (B) Budding of the virions from cell membrane (arrow). (C) Partially liberated virions (arrow). (D) Free virions in intracellular space. Scale bars 100 nm / 500 nm as indicated. Images provided by Richard Paley and Kelly Bateman of the Centre for Environment, Fisheries and Aquaculture Science (Cefas), UK. Crown copyright, Cefas 2023.

the development of any clinical signs associated with the disease state of the fish from which the virus was isolated initially.

FACTS SHEET

Disease: Unknown – associated with gill disease
Pathogen: Common carp paramyxovirus (CCPV)
Susceptible hosts: Common carp
Clinical signs: Gill necrosis – needs conformation
Mortality: Unknown
Temperature: Unknown
Prevention and treatment: Unknown

6.3.2 Common carp (koi carp) orthomyxovirus (CCOV)

Like common carp paramyxovirus, an orthomyxovirus has also been isolated in Europe and North America from koi with gill necrosis. Initial sequencing of the *PB1* gene showed significant 43% amino acid sequence identity with another fish orthomyxovirus, called infectious salmon anaemia virus (ISAV). The sequencing of the virus genome is ongoing, but preliminary results show that CCOV is indeed a member of the family *Orthomyxoviridae*. The virus replicates in CCB and KFC, inducing the rounding of cells and later causing cell detachment leading to focal lysis. The virus appears not to replicate in common carp gills (CCG) and EPC cell lines. Association with the disease is very weak, as the experimental infections could not replicate the development of any clinical signs.

FACTS SHEET

Disease: Unknown – associated with gill disease
Pathogen: Common carp orthomyxovirus (CCOV)
Susceptible hosts: Common carp
Clinical signs: Gill necrosis – needs confirmation
Mortality: Presumably none
Temperature: Unknown
Prevention and treatment: Unknown

6.3.3 Common carp birnavirus (CCBV)

A putative un-enveloped virus was isolated from the necrotic gills of common carp in Germany in the 1990s. The virus is another example of a non-virulent member of viriome for which subsequent experimental infections could not establish a link between the CCBV and the disease, nor replicate the development of any clinical signs associated with the disease state of the fish from which the virus was isolated initially.

FACTS SHEET

Disease: Unknown – associated with gill disease
Pathogen: Common carp birnavirus (CCBV)
Susceptible hosts: Common carp, crucian carp, Prussian carp, catla
Clinical signs: Gill necrosis – needs confirmation
Mortality: Presumably none
Temperature: Unknown Prevention and treatment: Unknown

6.3.4 Common carp picornavirus (CPV-1)

Common carp picornavirus (CPV-1) has been isolated from common carp that died after an accidental discharge of organic waste into a fishpond in Germany. The fish died without showing obvious clinical signs of an infectious disease. Subsequent experimental infections using intraperitoneal infection confirmed that the virus did not cause any clinical signs in young carp. However, CPV-1 can be isolated from various organs of infected fish using FHM cells. It appears that CPV-1 is presumably yet another non-virulent member of common carp viriome, which can be isolated from fish using cell lines without being the causative agent of a disease *in vivo*.

FACTS SHEET

Disease: Presumably none
Pathogen: Common carp picornavirus (CCPiV)
Susceptible hosts: Common carp
Clinical signs: Presumably none
Mortality: Presumably none
Temperature: Unknown
Prevention and treatment: Unknown

6.3.5 Common carp iridovirus (CCIV)

This virus was isolated from necrotic gills of common carp in the Soviet Union in the 1970s. Extensive investigations, which included multiple infection experiments with common carp, did not induce any clinical signs of a disease or mortality. The virus was also very effectively cleared in common carp maintained at 18–20°C, and subsequent passages *in vivo* did not enhance the virulence of the virus. Lower temperatures do allow the virus to persist in carp for 12–27 weeks after injection, although the virus is not very infective via cohabitation or bath infection. This most likely confirms that the virus is a non-virulent member of common carp viriome which effectively replicates in multiple cell lines without causing disease *in vivo*.

FACTS SHEET

Disease: Unknown – associated with gill disease
Pathogen: Common carp iridovirus (CCIV)
Susceptible hosts: Common carp
Clinical signs: None
Mortality: Most likely no mortality
Temperature: Unknown
Prevention and treatment: Unknown.

6.4 Selected References for Further Reading

CyHV-1

Crossland, N., Hawke, J., Del Piero, F., Sokolova, Y., Waltzek, T. *et al.* (2018) Investigation of a cyprinid herpesvirus 1 disease episode in a group of pond-reared koi. *Journal of Aquatic Animal Health* 30, 185–190.

Rahmati-Holasoo, H., Ahmadivand, S., Shokrpoor, S. and El-Matbouli, M. (2020) Detection of Carp pox virus (CyHV-1) from koi (*Cyprinus carpio* L.) in Iran; clinico-pathological and molecular characterization. *Molecular Cellular Probes* 54, 101668.

Sirri, R., Ciulli, S., Barbé, T., Volpe, E., Lazzari, M. *et al.* (2018) Detection of Cyprinid Herpesvirus 1 DNA in cutaneous squamous cell carcinoma of koi carp (*Cyprinus carpio*). *Veterinary Dermatology* 29, 60–e24.

Viadanna, P.H., Miller-Morgan, T., Peterson, T., Way, K., Stone, D.M. *et al.* (2017) Development of a PCR assay to detect cyprinid herpesvirus 1 in koi and common carp. *Diseases of Aquatic Organisms* 123, 19–27.

CyHV-3

Adamek, M., Matras, M., Rebl, A., Stachnik, M., Falco, A. *et al.* (2022) Don't Let It Get Under Your Skin! – Vaccination protects the skin barrier of common carp from disruption caused by Cyprinid Herpesvirus 3. *Frontiers in Immunology* 13, 787021.

Boutier, M., Ronsmans, M., Rakus, K., Jazowiecka-Rakus, J., Vancsok, C. *et al.* (2015) Cyprinid herpesvirus 3: An archetype of fish alloherpesviruses. *Advances in Virus Research* 93, 161–256.

Boutier, M., Gao, Y., Donohoe, O. and Vanderplasschen, A. (2019) Current knowledge and future prospects of vaccines against cyprinid herpesvirus 3 (CyHV-3). *Fish and Shellfish Immunology* 93, 531–541.

Negenborn, J., van der Marel, M.C., Ganter, M. and Steinhagen, D. (2015) Cyprinid herpesvirus-3 (CyHV-3) disturbs osmotic balance in carp (*Cyprinus carpio* L.) – A potential cause of mortality. *Veterinary Microbiology* 177, 280–288.

Rakus, K., Ouyang, P., Boutier, M., Ronsmans, M., Reschner, A. *et al.* (2013) Cyprinid herpesvirus 3: An interesting virus for applied and fundamental research. *Veterinary Research* 44, 85.

CEV

Adamek, M., Teitge, F., Baumann, I., Jung-Schroers, V., El Rahman, S.A. *et al.* (2021) Koi sleepy disease as a pathophysiological and immunological consequence of a branchial infection of common carp with carp edema virus. *Virulence* 12, 1855–1883.

Lewisch, E., Gorgoglione, B., Way, K. and El-Matbouli, M. (2015) Carp edema virus/Koi sleepy disease: An emerging disease in Central-East Europe. *Transboundary and Emerging Diseases* 62, 6–12.

Matras, M., Borzym, E., Stone, D., Way, K., Stachnik, M. *et al.* (2016) Carp edema virus in Polish aquaculture – evidence of significant sequence divergence and a new lineage in common carp *Cyprinus carpio* (L.). *Journal of Fish Diseases*. DOI: 10.1111/jfd.12518.

Wang, N., Zhang, Z., Jing, H., Zhang, M., Wu, S. *et al.* (2021) Development of a novel droplet digital PCR assay for the sensitive detection of carp edema virus. *Aquaculture* 545, 737162.

Way, K., Haenen, O., Stone, D., Adamek, M., Bergmann, S.M. *et al.* (2017) Emergence of Carp Edema Virus (CEV) and its significance to European common Carp and Koi *Cyprinus carpio*. *Disease of Aquatic Organisms* 126, 155–166.

Zawisza, M., Chadzinska, M., Steinhagen, D., Rakus, K. and Adamek, M. (2024) Gill disorders in fish: Lessons from poxvirus infections. *Reviews in Aquaculture* 16, 234–253.

Ziafati Kafi, Z., Najafi, H., Alishahi, M., Rahmati Holasoo, H., Mouloki, A. *et al.* (2022) Detection and phylogenetic analysis of carp edema virus in common carp (*Cyprinus carpio*) in Iran; 2020–2021. *Aquaculture* 558, 738381.

SVCV

Ahne, W., Bjorklund, H.V., Essbauer, S., Fijan, N., Kurath, G. *et al.* (2002) Spring viremia of carp (SVC). *Diseases of Aquatic Organisms* 52, 261–272.

Ashraf, U., Lu, Y., Lin, L., Yuan, J., Wang, M. *et al.* (2016) Spring viraemia of carp virus: Recent advances. *The Journal of General Virology* 97, 1037–1051.

Boonthai, T., Loch, T.P., Standish, I. and Faisal, M. (2017) Susceptibility of representative great lakes fish species to the North Carolina Strain of Spring Viremia of Carp Virus (SVCV). *Journal Aquatic Animal Health* 29, 214–224.

Cui, L.C., Guan, X.T., Liu, Z.M., Tian, C.Y. and Xu, Y.G. (2015) Recombinant lactobacillus expressing G protein of spring viremia of carp virus (SVCV) combined with ORF81 protein of koi herpesvirus (KHV): A promising way to induce protective immunity against SVCV and KHV infection in cyprinid fish via oral vaccination. *Vaccine* 33, 3092–3099.

Emmenegger, E.J., Bueren, E.K., Jia, P., Hendrix, N. and Liu, H. (2022) Comparative virulence of spring viremia of carp virus (SVCV) genotypes in two koi varieties. *Diseases of Aquatic Organisms* 148, 95–112.

CyHV-2

Ito, T. and Maeno, Y. (2014) Susceptibility of Japanese Cyprininae fish species to cyprinid herpesvirus 2 (CyHV-2). *Veterinary Microbiology* 169, 128–134.

Jeffery, K.R., Bateman, K., Bayley, A., Feist, S.W., Hulland, J. *et al.* (2007) Isolation of a cyprinid herpesvirus 2 from goldfish, *Carassius auratus* (L.), in the UK. *Journal of Fish Diseases* 30, 649–656.

Kurobe, T., Kurita, J., Haenen, O., Voorbergen-Laarman, M. and Ito, T. (2024) Mass mortality events associated with cyprinid herpesvirus 2 (CyHV-2) infection in wild Prussian carp *Carassius gibelio* in the Netherlands and molecular biology of virus strains. *Journal of Fish Diseases* 47, e13868.

Sun, Y., Xu, C., Wang, H., Qiao, G., Wang, Z. *et al.* (2023) An attenuated strain of cyprinid herpesvirus 2 as a vaccine candidate against herpesviral hematopoietic necrosis disease in gibel carp, *Carassius auratus gibelio. Fish and Shellfish Immunology* 138, 108826.

Wang, H., Xu, L. and Lu, L. (2016) Detection of cyprinid herpesvirus 2 in peripheral blood cells of silver crucian carp, *Carassius auratus gibelio* (Bloch), suggests its potential in viral diagnosis. *Journal of Fish Diseases* 39, 155–162.

Wei, C., Kakazu, T., Chuah, Q.Y., Tanaka, M., Kato, G. *et al.* (2020) Reactivation of cyprinid herpesvirus 2 (CyHV-2) in asymptomatic surviving goldfish *Carassius auratus* (L.) under immunosuppression. *Fish and Shellfish Immunology* 103, 302–309.

CaHV

Gao, F.X., Wang, Y., Zhang, Q.Y., Mou, C.Y., Li, Z. *et al.* (2017) Distinct herpesvirus resistances and immune responses of three gynogenetic clones of gibel carp revealed by comprehensive transcriptomes. *BMC Genomics* 18, 561.

Gui, L., Zhao, Y., Xu, D., Li, X., Luo, J. *et al.* (2022) Quick detection of *Carassius auratus* herpesvirus (CaHV) by recombinase-aid amplification lateral flow dipstick (RAA-LFD) method. *Frontiers in Cellular and Infection Microbiology* 12, 981911.

Ke, F. and Zhang, Q.-Y. (2023) Advances on genomes studies of large DNA viruses in aquaculture: A mini review. *Genomics* 115, 110720.

Wang, Z.H., Ke, F., Zhang, Q.Y. and Gui, J.F. (2021) Structural and Functional Diversity among Five RING Finger Proteins from *Carassius auratus* Herpesvirus (CaHV). *Viruses* 13(2), 254. DOI: 10.3390/v13020254.

Zeng, X.T., Chen, Z.Y., Deng, Y.S., Gui, J.F. and Zhang, Q.Y. (2016) Complete genome sequence and architecture of crucian carp *Carassius auratus* herpesvirus (CaHV). *Archives of Virology* 161, 3577–3581.

GCRV

Fan, Y., Rao, S., Zeng, L., Ma, J., Zhou, Y. *et al.* (2013) Identification and genomic characterization of a novel fish reovirus, Hubei grass carp disease reovirus, isolated in 2009 in China. *The Journal of General Virology* 94, 2266–2277.

Feng, Y., Shi, C., Ouyang, P., Huang, X., Chen, D. *et al.* (2022) The national surveillance study of grass carp reovirus in China reveals the spatial-temporal characteristics and potential risks. *Aquaculture* 547, 737449.

Gao, Y., Pei, C., Sun, X., Zhang, C., Li, L. *et al.* (2018) Plasmid pcDNA3.1-s11 constructed based on the S11 segment of grass carp reovirus as DNA vaccine provides immune protection. *Vaccine* 36, 3613–3621.

Liang, H.R., Li, Y.G., Zeng, W.W., Wang, Y.Y., Wang, Q. *et al.* (2014) Pathogenicity and tissue distribution of grass carp reovirus after intraperitoneal administration. *Virology Journal* 11, 178. DOI: 10.1186/1743-1422X-1111-1178.

Liang, H.R., Fu, X.Z., Li, N.Q., Liu, L.H., Lin, Q. *et al.* (2016) The distribution of different virulence grass carp reovirus strains in some neglected tissues. *Polish Journal of Veterinary Sciences* 19, 763–770.

Wu, M., Li, H., Chen, X., Jiang, Y. and Jiang, W. (2020) Studies on the clinical symptoms, virus distribution, and mRNA expression of several antiviral immunity-related genes in grass carp after infection with genotype II grass carp reovirus. *Archives of Virology* 165, 1599–1609.

GrCRV

Walker, P.J., Bigarré, L., Kurath, G., Dacheux, L. and Pallandre, L. (2022) Revised taxonomy of Rhabdoviruses infecting fish and marine mammals. *Animals* 12, 1363.

KIRV

George, M.R., John, K.R., Mansoor, M.M., Saravanakumar, R., Sundar, P. *et al.* (2015) Isolation and characterization of a ranavirus from koi, *Cyprinus carpio* L., experiencing mass mortalities in India. *Journal of Fish Diseases* 38, 389–403.

LCDV

Shahi, N., Mallik, S.K., Kala, K., Joshi, N., Patiyal, R.S. *et al.* (2020) Seasonal emergence of benign epidermal tumor in farm-reared adult grass carp (*Ctenopharyngodon idella*) caused by lymphocystis disease virus at Uttarakhand, India. *Aquaculture* 526, 735408.

CCPV / CCOV / CCBV

Granzow, H., Fichtner, D., Schütze, H., Lenk, M., Dresenkamp, B. *et al.* (2014) Isolation and partial characterisation of a novel virus from different carp species suffering from gill necrosis – ultrastructure and morphogenesis. *Journal of Fish Diseases* 37, 559–569.

Neukirch, M. and Kunz, U. (2001) Isolation and preliminary characterization of several viruses from koi (*Cyprinus carpio*) suffering gill necrosis and mortality. *Bulletin of the European Association of Fish Pathologists* 21, 125–135.

CPV-1

Lange, J., Groth, M., Fichtner, D., Granzow, H., Keller, B. *et al.* (2014) Virus isolate from carp: Genetic characterization reveals a novel picornavirus with two aphthovirus 2A-like sequences. *The Journal of General Virology* 95, 80–90.

CCIV

Shchelkunov, I.S. and Shchelkunova, T.I. (1990) Infectivity experiments with *Cyprinus carpio* iridovirus (CCIV), a virus unassociated with carp gill necrosis. *Journal of Fish Diseases* 13, 475–484.

Chapter 7

Fungal and Fungal-Like Pathogens

7.1 Introduction

Infections with organisms from the former eukaryote kingdom 'Fungi' are known from all freshwater fish species. In conventional classification, 'fungi' are considered as heterotrophic eukaryotes with osmotrophic nutrition. Most of the conventional 'fungi' are multicellular organisms, formed of a mycelium of septate or aseptate hyphae, which grow apically and have a saprophytic or parasitic mode of life. In addition, many 'fungi' form permanent symbiotic associations with algae (lichens) or plants (mycorrhizae). Ultrastructural research and the inclusion of molecular sequence data have extended phylogenetic analyses and changed substantially the understanding of the phylogeny of eukaryote organisms. During this process, the organisms grouped into the former kingdom 'Fungi' were recognised to have only morphological similarities, but no taxonomic affinities, and therefore are now grouped into different eukaryote supergroups. Organisms now regarded as true fungi are placed together with animals into the supergroup Opisthokonta, while fungal-like oomycete pathogens, such as *Saprolegnia* and *Aphanomyces*, are placed into the Stramenopiles, Alveolates, Rhizaria (SAR) supergroup, and now are referred to as Peronosporomycetes (Stramenopiles or Chromista).

In general, infections with fungal and fungal-like pathogens are increasing globally. These pathogens are widely regarded as opportunistic generalists and the majority of species infect fish from various species or even families. The life cycle of all fungal and fungal-like organisms includes proliferative asexual reproduction with the production of a large number of motile or non-motile spores, which ensure further dissemination or spreading to previously non-infected

 Carps: A Practical Approach to Disease Control and Health Management (eds David Hoole *et al.*)
DOI: 10.1079/9781836990024.0007

hosts. Opisthokont 'true' fungi produce non-motile spores with a chitinous cell wall, which is resilient and relatively long-lived and can survive unfavourable growth conditions. Fungal-like oomycetes produce biflagellate zoospores, which disperse the parasite among hosts. These zoospores typically encyst after a short period of motility. In some species, longer-lived secondary zoospores hatch from these cysts and distribute the pathogen further. In *Saprolegnia*, these secondary zoospores may undergo further cycles of encystment and zoospore formation. This process is called 'polyplanetism' and most likely provides several occasions to contact a new host. Many oomycete zoospores are chemotactic and are attracted by, for example, amino acids or carbohydrates.

The opisthokont fungi, which share the presence of beta glucan and chitin in the cell wall, primarily in the spore, are unicellular in form or grow as a mycelium, possess flattened mitochondrial cristae, and synthesise lysine by the amino adipic pathway. This group of organisms includes Ascomycota and Basidiomycota, as well as Chytridomycota and Zygomycota, which are both now considered as paraphyletic groups. In addition, Microsporidia was recognised as a member of the Opisthosporidia, a group of opisthokont intracellular parasites or parasitoids with an amoeboid vegetative stage and invasive spores with a chitin cell wall and specialised apparatus for penetration into the host cell. These Opisthosporidia are also regarded as members of fungi in some phylogenetic analyses, but the systematic position of this group of organisms is still a matter of debate.

Opisthokont fungi that infect fish are mainly found among the Ascomycota and Mucoromycota (part of the Zygomycota) and include organisms from the genera *Ochroconis*, *Exophiala*, *Phoma* and *Fusarium*, with the latter also known as crustacean pathogens. Infections with all these organisms are sporadic and originate often from the invasion of wounds. However, species that can cause infections in fish are generalists and cause infections in other organisms, including plants. Because of their thick-walled resilient spores, in particular the spores from the opportunistic Ascomycota and Mucoromycota may be dispersed in the water of aquaculture ponds in tropical or subtropical countries, and may infect fish kept under unfavourable conditions.

Various microsporidia are reported as fish pathogens and can be regarded as primary pathogens. In addition, organisms of the class Ichthyosporea, namely Dermocystidia and Ichthyophonidia, are also included in this chapter. These organisms, together with the parasite *Sphareothecum destrues*, form a group of opisthokonts close to the animal–fungal boundary.

7.2 Peronosporomycetes (Oomycetes, water moulds)

7.2.1 Saprolegniosis

Introduction

Infections with oomycetous water moulds are responsible for serious losses in fish populations in aquaculture and in natural stocks. Oomycetes are known as saprophytic opportunists, which can be responsible for disease outbreaks in plants and animals ranging from nematodes, insects and crustaceans to aquatic and terrestrial vertebrates, including humans. Fish are mainly infested by organisms of the genera *Aphanomyces*, *Saprolegnia* and *Achlya* from the Saprolegniaceae family. Whilst species such as *Aphanomyces invadans* and *Saprolegnia parasitica* are pathogenic to various fish species, others such as other *Saprolegnia* species or other members of the Saprolegniaceae family are considered as saprophytes, which colonise dead cells or tissue, and thus act as secondary pathogens.

The oomycetes grow in filamentous aseptated hyphae and form a mycelium, as in many opisthokont fungi. In addition, oomycetes and fungi share a similar nutrition mode and feed by secreting enzymes that degrade cellulose, lipids or proteins and, subsequently, the organisms absorb degradation products. In contrast to fungi, oomycetes are diploid and lack chitin in their cell wall, which instead contains other glucans, including cellulose. Oomycetes produce motile, flagellated zoospores by means of asexual reproduction; their life cycle also includes a sexual reproduction cycle during which vegetative hyphae form specialised gametangia, the oogonium and the antheridium, which give rise to oospores. Sexual structures are important for species identification, but in the fish pathogenic species *A. invadans* and *S. parasitica*, these sexual structures may be absent or suppressed when cultured *in vitro*.

Disease description

In most cases, saprolegniosis is a chronic disease of adult and juvenile fish, but can also occur in eggs, which develop cotton-wool-like tufts. In fish, the clinical picture of a *Saprolegnia* infection manifests as focal grey-whitish patches on the skin or fins (Fig. 7.1), also with a cotton-wool-like appearance (Fig. 7.2). This results from hyphal filaments, which extend into the water. Larger mycelia may entrap mud or algae and thus may have a brown or green colour. The distribution of these patches may be random, but certain parts of the body, such as head, back or peduncle, are particularly affected. In addition to skin, gills may also be infected, while internal organs are involved less frequently. In histology, *Saprolegnia* hyphae can be seen in sections of infected skin areas on the surface, penetrating the epidermis to the stratum spongiosum of the dermis, where it spreads laterally. In chronic infections, hyphae may penetrate from the dermis to the intermyotomal fasciae. While hyphae of true fungi are readily visible on standard histological sections in haematoxylin–eosin (H&E) staining, oomycete hyphae are much more difficult to differentiate from the tissue and would require a Grocott methylamine silver staining with H&E counterstaining.

Pathogen

Several species from the Saprolegniaceae family are parasites of fish. However, because of their saprophytic life habit, most of these species are opportunists and infect wounds or the skin of immunocompromised fish. The genera *Achlya*, *Aphanomyces* and *Saprolegnia* can be identified based on the morphology of zoosporangia and the development of zoospores. In *Saprolegnia*, club-shaped sporangia release motile primary zoospores, which encyst after a short period

Figure 7.1 *Saprolegnia* infection in carp. Common carp with a grey overlay of *Saprolegnia* hyphae on the skin of the back and the head. Courtesy of the Environment Agency, National Fisheries Laboratory, UK.

Figure 7.2 *Saprolegnia* infection in carp. Common carp with *Saprolegnia* hyphae of a cotton-wool-like appearance.

of swimming in a primary cyst. Primary cysts release reniform secondary zoospores, which constitute the main motile stage and maintain motility for up to several days in order to infect a host or produce a secondary cyst. *Achlya* sporangia are also club-shaped, but discharge amoeboid, immotile primary spores, which encyst in a primary cyst at the tip of the sporangium. Secondary motile zoospores are released from these primary cysts. *Aphanomyces* differs from *Achlya* and *Saprolegnia* by having narrower vegetative hyphae. The zoosporangium also has a similar diameter as the hyphae; it contains a single row of spores and discharges immotile primary spores like *Achlya*, which also encyst at the tip of the sporangium and subsequently release motile secondary zoospores.

The cell wall of secondary zoospores from *S. parasitica* is decorated with long hooked hairs, which most likely facilitate attachment to a new host. These hooked hairs are not observed on zoospores from other *Saprolegnia* species or from other genera, and therefore might explain a higher virulence of *S. parasitica* compared with other *Saprolegnia* species. However, infections with *S. parasitica*, as with other *Saprolegnia* or *Achlya* species, mainly occur in immunocompromised hosts, fish with skin damage such as wounds, or damaged mucosal skin barrier. *S. parasitica* is frequently reported from salmonids in temperate waters. Cyprinids such as common carp, tench and silver carp are more often infected with saprophytic *Saprolegnia* or *Achlya* species such as *S. diclina,* but may also carry *S. parasitica* infections. In tropical waters, saprophytic *Achlya* and *Saprolegina*, but also *S. parasitica*, infect grass carp, silver carp and common carp, as well as Indian major carps, in particular rohu (*Labeo rohita*).

Detection

Characteristic lesions for oomycete infections and the characteristic cotton-wool-like tufts can be easily recognised during clinical inspection, and hyphae are visible during microscopic observation of fresh preparations scraped off affected skin or gill areas. In addition, hyphae are visible in histological sections after appropriate staining. The identification of the pathogen is more difficult and requires *in vitro* cultivation of the oomycetes. For this, pieces of affected tissue are collected, washed with media containing antibiotics in order to prevent bacterial overgrowth and then cultivated on agar plates using aseptic technique. The media may be flooded with a thin layer of freshwater or seawater containing antibiotics. In addition, identification of oomycete species can be based on sequencing the ITS spacer region of ribosomal DNA. This is hampered because of a large number of mis-assigned sequences in DNA databases like GenBank.

Prevention and treatment

Oomycete infections are mainly reported from immunocompromised hosts or fish with skin lesions, including damage to the mucous layer of the skin. In particular, crowding of fish during storage, but also rough handling of fish during catching, sorting, transfer or transport, may induce skin lesions that allow oomycete infection. Therefore, improvement of water quality and management measures is required, to prevent oomycete outbreaks. Chemical treatment of acute infections is difficult to achieve. All infectious stages of *Saprolegnia* and *Achlya* can be successfully treated with malachite green, formalin or hydrogen peroxide; however, the use of these chemicals for the treatment of fish for human consumption is regulated by legal restrictions and is not allowed in many countries. In particular, malachite green has been recognised as a mutagen, teratogen and tumour-promoting agent in animals and humans, and therefore its use is banned in many countries. Treatment of water with ozone may effectively reduce the frequency of saprolegniosis in hatcheries, while the treatment of eggs with bicarbonate-buffered iodophores is used to destroy potential pathogens, including oomycetes.

FACTS SHEET

Disease: Saprolegniosis
Pathogens: Oomycete pathogens *Achlya* spp., *Saprolegnia* spp., *S. parasitica*
Susceptible hosts: *Carassius* species, including goldfish (*Carassius auratus*), common carp (*Cyprinus carpio*), grass carp (*Ctenopharyngodon idella*), silver carp (*Hypophthalmichthys molitrix*), rohu (*Labeo rohita*) and other cyprinids
Mortality: Low to moderate in susceptible and stressed populations
Temperature: 5–30°C
Prevention and treatment: Improvement of hygiene and keeping conditions; improvement of management during catching, sorting, transport, water disinfection with ozone. The use of chemicals like malachite green or formalin for the treatment of fish for human consumption is regulated by legal prescription and thus may not be used in many countries

7.2.2 Epidermal ulcerative syndrome (EUS)

Introduction

The disease syndrome called epidermal ulcerative syndrome (EUS) was observed independently in separate parts of the world. Initially a range of names were used to describe this syndrome, before it was recognised that this might be a single disease. The disease is characterised by the development of dermal ulcers and was previously called mycotic granulomatosis of ayu (*Plecoglossus altevelis*) in Japan, red spot disease in estuarine mullets (*Mugil cephalus*) in Australia, and ulcerative mycosis in Atlantic menhaden (*Brevoortia tyrannus*) in estuaries of the US east coast. It also affects snakehead (*Channa striata*), silver barb (*Barbodes gonionutus*), swamp eel (*Monopterus albus*) and rohu (*Labeo rohita*) in South and South-east Asia, as well as cichlids (including tilapia) in Africa. All these different disease syndromes were found to be associated with the presence of an infection with the invasive oomycete *Aphanomyces invadans*. After a long period of scientific controversies, it is now recognised that *A. invadans* is the causative agent of EUS.

Disease description

First clinical signs of EUS are small pinhead-sized red lesions in the skin of affected fish, mainly on the lateral surface, but also on the head, operculum or peduncle. With progression of the disease, localised swellings are visible, along with protruding scales, scale loss, skin erosion, reddened areas of the skin under the scales or even exposure of underlying musculature and ulceration. In fresh squashed preparations taken from the area around such lesions, aseptate hyphae can be observed in microscopic examination. The infection may be responsible for massive losses that occur within a short period of time.

Pathogen

The oomycete *Aphanomyces invadans* has been identified as the agent responsible for the lesions leading to EUS. This was confirmed by the isolation of the pathogen from an affected specimen, as well as by an induction of the disease by experimental infection of susceptible fish with zoospores. Infection of fish occurs when zoospores of *A. invadans* penetrate the skin of the host, germinate and form hyphae, which subsequently penetrate the skin dermis and may extend to the skeletal muscle. Granulomas may be formed around the hyphae. *Aphanomyces* hyphae are about 12–25 μm in diameter in infected tissue, but considerably narrower in culture. Sporangia are as narrow as vegetative hyphae and contain one row of primary spores. These spores are amoeboid and, after being released from the sporangium, form primary cysts at the tip of the sporangium. These primary cysts release motile biflagellate secondary zoospores, which are attracted to the skin of fish, where they germinate and start an infection by penetrating the skin. Outbreaks often occur during periods of a sudden drop in temperature, such as after a period of rainfall. These conditions favour sporulation of the pathogen and may delay the inflammatory response of the fish to the infection. The pathogen infects a wide range of fish species from various freshwater and estuarine habitats, including barbs, catfish, cichlids, gobies, snakeheads and gouramis, but also Indian carp such as catla (*Gibelion catla*), mrigal (*Cirrhinus mrigala*) or rohu (*Labeo rohita*), particularly when kept in the lower range of their permissive temperature. Common carp (*Cyprinus carpio*), crucian carp (*Carassius carassius*) and goldfish (*Carassius auratus*) have been experimentally infected, but appear less susceptible to EUS.

Detection

A presumptive diagnosis of EUS can be based on the presence of open dermal ulcers in the skin of a suspected specimen and the observation of aseptate hyphae in fresh preparations of the muscle around the lesions. However, skin lesions or ulcers may be associated with a variety of infective or non-infective causes, including parasites, bacteria, viruses, or, in the estuarine environment, the presence of toxic algae. Thus, the preliminary diagnosis has to be confirmed by the detection of a granulomatous inflammation around invasive hyphae in histological sections of the tissue around the lesions, or by an isolation of *A. invadans* from the affected tissue location. For isolation of the pathogen, a sample of the skin underlying a moderately raised dermal lesion is collected aseptically and inoculated into specialised agar plates with antibiotics in order to prevent bacterial growth. The cultures are inspected daily for emerging hyphae, which then are transferred to fresh agar plates in order to obtain a pure culture. In addition, the presence of *A. invadans* in a tissue sample can be detected by PCR techniques using primer sequences targeting a nucleotide sequence of the ITS regions of ribosomal DNA.

Prevention and treatment

Outbreaks of EUS in susceptible fish populations are associated with the presence of *A. invadans* zoospores in the affected water body. The pathogen is transferred to previously non-affected sites by flooding events, but also by the movement of contaminated water, fish or equipment. In order to prevent further spreading of the pathogen, contaminated river water should not be used for farms with susceptible species. Likewise, fish from contaminated or uncertain sites should not be used for stocking of farms or natural water bodies. For any movement of live fish, health certificates and quarantine regulations should be in place. Likewise, there should be regular monitoring of the health status of fish populations. In endemic areas, healthy husbandry is essential in fish farming. When outbreaks occur in small, closed waterbodies, liming of water and improvement of water quality, together with removal of infected fish, may reduce the severity of the outbreak. Likewise, drying out and liming of ponds after an outbreak may decrease the frequency and severity of further outbreaks. Adequate treatment of EUS and protective vaccines are not available, but dietary immunostimulants and vitamins may increase fish survival during an outbreak in susceptible fish populations.

FACTS SHEET

Disease: Epizootic ulcerative syndrome (EUS)
Pathogen: Oomycete *Aphanomyces invadans*
Susceptible hosts: Indian major carp, including catla (*Gibelion catla*), mrigal (*Cirrhinus mrigala*) and rohu (*Labeo rohita*). Goldfish (*Carassius auratus*), crucian carp (*Carassius carassius*) and common carp (*Cyprinus carpio*) proved less susceptible in controlled laboratory infections
Mortality: Moderate to high in susceptible and stressed populations
Temperature: 22–30°C
Prevention and treatment: Avoid movements of live fish for stocking, or feeding of aquaculture facilities with water from affected sites. Instead, strict quarantine and health monitoring programmes should be in place for movements of live fish. Removal of affected fish, drying out and liming of affected ponds will reduce frequency and severity of outbreaks. An effective treatment of the disease or a protective vaccine is not available

7.2.3 Ascomycosis and zygomycosis

Introduction

Ascomycota and Zygomycota represent a diverse group of fungal organisms that mainly grow as saprophytes on almost all organic substrates, and some species are also known as opportunistic pathogens of plants, animals and humans. Several species are true generalists and infect hosts from various species from a wide range of systematic groups. These Ascomycota and Zygomycota may also cause clinical infections in fish from aquaculture populations when these are kept under unfavourable conditions, such as in water or sediments with a high organic load and when the fish are challenged by low oxygen or high ammonia concentrations in the water. In addition, spores of these organisms can be introduced into fish populations via contaminated feed. Clinical cases from carp populations have been mainly reported from regions with a warm climate. The ascomycote or zygomycote pathogens have resilient and long-lived environmental stages, and therefore may be further spread by an increased global trade and the migration of invasive aquatic species. In particular, pathogens from the genera *Aspergillus*, *Mucor*, *Penicillium* and *Rhizopus* are recorded from clinically affected cyprinid fish.

Disease description

Clinical signs are non-specific and include haemorrhages in skin and operculum, eroded scales and fins, exophthalmos, and occasionally cataract formation.

Pathogens

The Ascomycota, *Aspergillus* and *Penicillium*, comprise a ramified, septated mycelium, which grows on the surface of the colonised substrate. The mycelium develops characteristic fruiting bodies, the conidia, which form a large number of asexual spores. The Zygomycota *Mucor* and *Rhizopus* grow in long aseptate hyphae within the substrate. In these organisms, only the sporangiophores, which carry spore-developing sporangia, are separated by a septum. In these sporangia, thousands of spores are formed by mitosis. In addition, sexual reproduction occurs by mating of thalli from opposite (+ or –) mating types, forming a zygospore.

Detection

To isolate these fungi from infected fish, biopsies are collected from infected organs of suspected specimens and inoculated on Sabouraud dextrose (SDA), malt extract (MEA), or potato dextrose (PDA) agars. The fungi grow in mycelia on these isolation media and form characteristic sporangiophores. Fungal genera may be identified based the morphology of the mycelium and the sporangiophores or zygospores, while sequencing of the gene encoding the 18S rRNA is required for species determination.

Prevention and treatment

Contaminated fish feed or other materials in aquaculture facilities are considered as sources for Ascomycota or Zygomycota spores. Poor water, in particular low oxygen, high ammonia and elevated organic load, are the most important factors favouring the growth of these fungi, and such unfavourable conditions might favour clinically apparent infections of fish by these opportunistic pathogens. Therefore, an improvement of hygiene and keeping conditions has to be included in each treatment plan and will be decisive for its success. Outbreaks in populations of ornamental fish, such as goldfish or koi, can be treated by baths with chemicals, such as formalin or malachite green. However, these substances cannot be used for fish for human consumption, because of legal restrictions in many countries.

FACTS SHEET

Disease: Ascomycosis and zygomycosis
Pathogens: Fungal pathogens like *Aspergillus, Mucor, Penicillium* and *Rhizopus*
Susceptible hosts: *Carassius* species, including goldfish (*Carassius auratus*), grass carp (*Ctenopharyngodon idella*), silver carp (*Hypophthalmichthys molitrix*), rohu (*Labeo rohita*) and other cyprinids
Mortality: Low to moderate in susceptible and stressed populations
Temperature: 25–30°C
Prevention and treatment: Improvement of hygiene and keeping conditions

7.3 Dermocystidiosis

7.3.1 Introduction

Dermocystidium infections manifest as cysts of variable shape in various organs of affected fish specimens. These cysts contain numerous spherical stages designated as spores, which have a large central refractive body and stain PAS-positive. Based on these morphological characteristics, several *Dermocystidium* species have been described from various hosts, including fish, amphibia and invertebrates. However, the identification of amphibian and invertebrate parasites is still debatable, and most likely species from the genus *Dermocystidium* infect only fish. Likewise, the systematic position of *Dermocystidium* is uncertain but, based on mtDNA evolution, these organisms are now placed in the class Ichthyosporea (former Mesomycetozoea), close to the animal–fungus boundary.

7.3.2 Disease description

In cyprinids, the disease is characterised by the development of cyst-like structures in fins, skin (Fig. 7.3), muscle (Fig. 7.4), gills and even the eye. The cysts may have a nodular, lenticular or vermiform appearance (Fig. 7.5) and an inflammation may be observed in the surrounding tissue. The lesions contain whitish aseptate and thick-walled hyphae, which are filled with mature or immature spores. The spores are globular and contain an eccentrically situated nucleus and a large refractive body (Fig. 7.6).

7.3.3 Pathogens

Several species are reported from different tissue locations of carp or goldfish. *Dermocystidium cyprini* manifests as cysts in gills of common carp (Fig. 7.7) while *D. koi* forms cysts, measuring up to 10 mm, in the skin of carp and goldfish. An undetermined *Dermocystitium*

Figure 7.3 *Dermocystidium* infection in carp. Nodular lesions in the skin and dorsal fin of a common carp (arrows). Courtesy of the Environment Agency, National Fisheries Laboratory, UK.

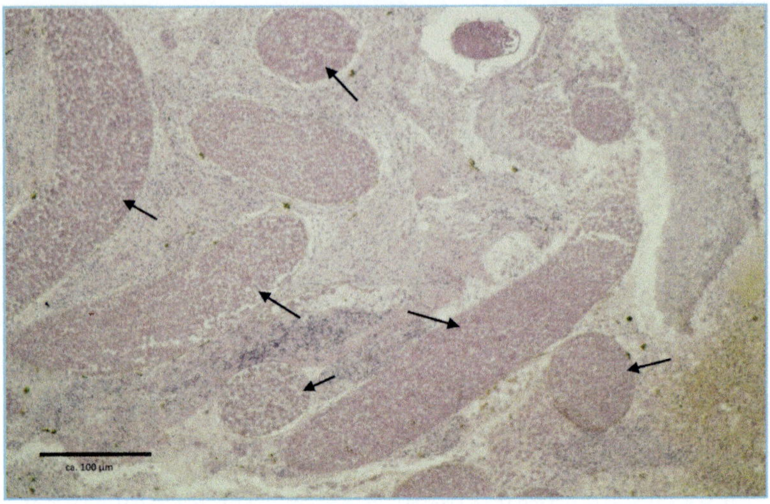

Figure 7.4 *Dermocystidium* infection in carp. Histological section of vermiform *Dermocystidium* cyst in the muscle of carp (arrows). HE staining. Indicative scale bar 100 µm.

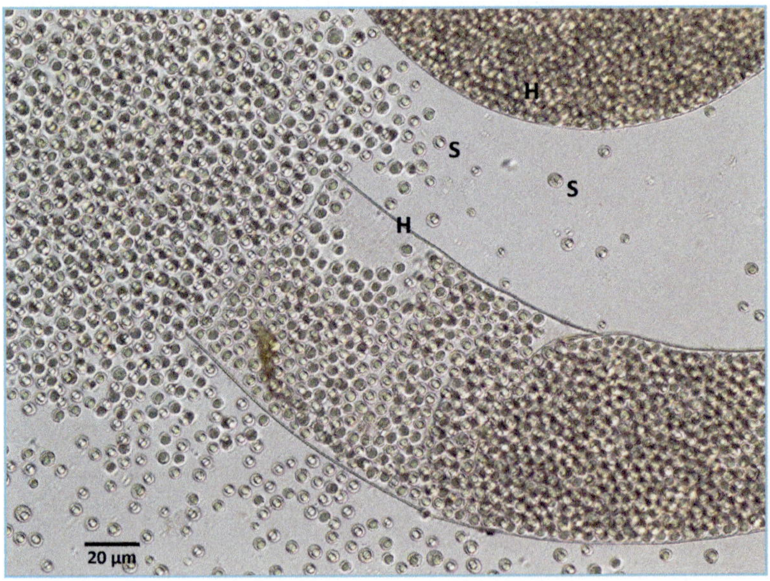

Figure 7.5 *Dermocystidium* infection in carp. Aseptate hyphae (H) from a *Dermocystidium* cyst with spores (S). Indicative scale bar 20 µm. Courtesy of the Environment Agency, National Fisheries Laboratory, UK.

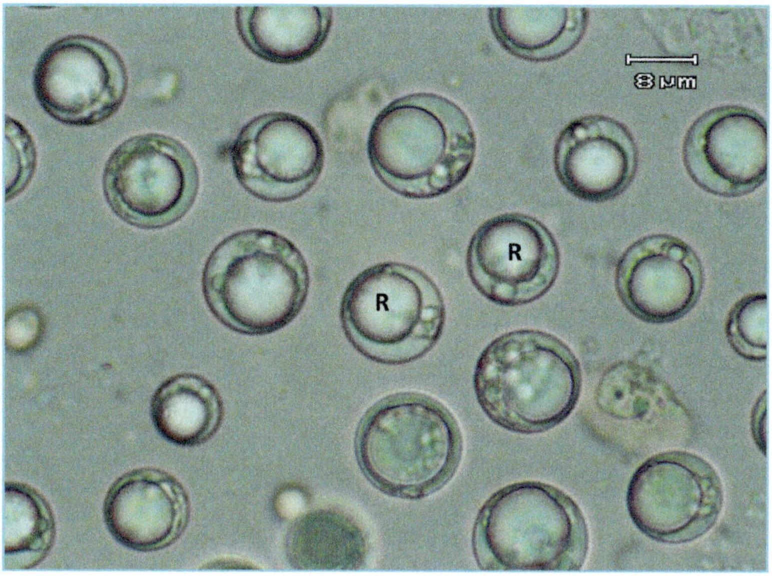

Figure 7.6 *Dermocystidium* infection in carp. *Dermocystidium* spores with a large refractive body (R). Indicative scale bar 8 µm. Courtesy of the Environment Agency, National Fisheries Laboratory, UK.

species forms 'nematode-like' cysts in skin and fins, and a further undetermined *Dermocystidium* species forms cysts with convoluting hyphae in the eye of common carp. In goldfish a *Dermocystidium* species induces whitish nodules on the head. Furthermore, a visceral infection with a *Dermocystidium* species has been detected in carp, goldfish and roach.

7.3.4 Detection

The tissue cysts are directly visible in infected tissue locations (Figs. 7.3 and 7.7) or in histological sections of affected tissue (Fig. 7.4). In fresh samples, numerous small globular spores with a prominent refractive body can be detected by microscopic observation at high magnification (Fig. 7.6). Species identification currently is based on morphological characteristics of cyst and spores; information on nucleotide sequences is urgently required.

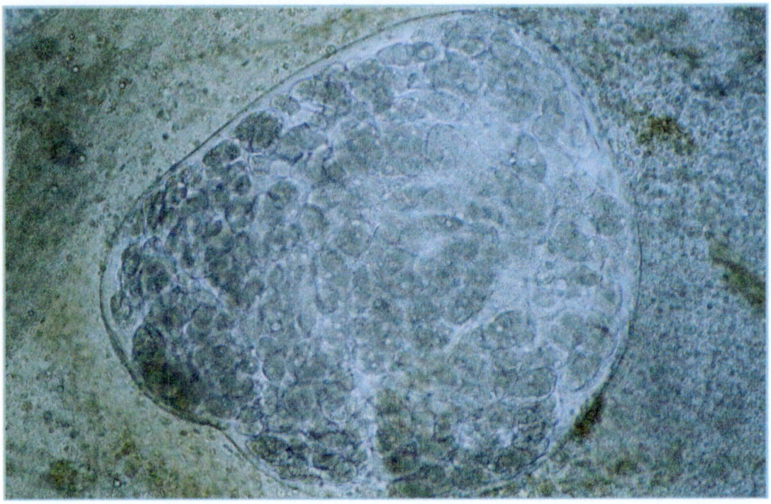

Figure 7.7 *Dermocystidium cyprini* cyst in gills of common carp.

7.3.5 Prevention and treatment

Treatment strategies for *Dermocystidium* are not established.

FACTS SHEET

Disease: Dermocystidiosis

Pathogens: *Dermocystidium cyprini* (gills of common carp), *D. erschowi* (skin of common carp), *D. koi* (skin, fins, muscle of common carp, goldfish), *Dermocystidium* sp. (head of common carp, goldfish), *Dermocystidium* sp. (fins of common carp), *Dermocystidium* sp. (eyes of crucian carp, koi).

Susceptible hosts: *Carassius* species, including goldfish (*Carassius auratus*) and crucian carp (*Carassius carassius*), common carp (*Cyprinus carpio*)

Mortality: None – too low in affected populations

Temperature: 5–25°C

Prevention and treatment: Treatment protocols are not established

7.4 Rosette Agent *Sphaerothecum destruens*

7.4.1 Introduction

With the introduction of the freshwater topmouth gudgeon, *Pseudorasbora parva*, from Chinese waters into Europe and Anatolia, a parasite, *Sphaerothecum destruens*, was co-transferred and now infects various cyprinid and non-cyprinid species from the native fish fauna. Recent observations revealed links between the presence of the invasive *P. parva*, and its parasite *S. destruens*, and the decline of several fish species from the endemic fauna in Europe and Turkey. In particular, the population of the shoaling sunbleak, *Leucaspius delinatus*, experienced a serious decline after contact with the parasite, but also other cyprinids, like carp and bream, are susceptible to infection by *S. destruens* and subsequent mortality. Hence, this parasite is considered a severe threat to several fish species endemic in the European fauna.

7.4.2 Disease description

Gross signs of the infection are hardly visible. While in some cases clinical signs might be completely absent, in other cases infected fish may show anaemia and have an enlarged liver, kidney and spleen. The disease may also be associated with distinct granulomas in visceral organs, but the infection may also be dispersed in various tissues with little host response.

7.4.3 Pathogen

Sphaerothecum destruens is an intracellular parasite of visceral organs, including liver and kidney of infected specimens. It forms characteristic clusters of spherical spores 2–6 µm in diameter, visible in stained smear preparations of infected organs, and therefore was initially termed the 'rosette agent'. In histological sections, the parasite stains eosinophilic with haematoxylin/eosin, PAS positive, basophilic with Giemsa stain, but is not acid-fast in Ziehl–Neelsen staining. The infection leads ultimately to cell death and in infected tissue, in particular in the liver, multifocal cellular necrosis is visible with aggregates of *S. destruens* spores in the cytoplasm of host cells or clusters of parasites released from lysed cells. In the nodular form of the disease, distinct granulomas enclose infected tissue areas. The parasite is released from infected fish with bile, urine or seminal and ovarian fluid during spawning. Shedding from gut epithelium, skin or gills is considered to contribute to the dispersal of the parasite into the environment. The parasite develops motile zoospores, which facilitate its spreading to naïve hosts. In infected fish, the organism grows slowly, but ultimately is responsible for death of hosts from susceptible species. While the invasive topmouth gudgeon is able to control the infection, various fish species endemic to the European fauna are susceptible to the disease, including common carp and bream.

7.4.4 Detection

The parasite can be detected in histological sections or smear preparations of affected tissue areas, in particular after Giemsa staining. In addition, PCR primer sequences are described for molecular biological analysis of tissue samples from suspected specimens. The parasite can be isolated from infected tissue by *in vitro* cultivation in cell cultures from salmonid or cyprinid origin (CHSE, FHM cells).

7.4.5 Prevention and treatment

Most likely, the parasite is introduced into natural waters or aquaculture facilities by infected hosts. In this context, the dispersal of the parasite by *P. parva* is of great importance. Therefore, to prevent the introduction of this parasite into a fish population, the exclusion of the invasive topmouth gudgeon and fish from infected stocks is required. Effective treatment protocols have not been developed.

FACTS SHEET

Disease: Rosette agent
Pathogen: *Sphaerothecum destruens*
Susceptible hosts: European cyprinids, including common carp (*Cyprinus carpio*)
Mortality: Moderate to high in populations of susceptible species (e.g. sunbleak, *Leucaspius delinatus*), low in common carp (*Cyprinus carpio*)
Temperature: 5–25°C
Prevention and treatment: Treatment protocols are not established

7.5 Microsporidiosis

7.5.1 Introduction

Microsporidia are obligate intracellular parasites of hosts from various animal phyla, but also from protists like Ciliophora, Apicomplexa or Amoebozoa. They infect a wide range of aquatic organisms, including various fish species such as several cyprinid species, which are infected by organisms from different microsporidian genera. The nervous system of the zebrafish *Danio rerio* is infected by *Pseudoloma neurophila*, the muscle of zebrafish by *Pleistophora hyphessobryconis*, and the ovaries of roach *Rutilus rutilus* and golden shiner *Neotropis cornutus* by *Pleistophora ovariae*. These infections may cause serious losses in affected populations. In contrast, there are only cursory reports of infections of cultured cyprinid species with microsporidia. *Pleistophora hyphessobryconis* infection was detected in goldfish (*Carassius auratus*), an undetermined *Pleistophora* species in catla (*Gibelion catla*), and an undetermined microsporidium in mrigal (*Cirrhinus mrigala*).

7.5.2 Disease description

Specimens heavily infected with *Pleistophora* may be lethargic; they may lose their colouration and white patches may appear under the skin. In goldfish and catla, scales may bristle, and haemorrhages or patches without scales may be visible. In heavily infected specimens, the parasite may spread to other organs, such as connective tissue of the ovaries, intestinal epithelium or skin. *Pleistophora* undergoes a merogonic multiplication in infected muscle cells and may occur as tumour-like masses in an advanced infection. Spores develop in a sporogonial plasmodium, which grows and subsequently is enveloped by a thick wall, forming a sporophorous vesicle (SPV). In this SPV, numerous ovoid spores develop with a size of 4 × 6 µm, a posterior vacuole and a coiled polar filament.

7.5.3 Pathogen

Pleistophora infects muscle cells, where multinucleate, rounded meronts develop (Fig. 7.8), which divide by plasmotomy. During subsequent sporogony, the sporont grows into a multinucleate sporogonial plasmodium, which is enveloped by a thick wall and forms a sporophorous vesicle (SPV). In the SPV, a multinucleate sporoplasma develops into uninucleate spores. The spores have an inner chitin and an outer proteinaceous wall, are lacking mitochondria, kinetosomes or cilia, but possess a specialised polar tube for host penetration, which is coiled in the posterior part of the spore with a characteristic number of turns. Spores are liberated from SPVs, pervade the body cavity and concentrate in the subcutaneous tissue from where they are discharged into the environment and spread to non-infected hosts. Heavy infections are lethal.

7.5.4 Detection

In heavily infected specimen, spores can be detected in skin scrapings or fresh preparations of affected tissue. Merogonic and sporogonic

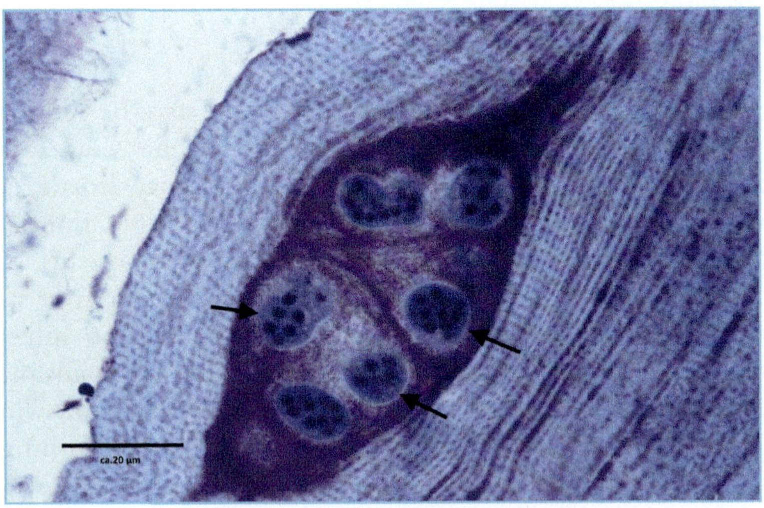

Figure 7.8 Histological section of *Pleistophora hyphessobryconis* merogonic developmental state (arrows) in muscle. Toluidin blue staining. Indicative scale bar 20 µm.

developmental stages can be detected in histological sections of affected tissues (Fig. 7.8). In addition, PCR protocols have been developed for the detection of several important microsporidian species, including *Pseudoloma neurophilia* in zebrafish. However, no PCR protocol exists for the specific detection of *Pleistophora hyphessobryconis*.

7.5.5 Prevention and treatment

Infections with microsporidia are introduced into an aquaculture facility by stocking with infected fish. Because infection might not be detected during an early phase of infection, care has to be taken before fish of unknown infection status are introduced into the facility, in particular in re-circulation systems. For treatment of infections with microsporidia, effective protocols are not established.

FACTS SHEET

Disease: Microsporidiosis
Pathogen: *Pleistophora* and other species
Susceptible host: Neon tetra (*Paracheirodon axelrodi*), goldfish (*Carassius auratus*) and other free-living cyprinids, Indian major carps
Mortality: Moderate to high in affected populations
Temperature: 5–25°C
Prevention: Quarantine and health monitoring protocols should be in place before stocking with susceptible species. Treatment protocols are not established

7.6 Selected References for Further Feading

Adl, S. M., Simpson, A. G. B., Lane, C. E., Lukes, J., Bass, D. *et al.* (2012) The revised classification of eukaryotes. *Journal of Eukaryotic Microbiology* 59, 429–493.

Alderman, D.J. (2008) Fungal diseases of fish. In: Eiras, J.C., Segner, H., Wahli, T. and Kapoor, B.G. (eds) *Fish Diseases.* Science Publishers, Enfield, New Hampshire, pp. 279–349.

Andreou, D., Gozlan, R.E., Stone, D., Martin, P., Bateman, K. *et al.* (2011) *Spherothecum destruens* pathology in cyprinids. *Diseases of Aquatic Organisms* 95, 145–151.

Andreou, D., Arkush, K.D., Guegan, J.-F. and Gozlan, R.E. (2012) Introduced pathogens and native freshwater biodiversity: A case study of *Sphaerothecum destruens*. *PLoS ONE* 7, e36998.

Bruno, D.W. and Wood, B.P. (1999) Saprolegnia and other Oomycetes. In: Woo, P.T.K. and Bruno, D.W. (eds) *Fish Diseases and Disorders: Viral Bacterial and Fungal Infections*, Vol. 3. CAB International, Wallingford, UK, pp. 599–659.

Cervinka, S., Vitovec, J., Lom, J., Hoska, J. and Kubu, F. (1974) Dermocystidiosis – a gill disease of the carp due to *Dermocystidium cyprini* n.sp. *Journal of Fish Biology* 6, 689–699.

Combe, M. and Gozlan, R.E. (2018) The rise of the rosette agent in Europe: An epidemiological enigma. *Transboundary and Emerging Diseases* 65, 1474–1481.

De La Bastide, P.Y., Leung, W.L. and Hintz, W.E. (2015) Species composition of the genus *Saprolegnia* in fin fish aquaculture environments, as determined by nucleotide sequence analysis of the nuclear rDNA ITS region. *Fungal Biology* 119, 27–43.

EFSA (2011) Scientific opinion on epizootic ulcerative syndrome. *EFSA Journal* 9, 2387.

FAO (2009) *What you need to know about epizootic ulcerative syndrome (EUS) – An extension brochure*. FAO, Rome, Italy.

Gozlan, R.E., Marshall, W.L., Lilje, O., Jessop, C.N., Gleason, F.H. *et al.* (2014) Current ecological understanding of fungal-like pathogens of fish: What lies beneath? *Frontiers in Microbiology* 5, 62.

Iqbal, Z., Sheikh, U. and Mughal, R. (2012) Fungal infections in some economically important freshwater fishes. *Pakistan Veterinary Journal* 32, 422–426.

Keeling, P. (2009) Five questions about microsporidia. *PLoS Pathogens* 5, e1000489.

Lom, J. and Dykova, I. (1992) *Protozoan Parasites of Fishes*. Elsevier, Amsterdam, The Netherlands.

Lom, J. and Nilsen, F. (2003) Fish microsporidia: Fine structural diversity and phylogeny. *International Journal for Parasitology* 33, 107–127.

Lom, J. (2008) Microsporidia. In: Eiras, J., Segner, H., Wahli, T. and Kapoor, B.G. (eds) *Fish Diseases*. Science Publishers, Enfield, New Jersey, pp. 351–395.

Pekkarinen, M., Lom, J., Murphy, C.A., Ragan, M.A. and Dykova, I. (2003) Phylogenetic position and ultrastructure of two *Dermocystidium* species (Ichthyosporea) from the common perch (*Perca fluviatilis*). *Acta Protozoologica* 42, 287–307.

Ruggiero, M.A., Gordon, D.P., Orrell, T.M., Bailly, N., Bourgoin, T. *et al.* (2015) A higher level of classification of all living organisms. *PLoS ONE* 10, e0119248.

Stentiford, G.D., Feist, S.W., Stone, D.M., Bateman, K.S. and Dunn, A.M. (2013) Microsporidia: Diverse, dynamic, and emergent pathogens in aquatic systems. *Trends in Parasitology* 29, 567–578.

Torruella, G., de Mondoza, A., Grau-Bove, X., Anto, M., Chaplin, M.A. *et al.* (2015) Phylogenomics reveals convergent evolution of lifestyles in close relatives of animals and fungi. *Current Biology* 25, 2404–2410.

Van den Berg, A.H., McLaggan, D., Dieguez-Uribeondo, J. and van West, P. (2013) The impact of the water moulds *Saprolegnia diclina* and *Saprolegnia parasitica* on natural ecosystems and the aquaculture industry. *Fungal Biology Reviews* 27, 33–42.

Van West, P. and Beakes, G.W. (2014) Animal pathogenic oomycetes. *Fungal Biology* 118, 525–526.

Yi, Z., Berney, C., Hartikainen, H., Mahamdallie, S., Gardner, M. *et al.* (2017) High-throughput sequencing of microbial eukaryotes in Lake Baikal reveals ecologically differentiated communities and novel evolutionary radiations. *FEMS Microbiology Ecology* 93, fix073.

Chapter 8

Bacteria

8.1 Introduction

Fish are continuously exposed to a variety of different bacteria in the water, leading to infections that are multifactorial. The total bacterial amount in pond or aquarium water is usually between 10^2 and 10^6 colony-forming units per millilitre of water, but can be even higher. The amount of bacteria present depends on many different factors, but is especially high when the organic load in the water is significant. Therefore, in aquaculture systems with higher stocking densities, large numbers of bacteria are present, and usually occur as planktonic stages, organised on all water-covered surfaces in biofilms, and colonise the skin and gills of fish. Additionally, the extensive intestinal microbiome of fish is in direct contact with the bacterial populations in the waterbody. Many bacteria occurring in fish-keeping systems are ubiquitous in the water and can multiply and survive without the fish as a host. Most of these bacteria belong to two different groups. There are autotrophic nitrifiers that oxidise ammonia and ammoniac to toxic nitrite and further to less toxic nitrate. Heterotrophic bacteria can utilise and degrade organic matter and, therefore, also play an important role in the aquatic environment. Nevertheless, heterotrophic bacteria have much shorter generation times compared with autotrophic bacteria and they can replicate very fast under conditions of high organic load in the water. Many species of heterotrophic bacteria are known as facultative pathogens for fish, whereas obligate pathogenic bacteria are very rare in fish-keeping systems.

The bacteria occurring in the water can usually be detected on the skin and gills of fish, including those of clinically healthy

 Carps: A Practical Approach to Disease Control and Health Management (eds David Hoole *et al.*)
DOI: 10.1079/9781836990024.0008

individuals. The manifestation of a disease depends on the pathogenicity or virulence of the pathogen, the environmental conditions, and the condition and susceptibility of the fish. Stress and non-optimal environmental conditions are, in most cases, involved in an outbreak of a bacterial infection. Both facultative pathogenic and obligate pathogenic bacteria can cause infections over a wide temperature range. The clinical signs of a bacterial infection are in most cases non-specific and may include skin ulcerations, gill damages, discolouration lethargy and ulcerative necrotic changes in the fins. As a bacterial infection progresses, it can become systemic and in these cases an increase in circumference in the abdominal area, oedema of the subcutaneous tissue, manifests itself in protruding scales. In addition, sometimes ocular changes, such as unilateral or bilateral exophthalmos, can be observed. Autopsy often reveals changes in the internal organs, such as swelling or haemorrhages, especially of spleen and kidneys. In addition, serous fluid is often found in the abdominal cavity.

Bacterial diseases can lead to high mortality and morbidity in a fish stock. Although knowledge about bacterial infections in fish has increased in recent decades, we still do not know nearly as much as we do about comparable diseases in mammals. Knowledge about the epidemiology of many pathogens and the management of bacterial diseases in fish stocks is sometimes incomplete. Also, by means of the standard methods used in routine diagnostic laboratories for species identification of bacteria, ubiquitous bacteria from aquatic environments cannot always be correctly identified, as some species are not included in the databases for comparison. Therefore, often no proper diagnosis of a bacterial infection can be made. Assessment of the significance of the detected bacteria is difficult, especially considering that the skin and gills of clinically healthy fish are always colonised by bacteria.

8.2 Pathogenesis

Bacterial infections in carp are, in most cases, secondary infections, and only a few bacterial species are considered as obligate pathogenic for these fish species. The main entry points for bacteria are the skin, the gills and the gastrointestinal tract. To induce signs of a disease, it is necessary for bacteria to be able to attach themselves to these tissues and eventually invade them. The first barrier to invading bacteria is the mucous layer that covers the fish's skin and gills, as well as lining the gastrointestinal tract, and most bacterial infections in fish are caused by colonisation of these mucosal surfaces with potentially pathogenic bacteria. A dysbacteriosis of the physiological microbiome on mucous membranes facilitates the colonisation of potentially pathogenic bacteria. The bacteria can adhere to the mucosa of the skin, gills and intestine, although the intensity of this adherence depends on the bacterial species and on the composition of the mucus. The adhesiveness of specific bacterial isolates seems to be based on different extracellular proteins that influence virulence. Various pathogenicity factors further contribute to the isolates' ability to damage host cells. In this way, extracellular toxins produced can be applied to host cells by means of secretion systems, or cells can be damaged by the secretion of proteases, lipases and nucleases. If the mucous layer is already damaged due to other causes, it is particularly easy for the bacteria to settle in these previously damaged areas. Circumstances that can lead to damage to the mucous layer are numerous and can include, for example, an infestation with parasites or increased levels of nitrogen compounds such as ammonia or nitrite in the water. Bacterial infections of the skin usually first manifest themselves as increased mucus formation and reddening of a localised area of the skin. Oedema of the scale pockets and subsequent loss of scales can also occur. This is followed by further tissue destruction, which can result in deep ulcerations that might also affect the underlying muscles. Bacterial infections of the gills usually initially manifest themselves as increased mucus production and swelling of the gill epithelium due to cell oedema. Subsequently, focal necrotic areas can appear in the gill epithelium. Bacterial infections of the intestinal tract are rarely observed in carp. They can manifest as reddening of the intestinal mucosa, mucosal oedema, and subsequent necrotic areas within the epithelia of the digestive tract. Subsequently septicaemia can occur, and carp can show typical clinical signs like ascites and water retention in the swim bladder, as well as bleedings and oedema in internal organs.

8.3 Diagnostics

Because aquatic bacteria can often occur as secondary pathogens, it is essential that a possible primary cause should always be clarified first if a bacterial infection is suspected. A range of other causes may occur and hence a thorough parasitological and virological examination, as well as the assessment of the environmental conditions, such as chemical water parameters, is recommended. When taking samples for a suspected bacterial infection, great care must be taken, especially in the case of skin changes. The sample should only be taken with a fine sterile swab from the edge of the alteration, since in the central area of damage there are usually ubiquitous heterotrophic bacteria and fungi to be found in the water, the detection of which makes it difficult to interpret the bacteriological test result.

The swabs should be processed as soon as possible in a laboratory with the appropriate equipment and the necessary knowledge. If the sample has to be sent to such a specialised laboratory, then it is recommended that it be transported in a cooled state as quickly as possible for diagnosing bacterial infectious agents.

The detection of bacterial infectious agents in fish differs in many areas from that in mammals and requires some relevant experience and appropriate equipment. Although normal culture media can be used to cultivate the bacteria, the incubation temperature for bacterial species from the aquatic environment must be well below the 37°C that is usual for the culture of bacteria from mammals. Some of the bacteria relevant to fish will only show growth on media at temperatures of 20°C or lower. It should be noted that such temperatures cannot be maintained with the usual incubators, which can heat but not cool, especially in summer. It is therefore essential to use a cooled incubator in which the desired temperature can be maintained regardless of the outside temperature. The samples from carp are cultivated in cooled incubators at a temperature that roughly corresponds to the water temperature of the waterbody from which the examined fish was obtained. Due to the lower temperature, the samples must be incubated for a longer period than samples from mammals or birds, normally for several days with daily checks. The difficulty here is the usually very good growth of motile aeromonads and other heterotrophic bacterial species, which under certain circumstances can overgrow slowly growing bacteria. Furthermore, different bacterial species sometimes have a very similar colony morphology, which makes it difficult to distinguish between them.

The identification of cultivated bacteria is currently carried out in diagnostic laboratories mostly by means of MALDI–TOF MS or by sequencing a section of the 16S rRNA gene. Biochemical identification methods are also sometimes used, such as analysis of specific enzymes and fermentation of sugars. A comparison of the result with entries in the corresponding databases is often difficult, since bacteria from the aquatic environment, which make up most of the bacteria detected in diseased fish, are sometimes not found or only occur in small numbers. In some cases, this makes it impossible to clearly identify the bacteria using routine diagnostics, and there is therefore an urgent need to optimise the methods and databases utilised. This is especially the case when using commercial biochemical test systems, as these tests are only conditionally suitable for diagnosing bacterial pathogens in fish.

The interpretation of the results can also prove difficult, as bacteria are always detected on the skin and gills, even in clinically healthy carp, and often from the internal organs. In addition, if more than one bacterial species is detected (which is usually the case) it is not always clear which of these species are causally involved in the disease. Processing of the sample can also cause some issues. For example, some bacterial species survive transport in a medium better than others, which means that relevant bacteria can either have died off completely during transport or only a few survive, which might be overgrown by other bacteria during incubation. Thus, primary pathogens are no longer detectable in some cases. This is particularly common in chronic diseases where the primary pathogen can no longer be detected. It must therefore be assumed that only part of the bacterial fauna involved in diseases in fish can be cultured. For these reasons, to interpret the results accurately, possible difficulties in sampling, cultivation and identification of the bacteria must be taken into account. If ubiquitous, mostly heterotrophic bacteria from the aquatic environment are detected together with parasitic infections or water quality problems, the parasitic infections should be treated first or the water quality optimised. Only after these have been eliminated as possible causes of ill health in fish should antibiotic therapy be considered. A decision on treatment strategies must be made on a case-by-case basis as to whether and which measures or treatments should be initiated. An antibiotic treatment should only be performed if the bacterial infection is clearly the reason for the clinical signs seen in carp.

This is especially important since bacteria in fish already show a very wide range of antibiotic resistances. Due to this, an antibiogram should always be carried out for bacteria from the aquatic environment, as suggested by the Clinical and Laboratory Standards Institute (CLSI). Unfortunately, very few specifications are available for the interpretation of the results. For example, breakpoints for determining minimum inhibitory concentrations (MIC) are not available for all bacteria of the genus *Aeromonas*. The same applies to various other bacteria from the aquatic environment. Therefore, evaluation criteria of other animal species or epidemiological cut-off values are used for the interpretation of the resistance tests.

8.4 Prophylaxis and Treatment

Vaccination against specific pathogens is, in many cases, the most effective way of prophylaxis. Nevertheless, in Europe, only a few commercially available vaccines against bacterial pathogens for fish are approved and none are approved for carp species. There are vaccines against *Yersinia ruckeri*, *Vibrio* species and *Flavobacterium* species for salmonids available that unfortunately cannot be easily used in carp. In individual cases, in the case of regular outbreaks of disease signs caused by one specific bacterial pathogen in an aquaculture system, the use of a herd-specific vaccine can be considered. Prophylactic measures such as tank hygiene, quarantine of new arrivals and the separation of individual batches of fish or crustaceans are recommended in any case. This includes maintaining good water quality, feeding the animals according to their needs, and regular cleaning and disinfection of the equipment. Feed additives such as probiotics, which can stimulate the immune system of the fish, can also be utilised. It should be noted that, when buying new fish, it is recommended to keep them in quarantine for several weeks. This makes sense, not only for reasons of possible bacterial infections, but also because of the possibility of the presence of other diseases.

Antibiotic treatment should be carefully considered in every case, since antibiotic agents affect not only the pathogenic bacteria, but also the bacteria of the natural microbiome. Nevertheless, after the outbreak of a clinically manifest bacterial infection with obligate or facultative pathogens, the only option left is antibiotic treatment. Food-producing fish, especially in recirculation systems, are usually treated with an antibiotic via the feed. This type of treatment can be relatively stress-free for fish, as the animals can remain in their familiar surroundings. In this case, it is unavoidable to treat the entire shoal, whereby care must be taken to ensure that the affected animals are also treated. If the general condition is disturbed, however, the affected fish may no longer eat any food. The fish can also be injected intramuscularly with an antibiotic, but this is not common in many countries for economic and work-related reasons, especially when large numbers of fish must be treated. Treatment by an application of antibiotics to the water of the aquaculture system is not recommended, due to the effects on the autotrophic bacteria and thus the chemical water quality. As moving the animals to quarantine tanks for antibiotic treatments is usually not possible, due to the large number of individuals, this form of application is not used regularly. In any case, antibiotic treatment of fish should be carried out for a sufficiently long period, usually 8–10 days, and with a sufficiently high dosage of the active substance. Both under-dose use and an insufficiently long therapy duration can lead to the development of resistant bacteria, which must be avoided. The legal regulations on reallocations and waiting times are different in different countries.

The specific antibiotic agents to be used and their dosages, together with the tolerability of different fish species, can vary greatly. The development of antimicrobial resistance (AMR) of bacterial species adds another complication in prescribing the actual antibiotic and its dose to be used. It is important to consult an appropriate professional before antibiotics are applied, as overuse or underuse of the antibiotic can lead to the development of AMR.

8.5 Gram-Negative Disease-Causing Bacteria

Most bacteria that can be isolated from fish and the aquatic environment are Gram-negative, i.e. they do not retain the crystal violet stain that is part of Gram staining, a method for bacterial identification. The reason for this is the structure of the cell wall of these bacteria. Different from Gram-positive bacteria, Gram-negative bacteria have an outer lipid membrane and only a thin peptidoglycan layer. This structure makes Gram-negative bacteria more resistant to disinfectants and multiple drugs, including many antibiotics.

8.5.1 Aeromonadaceae

Among the family Aeromonadaceae, bacteria of the genus *Aeromonas* are important pathogens for carp. The bacteria belong to the class Gammaproteobacteria, are Gram-negative and are ubiquitous in freshwater and brackish water habitats. Depending on the species, they are known as obligate or facultative pathogens of fish in general. A distinction is made between mesophilic and psychrophilic species within the genus *Aeromonas*. The only psychrophilic and the only immobile species within the genus, *Aeromonas salmonicida*, has been described as an obligate pathogen for fish. The subspecies *A. salmonicida* subsp. *salmonicida* is the causative agent of furunculosis in salmonids, while other subspecies can also lead to diseases in other fish species, like carp. Some *A. salmonicida* isolates only grow at temperatures below 20°C and therefore disease outbreaks are usually seen in cold water. All other species of the genus *Aeromonas* are mesophilic and motile. These species are considered facultative pathogens and often only lead to bacterial infections if the environmental conditions, pathogen pressure, health status and condition of the animals are suboptimal. Unlike *A. salmonicida*, motile aeromonads show growth over a very wide temperature range between 4°C and 37°C. By 2016, at least 31 species within the genus *Aeromonas* had been identified. Motile aeromonads can almost always be detected from the skin of clinically healthy freshwater fish. In clinically manifest infections caused by motile aeromonads, dermatitis, deep ulcerations of the skin, abdominal oedema, exophthalmos and septicaemia can occur. Motile aeromonads are not host-specific and diseases have been described in a variety of fish species. In addition, different species of motile aeromonads are considered zoonotic pathogens that can cause gastrointestinal infections and wound infections in humans. Various factors contribute to the virulence of bacteria of the genus *Aeromonas*, which is why it is considered multifactorial. The virulence of a specific isolate is mainly caused by the formation of S-layers and by surface polysaccharides, iron-binding systems, exotoxins and extracellular enzymes, secretion systems, fimbriae and other adhesins, and flagella. Exotoxins secreted by the *Aeromonas* genus include aerolysins, nucleases, lipases and proteases. In particular, the secretion of aerolysins, which are cytolytic pore-forming toxins specific to bacteria of the genus *Aeromonas* in particular *Aeromonas hydrophila*, appear to play a crucial role in the virulence of mesophilic species of the genus *Aeromonas*. In the pathogenesis of *A. salmonicida* subsp. *salmonicida*, the lipases H3 and GCAT (glycerophospholipid-cholesterol acetyltransferase) in particular seem to be decisive. Overall, lateral flagella, the cytolytic enterotoxins Ast and Alt, aerolysins, haemolysins and GCAT are considered to be the five most important pathogenicity factors of bacteria of the *Aeromonas* genus.

Aeromonas hydrophila

AETIOLOGY. *A. hydrophila* is a Gram-negative, facultative anaerobic, non-sporulating, rod- to cocci-shaped bacterium. Different subspecies of this species are described, such as *A. hydrophila* subsp. *hydrophila*, *A. hydrophila* subsp. *dharkensis* and *A. hydrophila* subsp. *ranae*. The former subspecies *A. hydrophila* subsp. *proteolytica* was reclassified and now is known as *Vibrio proteolyticus*. *A. hydrophila* subsp. *ranae* can induce infections in frogs and *A. hydrophila* subsp. *dharkensis* has been isolated from diseased children. *A. hydrophila* subsp. *hydrophila* is an infectious agent for fish that causes a disease known as motile *Aeromonas* septicaemia (MAS) or bacterial haemorrhagic septicaemia. These bacteria, together with *Aeromonas sobria*, are one of the causative agents in the so-called red spot discase in farmed grass carp (*Ctenopharyngodon idella*). The bacteria are equipped with a polar flagellum that enables them to move actively. *A. hydrophila* is ubiquitous in aquatic ecosystems. In intensively managed fishponds, more than a thousand bacteria of this species can occur per millilitre of water. Although the bacteria are more common in freshwater, they also occur in seawater. All freshwater fish can contract an infection, and hence the bacterium is distributed worldwide and can often be found in large numbers, particularly associated with heavily organically polluted water bodies. *A. hydrophila* can also cause diseases in many other vertebrates, including amphibians and mammals, and is thus a zoonotic agent. In humans, gastrointestinal and systemic infections have been described. However, since the bacterium is ubiquitous, and people are regularly exposed to it, the risk of zoonosis can be assessed as relatively low.

A. hydrophila, which can be part of the microbiome of aquatic organisms, is very often involved as a secondary pathogen in bacterial infections of fish. At water temperatures above 8°C and especially in spring and summer, when the water temperature increases, disease outbreaks can occur. The highest mortality rates for goldfish were detected at water temperatures between 17°C and 25°C.

Several factors appear to determine the virulence of different isolates of *A. hydrophila*. Virulence factors like exotoxins such as enterotoxin, haemolysin, lipases and proteases have been described, and the formation of an S-layer and an increased production of elastase are more prominent in virulent isolates.

PATHOGENESIS. Fish can become infected via the water, especially when the skin and gills have been damaged. The environment, in particular high temperatures, organic pollution and lack of oxygen, together with the immune status of the animal and possible primary diseases, play a role in the pathogenesis. Skin lesions caused by an infestation with ectoparasites or due to suboptimal water values are very often secondarily colonised by *A. hydrophila*. In the pathogenesis of virulent strains, extracellular products of the bacteria seem to play a crucial role and can induce high mortalities in only a few hours. In these cases, cell death occurs mainly in the kidney, primarily due to cytotoxic and haemolytic activities of the extracellular products.

CLINICAL SIGNS OF THE DISEASE. In carp species, infections with *A. hydrophila* can be peracute to chronic, with latent infections also occurring. In peracute cases, the animals usually show no clinical symptoms, but are dying very rapidly. Ulcerative skin lesions, necrosis of the fins and septicaemia occur in acute cases (Fig. 8.1). The skin lesions can be superficial to deep. Large, long-standing ulcerations that can extend into the muscles and abdominal oedema can be observed in the chronic form. In some cases, deep ulcerations lead to septicaemia, which may be characterised by exophthalmos and ascites. Oedema in the scale pouches, abscesses, haemorrhages in the gills and exophthalmos also occur regularly. In systemic forms, anorexia and darkening of the body can usually be observed. Mortality can be very high in a population.

PATHOLOGY. In dissections, oedema of the internal organs, haemorrhages and diffuse necrosis are common. Necrosis of the intestinal mucosa with detachment of the epithelium can often be seen histologically.

DIAGNOSIS. *A. hydrophila* survives well in water and grows easily on common culture media, such as CSB agar (Fig. 8.2). The bacterium can be cultivated within a wide range of temperatures and grows even at temperatures of 37°C. For these reasons, the bacterium can very often be isolated from diseased carp. Of course, diagnostic samples should still be incubated at lower temperatures and for a longer period of time in order to be able to identify possible other bacteria that may be present. Identification of certain species of motile aeromonads is possible for most species by MALDI–TOF

Figure 8.1 Carp showing external clinical signs of an infection with motile aeromonads. (A) An extensive skin ulceration and necrosis. (B) Dorsal bacterial lesion. (C) Deformed mouth of common carp due to bacterial infection of the lower jaw. (A) and (B) courtesy of the Environment Agency, National Fisheries Laboratory, UK.

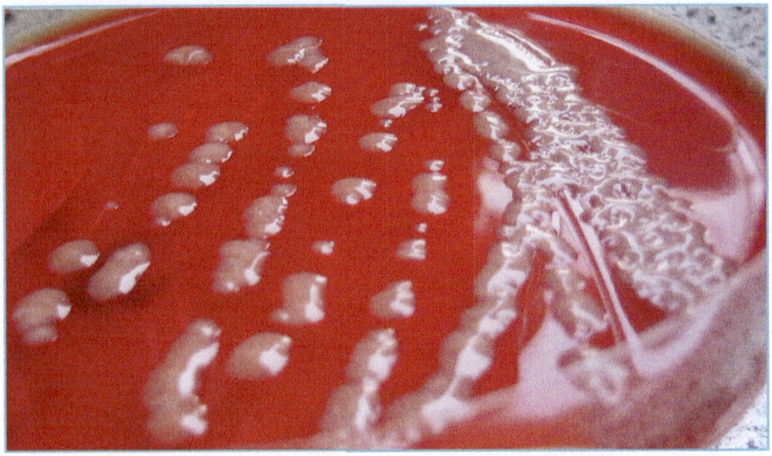

Figure 8.2 Colonies of *A. hydrophila* growing on Columbia sheep blood agar.

MS or by sequencing specific housekeeping genes, like the Gyrase B gene. Biochemical methods and 16S rRNA sequencing, however, regularly only secure genus identification; species identification is particularly difficult. After having a diagnosis, it is especially important to interpret the results correctly. As already mentioned, the bacterium is often found on the skin and the gills of clinically healthy fish. Therefore, only the detection of a significant number of these bacteria in connection with the clinical signs is meaningful. As *A. hydrophila* can also overgrow other bacterial species due to its very rapid growth, the highest level of significance is obtained when examining several fish in a population and preferably taking samples of the kidneys.

TREATMENT. Infections with *A. hydrophila* are usually secondary. Therefore, the treatment of possible parasitic infections and an optimisation of the environmental conditions are necessary in most cases. If an antibiotic treatment is necessary, whenever possible the antibiotic substance should be chosen according to the results of an antibiogram, as *A. hydrophila* often shows multiple resistances.

FACTS SHEET

Disease: Motile *Aeromonas* septicaemia (MAS)/Bacterial haemorrhagic septicaemia
Pathogen: *Aeromonas hydrophila*
Susceptible hosts: All carp species
Clinical signs: Skin alterations and ulcerations, oedema in the scale pouches, exophthalmos
Pathology: Oedema, haemorrhages, and necrosis in the internal organs, ascites
Mortality: Can be high, especially in warm water temperatures
Temperature: Growing between 4°C and 37°C; clinical disease signs mostly at warmer temperatures
Prevention and treatment: Prevention is by optimisation of environmental conditions, reduction of ectoparasites. Treatment is dependent on the result of the antibiogram

Aeromonas sobria

AETIOLOGY. *Aeromonas sobria*, a Gram-negative, motile, rod-shaped bacterium, can be detected in diseased carp at about the same frequency as *A. hydrophila*. These two bacteria cause similar clinical symptoms and have similar characteristics. Together with *A. hydrophila*, *A. sobria* is the causative agent of red spot disease in grass carp (*Ctenopharyngodon idella*). The bacterium can also cause clinical signs of a disease in other carp species, including common carp (*Cyprinus carpio*), silver carp (*Hypophthalmichthys molitrix*) and rohu (*Labeo rohita*).

PATHOGENESIS. *A. sobria* usually causes secondary infections in carp species and can be isolated in polluted waters and in nutrient-rich freshwater. Poor management practices, as well as parasitic and other infections, are also associated with clinical infections. Fish normally become infected through the water.

CLINICAL SIGNS OF THE DISEASE. Affected fish show skin lesions, haemorrhagic septicaemia, or fin rot. Symptoms are very similar to those of *A. hydrophila* infections (see Fig. 8.1). In rohu, loss of the escape reflex and skin darkness associated with skin haemorrhages have been described.

PATHOLOGY. As in infections with other motile aeromonads, oedema, haemorrhages and diffuse necrosis in the internal organs are often associated with infections with *A. sobria*. Changes of the gills are regularly seen, and haemorrhagic gill epithelia and epithelial hyperplasia can be detected. Degenerative changes of the epithelium and hypertrophy of epithelial cells have also been described.

DIAGNOSIS. See *Aeromonas hydrophila.*

TREATMENT. Environmental optimisation is advisable before treating with antibiotics. Since resistance to different groups of active substances occurs, antibiotic therapy should only be given after a resistance test has been undertaken.

FACTS SHEET

Disease: Motile *Aeromonas* septicaemia (MAS); Motile *Aeromonas* infection (MAI); Red spot disease in grass carp
Pathogen: *Aeromonas sobria*
Susceptible hosts: All carp species
Clinical signs: Skin lesions, haemorrhagic septicaemia, fin rot
Pathology: Oedema, haemorrhages, diffuse necrosis in the internal organs, gill changes
Mortality: Can be high, especially in warm water temperatures
Temperature: Growing between 4°C and 37°C; clinical disease signs mostly at warmer temperatures
Prevention and treatment: Prevention is by optimisation of environmental conditions, reduction of ectoparasites. Treatment is dependent on the result of the antibiogram

Aeromonas salmonicida

AETIOLOGY. *Aeromonas salmonicida* is a Gram-negative, non-spore-forming, facultative anaerobic, rod-shaped bacterium. Unlike other aeromonads, it is immobile. It is considered as an obligate pathogenic bacterium for fish, causing different clinical signs in a variety of fish species. Outbreaks of diseases are stress-related and can lead to high mortalities, although the bacterium has a limited survival rate when not associated with a fish. *A. salmonicida* was first described in 1894 after an outbreak of the disease in trout in Germany. The bacterium is also widespread in carp almost worldwide, as shown by reports from Israel, Finland and Japan. It causes disease in freshwater, brackish and saltwater fish. *A. salmonicida* is perhaps best known as a bacterial pathogen of salmonids. However, five subspecies of this genus are confirmed that can be divided into the typical strain *A. salmonicida* subsp. *salmonicida*, the causative agent for furunculosis in salmonids, and four atypical subspecies, including *A. salmonicida* subsp. *smithia*, *A. salmonicida* subsp. *achromogenes*, *A. salmonicida* subsp. *masoucida* and *A. salmonicida* subsp. *pectinolytica*. High morbidity and mortality caused by *A. salmonicida* infections are mainly due to atypical subspecies of the bacterium, which are found in different non-salmonid species. In cyprinids, *A. salmonicida* infections have been reported in common carp, crucian carp (*Carassius carassius*) and largemouth bronze gudgeon (*Coreius guichenoti*). In studies, 76.9% of the tested *A. salmonicida* isolates were pathogenic for carp species. Infections with *A. salmonicida* also affect ornamental koi carp and goldfish (*Carassius auratus*), and are believed to be the main pathogenic agent in 'koi ulcer disease' (KUD) and 'goldfish ulcer syndrome', respectively. In common carp, infections with *A. salmonicida* are known as 'carp erythrodermatitis', 'carp furunculosis' or 'infectious ascites', resulting from haemorrhages in the skin. In cyprinid species especially, infections with atypical subspecies have been diagnosed; however, *A. salmonicida* subsp. *salmonicida* was recently detected in clinically diseased crucian carp. Even if the mortality remains low in a farm, affected fish cannot be sold as food or for stocking, because of the severe skin damage. The disease is therefore of great economic importance.

PATHOGENESIS. Fish become infected with *A. salmonicida* via the water. The disease can be associated with infestation with parasites that damage the mucous membrane. In particular, damage caused by infections with carp lice (*Argulus foliacaeus*) or fish leeches (*Piscicola geometra*) are considered the entry point for the pathogen. The bacterium can remain infectious in freshwater for about 17 days. Bacterially infected skin lesions tend to be the main source of infections for other fish within the waterbody. Bacterium virulence factors, which vary between strains, include leucocidin, which leads to leucopenia, and proteases. The production of several exotoxins and exoproteases has also been demonstrated in various strains of *A. salmonicida*.

CLINICAL SIGNS OF THE DISEASE. Atypical subspecies of *A. salmonicida* usually cause skin ulcerations without septicaemia (Fig. 8.3). The skin alterations range from whitish changes to deep ulcerations, which can also affect the underlying muscles and bone tissues. In many cases, individual scale pockets initially appear inflamed. Necrosis of the epidermis, the dermis and finally the underlying tissue develops. Bleeding at the base of the fins, in the mouth or orbit area can also occur. Further clinical signs can include exophthalmos, haemorrhage in the gills, severe bleeding on the body of the infected fish and anaemia. The infection usually does not become systemic until it is advanced, if at all. Other secondary pathogens, such as motile aeromonads and pseudomonads, can also attack the already damaged skin. Infections with atypical *A. salmonicida* subspecies in *Carassius* spp. are known to lead to necrosis and ulceration of the skin and to degeneration of the muscles. Erythrodermatitis can occur from subacute to chronic stages and can be detected at water temperatures greater than 4°C. Water temperatures appear to play

Figure 8.3 Carp showing a typical circular skin ulceration with a central necrotic area surrounded by inflamed reactive area caused by *A. salmonicida* infection. Courtesy of the Environment Agency, National Fisheries Laboratory, UK.

an important role in the expression of the disease, which usually occurs above 12°C. The higher the water temperature, the clearer the symptoms become. Although morbidity can be very high, mortality is rather low to moderate. Different carp species seem to have different susceptibility to infections and younger animals are more susceptible than older ones.

PATHOLOGY. In dissections, especially of the intestine, liver, spleen and kidney, varying degrees of inflammatory damage can be seen. Hyperplasia of goblet cells of the intestinal mucosal epithelium and degeneration and necrosis of the renal tubular epithelium cells have been described in infected crucian carp (*Carassius carassius*). In histopathological examinations, multifocal necrosis and infiltration of inflammatory cells in gill, liver, kidney and intestine can be detected in diseased fish.

DIAGNOSIS. The diagnosis is often difficult, due to the comparatively unstable pathogens and the special requirements for cultivation, especially in the case of skin ulcers. Typical subspecies of *A. salmonicida* often cannot be grown at temperatures above 18°C, whereas atypical isolates show good growth on agar plates at higher temperatures up to 28°C. Blood agar (CSB) can be used for culture. Typical isolates of the bacterium form a brown pigment on culture media like TSA, while atypical isolates only form very late or not at all. The colonies grow slowly, and therefore the incubation period should be at least 5–7 days. Since the bacterium can very quickly be overgrown by other bacterial species, such as *A. hydrophila*, the culture medium should be examined closely every day. Atypical *A. salmonicida* isolates grow in small grey colonies. It is recommended that samples are taken from several individual fish, especially from newly infected animals. In the case of acute infections, there is a greater probability that secondary infections have not yet occurred and that the pathogen can be isolated. Overall, false-negative findings are not uncommon, due to the problems mentioned. Identification of *A. salmonicida* is possible by MALDI–TOF MS, sequencing of the Gyrase B gene and, although limited, also by sequencing of the 16S rRNA gene.

TREATMENT. Antibiotic therapy can be carried out after preparation of an antibiogram. Local treatment of the ulcerations is also helpful. Since the skin changes are very quickly colonised by secondary pathogens, environmental optimisation must be carried out in all cases. After an outbreak of an infection of *A. salmonicida*, the animals should be closely observed and optimal environmental conditions maintained, otherwise carrier fish can cause re-emergence of disease outbreaks. The disease has a relatively good tendency to heal under optimal conditions and appropriate hygiene practice. Commercial vaccines are not available. Herd-specific vaccines, accompanied by good hygiene practices, are very well suited to reducing the infection pressure.

FACTS SHEET

Disease: Carp erythrodermatitis, Carp furunculosis, Infectious ascites, Koi ulcer disease, Goldfish ulcer syndrome
Pathogen: *Aeromonas salmonicida* (typical subspecies *salmonicida* and different atypical subspecies)
Susceptible hosts: *Cyprinus carpio*, *Carassius* spp.; other carp species may also be infected
Clinical signs: Skin ulcerations most prominent, exophthalmos, haemorrhages in the gills
Pathology: Inflammation, degeneration, and necrosis of internal organs
Mortality: Mostly moderate to low
Temperature: Mainly above 12°C
Prevention and treatment: Prevention is by optimisation of environmental conditions, reduction of ectoparasites, vaccination with herd-specific vaccines. Treatment is dependent on the result of the antibiogram

Aeromonas veronii

AETIOLOGY. *Aeromonas veronii* is a Gram-negative, motile, rod-shaped bacterium, which causes disease symptoms that are very similar to those caused by other motile aeromonads. Severe disease outbreaks due to infections with *A. veronii* have been described in crucian carp (*Carassius gibelio*) in China and in common carp in Korea. The bacteria can also cause clinical signs of a disease in other carp species, for example in grass carp and rohu.

PATHOGENESIS. Like other motile aeromonad species, *A. veronii* often causes secondary infections in carp. Fish usually become infected through the water, as the bacteria are ubiquitous in freshwater environments. Six virulence genes related to pathogenicity have been detected after disease outbreaks in *A. veronii*. These factors included aerolysin, cytotonic enterotoxins, elastase, cholesterol acyltransferase, lipase and serine protease.

CLINICAL SIGNS OF THE DISEASE. Diseased fish show skin lesions, which can comprise reddish foci, abdominal distension and fin rot (see Fig. 8.1). In affected common carp in Korea, the mortality rates reached up to 2% of the total fish in a farm per day. In experimental infections the lethal dosage of an *A. veronii* isolate was determined as 1.31×10^7 CFU/ml.

PATHOLOGY. Enteritis, liver congestion, enlarged spleen and kidney, and a general infiltration of inflammatory cells can be seen in infected carp. Histologically, alterations of the liver tissue, such as degeneration of hepatocytes and congestion in sinusoids, have been detected. Additionally, haemorrhages in the spleen, necrosis of the renal tubules and glomerular destruction in the kidney, necrosis and haemorrhages in the intestine, and haemorrhages in the muscle have also been observed.

DIAGNOSIS. See *Aeromonas hydrophila*.

TREATMENT. As in infections with other motile aeromonads, environmental optimisation is necessary before treating with antibiotics. An antibiotic therapy should only be given after a resistance test.

FACTS SHEET

Disease: Motile *Aeromonas* septicaemia (MAS), Motile *Aeromonas* infection (MAI)
Pathogen: *Aeromonas veronii* (and other motile aeromonads)
Susceptible hosts: Crucian carp, common carp, grass carp, rohu, other carp species
Clinical signs: Skin lesions, haemorrhages, fin rot
Pathology: Haemorrhages, necrosis in the internal organs
Mortality: Can be high
Temperature: Clinical disease signs mostly at warmer temperatures
Prevention and treatment: Prevention is by optimisation of environmental conditions, reduction of ectoparasites. Treatment is dependent on the result of the antibiogram

Other motile aeromonads

Various other species of motile aeromonads might induce disease signs in carp species under certain circumstances. Among these species are *A. allosaccharophila*, *A. aquariorum* (syn. *A. dharkensis*), *A. bestiarum*, *A. bivalvium*, *A. caviae* (syn. *A. punctata; A. culicola*), *A. dharkensis, A. diversa*, *A. encheleia*, *A. enteropelogenes* (syn. *A. trota*), *A. fluvialis*, *A. ichtyosmia*, *A. jandaei*, *A. media*, *A. mollus-corum*, *A. piscicola*, *A. popoffii*, *A. rivuli*, *A. sanarellii*, *A. schubertii, A. sharmana*, *A. simiae, A. taiwanensis, A. tecta* and *A. trota* (syn. *A. enteropelogenes*). Especially *A. caviae* could be detected in high frequencies in the Indian major carp catla (*Gibelion catla*), rohu (*Labeo rohita*) and mrigal (*Cirrhinus mrigala*).

8.5.2 Chlamydia-like bacteria

Aetiology

The disease epitheliocystis has been reported in more than 90 freshwater and marine fish species, including several species of carp. The aetiological agents are chlamydia-like organisms that belong to the order Chlamydiales. Molecular biological examinations reveal that these organisms are related to *Neochlamydia* spp., *Protochlamydia* spp. and *Piscichlamydia* spp. Chlamydia are small, coccoid, Gram-negative microorganisms that only occur intracellularly. They go through a special reproductive cycle in the host cell, in which small elementary bodies transform into larger initial bodies, which then divide and become elementary bodies again. These pathogens have a high pathogenicity and can lead to large losses in a fish stock.

Pathogenesis

As the pathogen cannot be cultured and infection studies are difficult to process, firm data on the pathogenesis is not available.

Clinic signs of the disease

Infected fish may be sluggish and exhibit swimming disorders. Hyperplasia of the epithelium of gills and skin can occur and respiration difficulties arise due to gill damage. In severe cases large mortalities can occur in the fish stock.

Pathology

White cysts of 20–400 μm diameter can be found in the gills and in the skin of affected fish. Cells of the gill and skin epithelium are enlarged due to the proliferation of the bacteria. Hypertrophy of the gill epithelium and fusion of the secondary lamellae can also be seen (Fig. 8.4).

Diagnosis

The clinical signs are the first indication of epitheliocystis. Histological examination reveals eosinophilic inclusions in the gills, which enclose chlamydia-like structures. Although cultivation of the

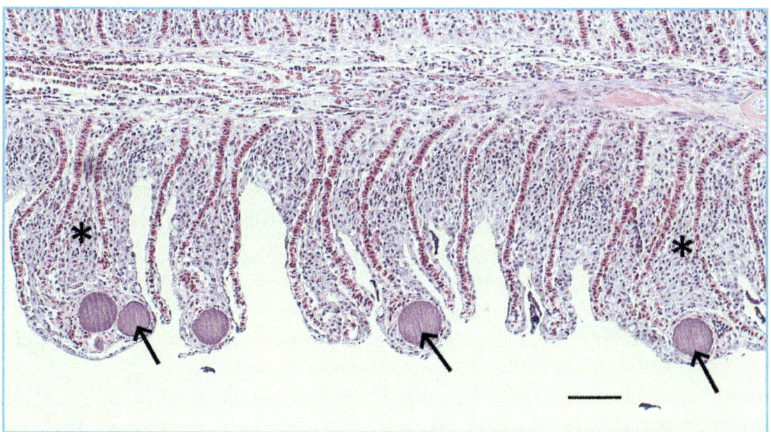

Figure 8.4 Histopathological section of carp gill showing infection with chlamydia-like organism (arrow). Note hypertrophy of the gill epithelium and fusion of the secondary lamellae (*). Indicative scale bar 100 μm. Courtesy of the Environment Agency, National Fisheries Laboratory, UK.

pathogens has not been achieved, molecular biological examinations for chlamydia-like organisms can be performed.

Treatment

Therapy is not yet possible. In some cases, an improvement in fish health was observed with antibiotic treatment. Prophylactic measures are therefore crucial to avoid introducing the infection, and newly acquired fish should be placed in a quarantine tank for at least 3 weeks.

FACTS SHEET

Disease: Epitheliocystis
Pathogen: Chlamydia-like organisms
Susceptible hosts: *Cyprinus carpio, Carassius gibelio, Carassius auratus, Ctenopharyngodon idella, Labeo rohita,* most probably also other carp species
Clinical signs: Shortness of breath, cysts on skin and gills
Pathology: Cysts on skin and gills
Mortality: Can be high, especially when fish are kept in high stocking densities and in juveniles
Temperature: Large temperature range
Prevention and treatment: Avoidance of introducing the pathogen with new fish. Antibiotic treatments could be tried, although success is questionable

8.5.3 Enterobacteriaceae

The family Enterobacteriaceae consist of more than 100 different bacterial species, many of which are commonly found in the gut of humans or animals. The Enterobacteriaceae comprise not only well-known pathogens for humans, but also important fish pathogens, such as *Citrobacter*, *Edwardsiella* and *Plesiomonas.*

Citrobacter spp.

AETIOLOGY. *Citrobacte*r spp. are Gram-negative, motile rod-shaped bacteria commonly found in soil, water and sewage, and from diseased or clinically healthy animals. Some species of *Citrobacter* can act as opportunistic pathogens in carp, and *Citrobacter freundii* tends to be the most abundant species. The pathogens occur in both freshwater and seawater, particularly when these environments are heavily polluted with organic materials. Different carp species, as well as other fish species, are infected by *Citrobacter* spp., among them common carp, crucian carp and grass carp. *Citrobacter freundii* plays also an important role as an infectious agent in ornamental fish, such as koi carp.

PATHOGENESIS. Stress is one of the known predisposing factors leading to disease outbreaks due to *Citrobacter* spp. in fish. Changes of the environmental conditions of the waterbody, such as drop in temperature, very high organic loads, low oxygen levels or high amounts of ammoniac, are especially important in supporting infections. Fish usually become infected through the bacteria that are present in high numbers in the water. In experimental infections with common carp and grass carp, *Citrobacter* had only a low to moderate virulence and the haemolysins present seemed not to be responsible for the pathogenesis of the infection.

CLINICAL SIGNS OF THE DISEASE. Infected fish show uncoordinated swimming movements, loss of appetite and skin changes, such as superficial reddening (Fig. 8.5). Bleeding can also occur in the eyes and at the bases of the fins. Carp fry and fingerlings are especially prone to the development of a septicaemia and the losses in a farm

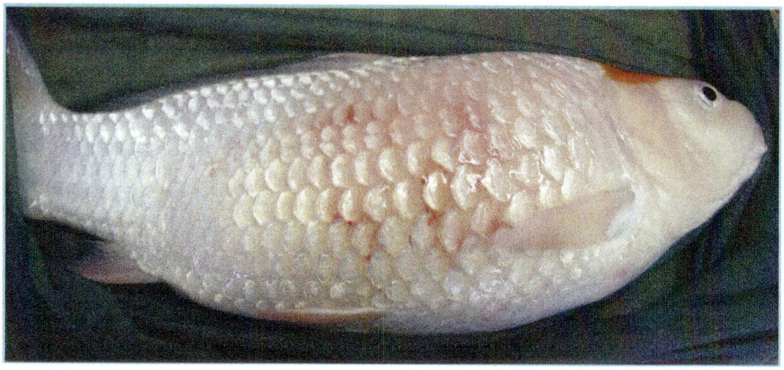

Figure 8.5 Koi carp with superficial reddening of the skin related to infection with *Citrobacter* sp.

can be high. In grass carp, gastroenteritis due to *Citrobacter* spp. infections is an increasing problem.

PATHOLOGY. A reddening of the spleen, a pale liver, haemorrhages in the peritoneum, and swelling and granulomas in the kidney are described in diseased carp.

DIAGNOSIS. Cultivation succeeds without any problems on CSB-agar plates. The identification of the pathogen can be made by 16S rRNA sequencing.

TREATMENT. Since the bacterium occurs particularly in water with a high organic level, environmental optimisation, in particular optimisation of the hygienic conditions, is necessary. Stress is one of the main factors for disease outbreaks; therefore, a reduction in the stressful conditions should be implemented as soon as possible to avoid progression of the infection on a farm. If possible, treatment of the affected fish with an antibiotic should only be carried out after an antibiogram, since some resistances for *Citrobacter* spp. have been described.

FACTS SHEET:

Disease: Infection with *Citrobacter* spp.
Pathogen: *Citrobacter* spp., especially *Citrobacter freundii*
Susceptible hosts: Common carp, crucian carp, grass carp and other carp species
Clinical signs: Uncoordinated swimming movements, skin changes, septicaemia
Pathology: Haemorrhages, swollen kidney, granulomas
Mortality: Can be high in fry and fingerlings
Temperature: Disease outbreaks especially in summer at higher water temperatures
Prevention and treatment: Prevention is by optimisation of environmental conditions, reduction of stress. Treatment is dependent on the result of the antibiogram

Edwardsiella tarda

AETIOLOGY. *Edwardsiella tarda* is a Gram-negative, rod-shaped bacterium with a peritrichous flagellum and is a facultative anaerobe. *E. tarda* can be found in many aquatic environments, such as ponds, lakes and rivers, but also occurs in soil. There are four serotypes

(A, B, C and D), with serotype A being primarily involved in disease outbreaks. There is some doubt as to whether *E. tarda* should be considered as a primary or secondary pathogen, as the bacterium appears to be part of the normal skin microflora, at least in catfish (*Silurus glanis*). The bacterium was first described in 1965 and is causing problems worldwide, particularly in the USA and Japan. *E. tarda* is primarily a pathogen of warmwater fish, particularly catfish and eels, but has also been isolated from many other freshwater fish, such as common carp, goldfish and grass carp following disease outbreaks. *E. tarda* has a zoonotic potential and humans can also become infected through food. Infections can result in serious intestinal infections, meningitis, liver abscesses and wound infections.

PATHOGENESIS. Fish become infected with bacteria from the environment, which remain infectious in the pond water for 76 days. There is a correlation between *E. tarda* pathogenicity and water temperature. Diseases occur above a water temperature of 10°C, but especially at higher temperatures up to 30°C or even higher. The pathogenicity of the bacteria is highest at water temperatures between 20°C and 25°C. Strong organic pollution of the water, high copper concentrations and stress also seem to play a role in pathogenicity.

CLINICAL SIGNS OF THE DISEASE. Depending on the stage of the disease, the symptoms range from small skin lesions to abscesses and necrosis of the skin. Hyperplasia of the epithelium and necrosis along the lateral line and on the skin and gills can occur. Diseased fish are lethargic, congregate on the surface and exhibit spiral swimming movements. The lateral skin lesions develop into large abscesses in the muscles or at the base of the caudal fin. The abscesses may be filled with gas and necrotic tissue, and have a foul odour. Although *E. tarda* does not possess endotoxins, the exotoxins seem to be responsible for the abscess. Infected carp on farms also show haemorrhages on the body and fins. In contrast, experimental infections with *E. tarda* did not cause any disease signs in common carp, Prussian carp (*Carassius gibelio*) and tench (*Tinca tinca*).

PATHOLOGY. The dissection may show whitish nodules in the gills, kidneys, liver, spleen and intestines.

DIAGNOSIS. *E. tarda* can be cultivated on simple culture media such as trypticase soy agar. The bacteria show a good growth at 37°C, but will also grow after 2–4 days at an incubation temperature of 25°C. The colonies appear grey-transparent, small and round. Identification is possible by 16S rRNA sequencing or MALDI–TOF MS.

TREATMENT. Since environmental conditions play a role in the pathogenesis of *E. tarda*, there is a need to optimise the keeping conditions. Many resistances against antibiotic agents are described in *E. tarda* and therefore a treatment should only be carried out after an antibiogram if possible.

FACTS SHEET

Disease: Infection with *Edwardsiella tarda*
Pathogen: *Edwardsiella tarda*
Susceptible hosts: Sometimes common carp, might be detected in other carp species as well
Clinical signs: Skin haemorrhages
Pathology: White nodules in the internal organs
Mortality: Can be high when the environmental conditions are suboptimal
Temperature: Disease outbreaks especially at 20–25°C
Prevention and treatment: Prevention is by optimisation of environmental conditions. Treatment is dependent on the result of the antibiogram

Escherichia vulneris

AETIOLOGY. *Escherichia vulneris* is a Gram-negative, rod-shaped bacterium belonging to the family Enterobacteriaceae. There are reports on bacterial infections due to *E. vulneris* in different fish species, including common carp.

PATHOGENESIS. Uptake of the bacteria is probably from the water and could lead to general bacterial infections.

CLINICAL SIGNS OF THE DISEASE. In infected fish, haemorrhagic lesions in the skin and exophthalmos are observed.

PATHOLOGY. A range of alterations of the liver, and a bloody exudate in the intestinal tract can be seen in infected fish.

DIAGNOSIS. The bacteria can be grown on common nutrient media such as blood agar, even at temperatures of 37°C. Identification of the species is possible utilising biochemical tests or by 16S rRNA sequencing.

TREATMENT. Antibiotic treatment of the affected fish is feasible, but should only be carried out after an antibiogram.

FACTS SHEET

Disease: Infection with *Escherichia vulneris*
Pathogen: *Escherichia vulneris*
Susceptible hosts: Common carp, most likely also other carp species
Clinical signs: Haemorrhages in the skin, exophthalmos
Pathology: Liver alterations, bloody exudate in the intestine
Mortality: Unknown
Temperature: Higher water temperatures
Prevention and treatment: Prevention is by optimisation of environmental conditions. Treatment is dependent on the result of the antibiogram

Plesiomonas shigelloides

AETIOLOGY. *Plesiomonas shigelloides* is a Gram-negative, motile, rod-shaped, facultative anaerobic bacterium. The bacteria are present in the aquatic environment of freshwater and brackish water habitats, especially in tropical and subtropical regions. *P. shigelloides* also occurs as part of the intestinal microbiome of fish. The bacterium has been isolated from different diseased fish species, including silver carp and grass carp, and has a zoonotic potential causing gastroenteritis in humans.

PATHOGENESIS. There is a correlation between disease outbreaks and rising water temperatures. Strong organic pollution of the water also favours the development of clinically manifest infections. *P. shigelloides* is found in the gastrointestinal tract of warmwater fish, where it can serve as a reservoir for infection.

CLINICAL SIGNS OF THE DISEASE. Diseased fish appear emaciated and show reddening of the skin, particularly around the anus. In silver carp 60% mortality has been described after a disease outbreak.

PATHOLOGY. Petechial haemorrhages in the muscles and ascites may occur. Kidney alterations such as tubular necrosis, and in gills hyperplasia of the lamellae, clubbing and swelling have been seen in diseased silver carp.

DIAGNOSIS. Cultivation is possible at temperatures between 22°C and 37°C on trypticase soy agar where *P. shigelloides* grows in round, whitish colonies. Identification is possible for example by 16S rRNA sequencing.

TREATMENT. Since the bacterium occurs particularly in water with a high level of organic pollution, environmental optimisation is necessary. If possible, treatment of the affected fish with an antibiotic should only be carried out after an antibiogram.

FACTS SHEET

Disease: Infection with *Plesiomonas shigelloides*
Pathogen: *Plesiomonas shigelloides*
Susceptible hosts: Silver carp, grass carp, common carp, most likely also other carp species
Clinical signs: Haemorrhages in the skin, exophthalmos
Pathology: Gill alterations, tubulus necrosis in the kidney
Mortality: Up to 60% in silver carp
Temperature: Higher water temperatures
Prevention and treatment: Prevention is by optimisation of environmental conditions. Treatment is dependent on the result of the antibiogram

Other Enterobacteriaceae

Other bacteria from the family Enterobacteriaceae also occur as facultative pathogenic pathogens in fish. Contamination of the storage or transport water plays an important role in this context. Investigations into the transport water of warmwater ornamental fish revealed a high level of contamination with so-called faecal germs which, in conjunction with other stress factors, could trigger disease symptoms in the fish. If these infectious agents are isolated in connection with diseases in fish, a critical review of the environmental conditions, especially the hygienic conditions, is essential. Optimisation will be necessary in most cases, otherwise problems will arise again and again. If the fish continue to show symptoms of illness despite optimal environmental conditions, antibiotic treatment can be carried out after an antibiogram.

8.5.4 Flavobacteriaceae

Flavobacteriaceae represent a group of yellow-pigmented bacteria, which include the genera *Cytophaga*, *Flavobacterium*, *Flexibacter*, *Myxobacterium*, *Myxococcus* and *Sporocytophaga*. The bacteria are Gram-negative and rod-shaped, and can be identified in many fish species associated with disease outbreaks, although not all bacterial species within this family are pathogenic to fish. All bacteria belonging to the Flavobacteriaceae show rapid growth. Depending on the genus present, the bacteria can cause disease symptoms in both freshwater and saltwater fish and in warm- and coldwater fish species. *Flavobacterium columnare*, *Flavobacterium psychrophilum* and *Flavobacterium branchiophilum* are frequently isolated from carp species. Infections with Flavobacteriaceae are one of the causes of bacterial gill swelling. Different genera can occur as the triggering agent. It should be noted, however, that bacterial gill swelling is probably induced not only by these bacteria, but also by non-optimal environmental conditions. Bacterial gill diseases occur in various fish species and at different water temperatures. Swelling of the secondary lamellae of the gills leads to reduced oxygen uptake, resulting in respiratory difficulties, such as gasping for air and rapid movements of the opercula. It is striking that often antibiotic therapy for an infection with Flavobacteriaceae, despite the effectiveness of the antibiotics *in vitro*, usually does not lead to a reduction in the clinical symptoms in the fish. Therefore, in cases of disease outbreaks the most important measure is the improvement of the environmental conditions, in particular the optimisation of the chemical and microbial water parameters, and a reduction of the population density.

Flavobacterium branchiophilum

AETIOLOGY. *Flavobacterium branchiophilum* is a Gram-negative, filamentous, non-gliding bacterium. The bacterium is one of the causative agents of bacterial gill disease (BGD), which is primarily an infectious disease of salmonids but also occurs in other freshwater fish.

PATHOGENESIS. *F. branchiophilum* is mainly a surface pathogen that infects the mucosal surface of the gills. In naturally infected fish the bacteria cannot be isolated from kidney or spleen, and therefore a distribution of the bacteria via the bloodstream and the development of a septicaemia seems to be unlikely. Fish are infected by bacteria from the water that probably get entrance to the fish by colonising the gill epithelium. Outbreaks of disease can occur at different water temperatures and are triggered by suboptimal environmental conditions. Disease outbreaks are known from fish from aquaculture facilities, whereas wild fish are usually not affected.

CLINICAL SIGNS OF THE DISEASE. Primarily, fish infected with *F. branchiophilum* show gill changes (Fig. 8.6). The hyperplasia of the

Figure 8.6 Common carp with operculum removed, showing gill swelling and hyperplasia related to *F. branchiophilum* infection.

gill epithelium usually starts at the lamellae tips and extends to the base of the gills. The mucus covering the gill epithelium often has a yellow-orange colour due to the bacteria it contains. As the distance between blood and water is enlarged due to the swelling, the fish usually show signs of oxygen deficiency and might also have problems with ammonia excretion. Infected fish gasp for air, and show rapid movements or abduction of the opercula. *F. branchiophilum* is identified as an important secondary pathogen in common carp (*Cyprinus carpio*) after infections with carp oedema virus (CEV).

PATHOLOGY. The described gill changes above can be detected histopathologically.

DIAGNOSIS. *Flavobacterium branchiophilum* can be detected microscopically in samples taken from the gill epithelium. The bacteria are seen as long, thin rods connected to the gill epithelium. On culture media *F. branchiophilum* grows in yellow, translucent colonies. Although some strains of the bacterium show good growing on standard agars, such as TSA and CSB agar, it is always recommended that a specific medium for *Flavobacterium* spp., such as Anacker–Ordal agar, Shie agar, or TYES agar, is used. A modified Anacker–Ordal agar with supplementation of different sugars has been shown to be very useful in the cultivation of *Flavobacterium* spp. On the selective media the growth of other bacterial species, like motile aeromonads or pseudomonads, is reduced and an identification of the Flavobacteria is much easier. The bacterium grows at temperatures between 5°C and 30°C, depending on the geographical region where it is isolated. While isolates from Japan can be grown at 30°C, isolates from the USA do not grow at these temperatures. Cultivation at 37°C is not possible with any isolate of *F. branchiophilum*, and therefore incubation at 18°C for 5 days under strictly aerobic conditions is recommended. The identification of the grown isolates is possible by 16S rRNA sequencing.

TREATMENT. A bath treatment with 1–5% sodium chloride for a few minutes can reduce the number of bacteria on the gills and the disease signs in some cases. Depending on the fish species treated, the application of salt should be carried out slowly and it should be considered that not all fish species are tolerant of such high salt concentrations. An antibiotic therapy can be attempted after an antibiogram, but this does not always cause any improvement in fish health. Optimising the environmental conditions is advisable in all cases.

FACTS SHEET

Disease: Bacterial gill disease (BGD)
Pathogen: *Flavobacterium branchiophilum*
Susceptible hosts: All carp species
Clinical signs: Hyperplasia of the gill epithelium, gasping for air
Pathology: Gill alterations
Mortality: Low to high, depending on the environmental conditions and the age of the affected fish
Temperature: Disease outbreaks can occur between 5 and 30°C
Prevention and treatment: Prevention by optimisation of environmental conditions. Treatment can be achieved with a salt bath and antibiotic application after antibiogram

Flavobacterium columnare

AETIOLOGY. *Flavobacterium columnare* is a Gram-negative, rod-shaped bacterium that can move by gliding. It forms characteristic columns in the mucus in infected fish that can be easily recognised under a light microscope. The bacterium is the causative agent of columnaris disease, which can affect many freshwater warmwater fish, including different carp species such as common carp, grass carp and goldfish (*Carassius auratus*). Outside the fish, the bacterium survives for a long time in warm, organically contaminated water. The pathogen can also occur on the skin of healthy fish. For this reason, the sole detection of the bacterium without the presence of clinical symptoms of the disease does not indicate anything about whether the bacterium plays a role as a causal problem.

PATHOGENESIS. The disease occurs at warm water temperatures and, in general, infections do not occur below 12°C water temperature. *F. columnare* adheres first to the tips of the gill filaments, then the bacteria colonise the middle part and the base of the filaments.

In common carp it has been shown that often the whole filament is colonised when the infection progresses. Also, lysis of the cartilaginous tissue of the gills has been recognised in carp. The severity of the disease and the mortality rates depend on the virulence of the strain of *F. columnare* involved and the condition of the affected fish. Non-optimal environmental conditions can trigger the onset of disease.

CLINICAL SIGNS OF THE DISEASE. The infection predominates in the gills, but can also affect the skin and muscles. Infections with *F. columnare* can also become systemic. Usually, the first signs of the disease are small lesions on the skin that appear as colour changes. These lesions are usually located at the base of the dorsal fin or the pelvic fins. The lesions can spread and become larger, especially in the back area, which has given the disease in rainbow trout the colloquial name 'saddleback disease'. The centre of the lesion may become necrotic and deep muscle ulceration may develop. Yellow to orange-coloured areas can often be seen at the edges of the ulcerations. These arise due to the pigment formation in the bacteria, which primarily accumulate in these peripheral areas and which can be seen under the microscope as columnar formations. Changes in the gills are often expressed in hyperplasia of the gill epithelium. Initially, the gill tips are thickened, then the changes extend to the base of the gills. The mucus on the gills often has a yellow-orange colour due to the bacterial pigmentation. Like in infections with other *Flavobacterium* species, the distance between water and blood in the gills is enlarged and the function of the gills is restricted. Therefore, the fish gasp for air and show higher ventilation rates, and the opercula are often abducted.

PATHOLOGY. Ulcerations of the skin and the muscles can be detected, as well as the gill changes described above.

DIAGNOSIS. The first indications of an infection with *F. columnare* are obtained from the microscopic examination of swabs from the mucosa of the skin and the gills of the changed areas, in which accumulations of the bacteria can be seen as columnar structures.

The characteristic gliding movement of the bacteria, which can be recognised microscopically, also gives an indication of *F. columnare*. The bacteria can be cultured at incubation temperatures between 4°C and 30°C on selective media for Flavobacteria, such as Anacker–Ordal agar (see also diagnosis of *F. branchiophilum*), under strictly aerobic conditions. Some isolates of *F. columnare* also grow at 37°C and as yellow-green colonies. Species identification is possible by 16S rRNA sequencing.

TREATMENT. Reducing the water temperature can prevent further progression of the disease. Antibiotic therapy according to the results of the antibiogram makes sense in the presence of skin ulcerations, although optimisation of the husbandry conditions is necessary in all cases. The success of the treatment is questionable in the case of advanced damage to the gill structure.

FACTS SHEET

Disease: Columnaris disease
Pathogen: *Flavobacterium columnare*
Susceptible hosts: All carp species
Clinical signs: Gill hyperplasia, colour changes of the skin, skin ulcerations, muscle ulcerations may occur
Pathology: Gill hyperplasia
Mortality: Can be high when fish are infected with a virulent strain
Temperature: Water temperatures above 12°C
Prevention and treatment: Prevention by optimisation of environmental conditions. Treatment utilising a reduction of the water temperature or antibiotic application after antibiogram

Flavobacterium psychrophilum

AETIOLOGY. *Flavobacterium psychrophilum* is a Gram-negative, rod-shaped bacterium that shows gliding movements and can be ubiquitously found in the aquatic environment, especially in freshwater. *F. psychrophilum* is the causative agent of bacterial coldwater disease (BCWD) and rainbow-trout fry syndrome (RTFS) in salmonids and plays only a minor role as an infectious agent in carp species. Nevertheless, *F. psychrophilum* was detected in common carp, goldfish and tench, although in much lower frequencies than in salmonids.

PATHOGENESIS. *F. psychrophilum* is ubiquitous in the aquatic environment and can survive for a month to years without a fish as a host. Small lesions in the skin seem to be the main entrance route of the bacteria, which have an affinity especially for the tissue of the lower jaw, fins and the caudal peduncle. The development of clinical signs can be triggered by infections with ectoparasites and suboptimal chemical water quality. After infection *F. psychrophilum* excretes psychrophilic protease and spreads through connective tissue, and expands further into the musculature. *F. psychrophilum* causes disease signs only in cold water temperatures between 4°C and 12°C.

CLINICAL SIGNS OF THE DISEASE. Affected fish show necrosis of the skin and muscles, particularly on the fins and the caudal peduncle. When the disease progresses, the muscles can be damaged revealing the bone, and disorders of the central nervous system and deformations of the skeleton can be observed as late effects.

PATHOLOGY. Necrosis of the skin and the muscle can be seen.

DIAGNOSIS. *Flavobacterium psychrophilum* grows on selective agar for Flavobacteria (see also *F. branchiophilum*) in the form of yellow-pigmented colonies (Fig. 8.7). Incubation must be strictly aerobic. The incubation temperature should be between 4°C and 23°C, and ideally at 20°C as growth stops at 30°C. An identification of the species is possible by 16S rRNA sequencing.

TREATMENT. Antibiotic therapy is unsuccessful in many cases. Increasing the water temperature can prevent further progression of the infection, but is usually not feasible in practice. For this reason, the most important measure is to optimise the husbandry conditions, with particular emphasis on water hygiene.

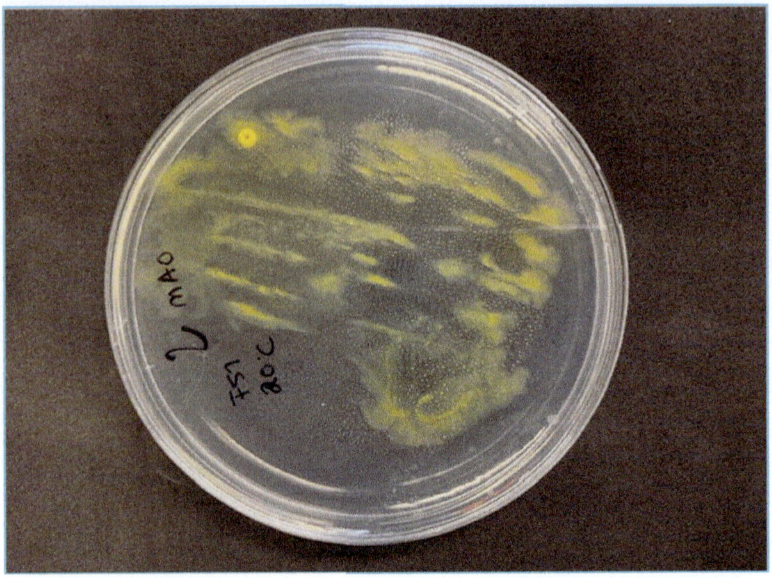

Figure 8.7 Flavobacterial growth on Anacker–Ordal agar.

FACTS SHEET

Disease: Bacterial cold-water disease (BCWD), rainbow-trout fry syndrome (RTFS)
Pathogen: *Flavobacterium psychrophilum*
Susceptible hosts: Salmonids; common carp and goldfish can get infected
Clinical signs: Necrosis of the skin and muscles
Pathology: Necrosis of the skin and muscles
Mortality: Can be high when fish are infected with a virulent strain
Temperature: 4–12°C
Prevention and treatment: Prevention by optimisation of environmental conditions. Treatment through an increase in the water temperature, antibiotic treatment after antibiogram

8.5.5 Francisellaceae

Francisella noatunensis subsp. *orientalis*

AETIOLOGY. *Francisella noatunensis* is a Gram-negative, coccoid to rod-shaped, non-motile and facultative intracellular, strictly aerobic bacterium. Although *F. noatunensis* subsp. *orientalis* is a well-known pathogen in tilapia, infections have also been described in common carp (*Cyprinus carpio*).

PATHOGENESIS. Fish are most likely infected by bacteria from the surrounding environment.

CLINICAL SIGNS OF THE DISEASE. Typical disease signs are haemorrhages and ulcerations in the skin. In the gills pale white nodules can be detected. However, in some infected common carp external signs of the disease have not been seen.

PATHOLOGY. White nodules (granulomas) in the spleen and other internal organs have been found (Fig. 8.8). As pathological alterations can look like granulomas caused by mycobacteria (mycobacteriosis), it is important that differential diagnosis between the two pathogens is carried out.

DIAGNOSIS. *F. noatunensis* subsp. *orientalis* can be cultured on either modified Thayer–Martin agar (MTM) or on cysteine heart

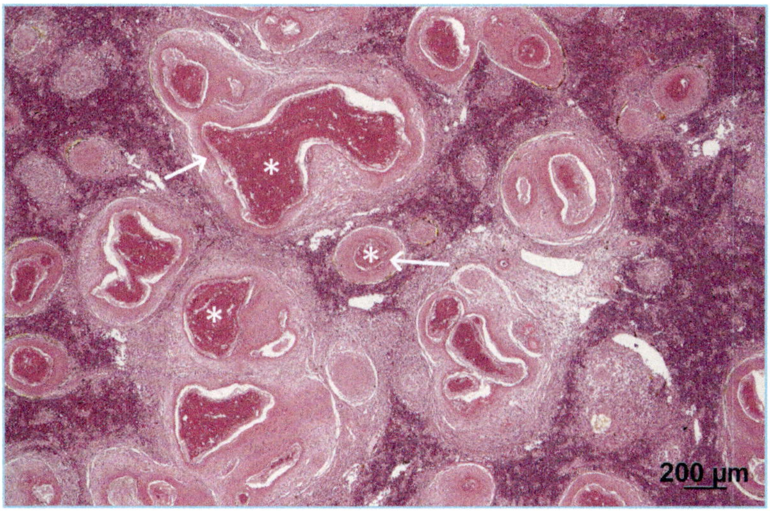

Figure 8.8 Histological section of common carp spleen showing multiple granulomas (*) encapsulated by connective tissue (arrow). Scale bar 200 µm. Images and interpretation provided by John Bignel of the Centre for Environment, Fisheries and Aquaculture Science (Cefas), UK. Crown copyright, Cefas 2022.

agar supplemented with 2% bovine haemoglobin (CHAB) at 28°C. Identification of the species is possible by molecular biological methods such as 16S rRNA sequencing.

TREATMENT. An antibiotic treatment according to the results of an antibiogram can be performed.

FACTS SHEET

Disease: Piscine francisellosis
Pathogen: *Francisella noatunensis* subsp. *orientalis*
Susceptible hosts: Fresh and marine warmwater fish species, including common carp
Clinical signs: Haemorrhages of the skin, ulcerations, pale gills with white nodules
Pathology: Granulomas in the internal organs
Mortality: High in tilapia, probably very low in common carp
Temperature: Water temperatures less than 25°C
Prevention and treatment: Prevention is by optimisation of environmental conditions. Treatment is dependent on the result of the antibiogram

8.5.6 Moraxellaceae

The family Moraxellaceae includes the genera *Acinetobacter* and *Moraxella*. Since *Acinetobacter* can regularly be detected in carp species, it warrants some consideration.

Acinetobacter spp.

AETIOLOGY. *Acinetobacter* spp. are Gram-negative, non-motile, rod-shaped, non-fermentative, facultative anaerobic bacteria. The bacteria are widely distributed in nature and occur in soil and water. *Acinetobacter* spp. can be regularly isolated from damaged tissues of common carp, and especially the species *Acinetobacter johnsonii* and *Acinetobacter lwoffii* appear to play an important role in disease. *Acinetobacter baumannii* was isolated from Indian major carp in a case of mortality.

PATHOGENESIS. *Acinetobacter* spp. are commonly found in freshwater and seawater, and high levels can be detected in water that is heavily organically polluted. The bacterium can also be found on the skin, the gills and in the intestines of clinically healthy fish.

Damage to the mucous membrane can therefore lead to bacterial colonisation in the underlying tissue and subsequently to a clinically manifest infection. The pathogenicity of *Acinetobacter* spp. is possibly caused by endotoxins.

CLINICAL SIGNS OF THE DISEASE. Affected fish may have hyperaemia of the superficial skin blood vessels and haemorrhages in the scale pockets, oedema and ulceration.

PATHOLOGY. Autopsy may reveal a range of changes in the kidneys, liver and spleen, and haemorrhages in the swim bladder and the peritoneum.

DIAGNOSIS. *Acinetobacter* spp. grow at incubation temperatures between 20°C and 37°C on standard agar plates in round, transparent, slimy colonies. An identification of the species is possible by 16S rRNA sequencing.

TREATMENT. Environmental optimisation must take place in all cases. If this is not sufficient, treatment can be carried out using an antibiotic based on the antibiogram.

FACTS SHEET

Disease: Infection with *Acinetobacter* sp.
Pathogen: *Acinetobacter* spp., mainly *A. johnsonii*, *A. lwoffii* and *A. baumannii*
Susceptible hosts: Common carp, Indian major carp, probably also other carp species
Clinical signs: Haemorrhages, oedema and ulcerations of the skin
Pathology: Changes in kidney, liver, spleen, and haemorrhages in the swim bladder and the peritoneum
Mortality: Not known
Temperature: Around 20°C
Prevention and treatment: Prevention is by optimisation of environmental conditions. Treatment is dependent on the result of the antibiogram

8.5.7 Pseudomonaceae

Pseudomonads are ubiquitous in aquatic ecosystems and can be found in both freshwater and saltwater. The bacteria are also part

Figure 8.9 Koi carp with extensive reddening of the skin and fins related to infection with *Pseudomonas* sp.

of the physiological microbiome of many fish species. However, *Pseudomonas* spp. have been detected in bacterial infections in fish, whereby especially *Pseudomonas fluorescens* can cause disease. Pseudomonads are primarily considered to be secondary pathogens that cause clinical signs as a result of another infection or poor environmental conditions (Fig. 8.9).

Pseudomonas fluorescens

AETIOLOGY. *Pseudomonas fluorescens*, a Gram-negative, motile, rod-shaped bacterium, causes disease in many fish species, including goldfish, crucian carp, Prussian carp (*Carassius gibelio*), silver carp (*Hypophthalmichthys molitrix*), bighead carp and common carp. The bacterium can produce the fluorescent dye fluorescein. *P. fluorescens* is also known as a zoonotic agent and can cause wound infections in humans.

PATHOGENESIS. *P. fluorescens* is one of the most common bacteria found in freshwater and fish become infected via this source. Stress, particularly falling water temperatures, can contribute to the onset of the disease.

CLINICAL SIGNS OF THE DISEASE. Skin ulcerations and large lesions can be seen in infected fish, and sometimes the gills may also be infected, resulting in clinical signs similar to those found in bacterial gill disease. Infections with *P. fluorescens* can lead to acute and chronic septicaemia. Such bacterial septicaemia usually manifests as fluid accumulation in the abdominal cavity and petechial bleeding in the gills, kidneys, liver and intestines. Septicaemia is often associated with stress and suboptimal housing conditions. Other clinical signs include fin rot and petechiae that occur in the mouth area.

PATHOLOGY. White nodules in the spleen and abscesses in the swim bladder have been described in infected fish.

DIAGNOSIS. *P. fluorescens* grows well on common culture media and on selective media for pseudomonads, such as Pseudomonas F agar under obligate aerobic conditions. The bacteria grow as small grey colonies on the culture medium. *P. fluorescens* growth is temperature dependent and bacterial growth does not occur at 42°C; ideal temperatures are between 22°C and 28°C.

TREATMENT. It is important to avoid any stress for the infected fish. Antibiotic therapy is often difficult due to the variety of resistances, but if applied this should only be carried out after an antibiogram has been made.

FACTS SHEET

Disease: Infection with *Pseudomonas fluorescens*
Pathogen: *Pseudomonas fluorescens*
Susceptible hosts: Probably all carp species can be affected
Clinical signs: Skin ulcerations, gill swelling, septicaemia
Pathology: Alterations in spleen and swim bladder can occur
Mortality: Can be up to 100%, depending on the environmental conditions
Temperature: Disease outbreaks mainly between 15°C and 20°C; may be disease outbreaks also at lower water temperatures
Prevention and treatment: Prevention is by optimisation of environmental conditions. Treatment is dependent on the result of the antibiogram

Pseudomonas putida

AETIOLOGY. *Pseudomonas putida* is a Gram-negative, motile, rod-shaped bacterium found regularly in soil and water. *P. putida* is known as a very important bacterial species in the environment and as it can control pollution in the water it is used for bioremediation. The bacterium can be found regularly as part of the physiological microbiome in fish, but also infects the skin of common carp.

PATHOGENESIS. Infections of common carp with *P. putida* that are present in the water are usually associated with high stocking density and poor chemical water quality. Therefore, *P. putida* seems to act as a secondary pathogen.

CLINICAL SIGNS OF THE DISEASE. Affected common carp exhibit haemorrhages on the skin and red patches on the opercula and on the lower abdomen.

PATHOLOGY. The skin alterations such as haemorrhages can be detected histologically. Other pathological signs are dependent on the primary cause of the disease.

DIAGNOSIS. Whilst *P. putida* is easy to cultivate on common culture media, for example blood agar, special culture media for pseudomonas are also available, for example Pseudomonas F agar. Bacteria grow under obligate aerobic conditions at temperatures between 25°C and 30°C as small grey colonies.

TREATMENT. Improvement of environmental conditions is essential in prevention of a progression of the infection. If necessary, treatments with antibiotics should only be carried out after an antibiogram has been made.

FACTS SHEET

Disease: Infection with *Pseudomonas putida*
Pathogen: *Pseudomonas putida*
Susceptible hosts: Common carp; probably also in other carp species
Clinical signs: Haemorrhages on the skin

Pathology: Haemorrhages on the skin
Mortality: Not known
Temperature: 25–30°C; but disease outbreaks also occur at lower water temperatures
Prevention and treatment: Prevention is by optimisation of environmental conditions. Treatment is dependent on the result of the antibiogram

8.5.8 Shewanellaceae

The sole genus within the family Shewanellaceae is *Shewanella*. The bacteria are originally marine organisms, but can be detected in freshwater habitats as well.

Shewanella spp.

AETIOLOGY. *Shewanella* spp. are Gram-negative, motile, rod-shaped, facultative anaerobic bacteria. The bacteria are widespread especially in the marine habitat, but also occur in the freshwater environment where they play an important role by converting organic compounds. *Shewanella* spp. belong to the physiological microbiome of fish from temperate zones. Over 60 species within the genus *Shewanella* have been described, but especially *Shewanella algae*, *Shewanella baltica*, *Shewanella oneidensis*, *Shewanella putrefaciens*, *Shewanella seohaensis* and *Shewanella xiamenensis* have been isolated from fish. Recently a novel species of the genus *Shewanella* was found in crucian carp and the name *Shewanella carassii* sp. nov. has been proposed. *Shewanella* spp. have a low zoonotic potential, although some cases of wound and soft tissue infections have been described in humans after contact with contaminated water, injuries by aquatic animals from contaminated water or consumption of raw fish. *Shewanella* spp. are not host specific and outbreaks of diseases occur in fish from different orders, including carp. Disease outbreaks can especially be seen at water temperatures between 7°C and 10°C, but also at higher temperatures. *Shewanella* spp. are considered to be opportunistic pathogens for carp species, and in most disease outbreaks additional bacterial species or parasites are involved. *Shewanella* spp. are also regularly isolated from the skin of clinically healthy carp.

PATHOGENESIS. *Shewanella* spp. are particularly prominent in fish that are already weakened by other infections. Carp seem to become infected usually through the water, as the bacteria are widely distributed in freshwater environments. As the progression of the disease differs greatly between fish from different farms, it has been suggested that pathogenicity of *Shewanella* spp. is isolate-dependent.

CLINICAL SIGNS OF THE DISEASE. Ulcerations, necrotic lesions and haemorrhages of the skin, colour changes, exophthalmos, and fin damage can often be seen in infected carp. Also, unspecific disease signs, such as lethargy, anorexia and darkening of the skin have been described.

PATHOLOGY. Haemorrhages in the internal organs such as the swim bladder, gut and gonads are often seen in infected fish. Oedema of the kidneys and hypertrophy of the spleen can also occur. In carp farms the mortality rate due to infections with *Shewanella* spp. is usually low, but in experimental infections mortality rates of up to 50% in common carp occur which are dependent on the bacterial isolate used.

DIAGNOSIS. The bacteria grow on CSB agar in typical large, brownish, shiny colonies. The genus *Shewanella* can be identified by biochemical examinations of the colonies, but with these methods an exact species identification is often not possible. The species can be identified by sequencing a part of the 16S rRNA gene if a secure diagnosis is needed.

TREATMENT. In isolates of *Shewanella* spp. multi-resistances against antibiotic agents can be regularly found. Therefore, an antibiogram is always recommended before any antibiotic treatment. As the bacteria are often acting as secondary pathogens, the primary pathogen should be treated first and, if necessary, the environmental conditions should be optimised.

FACTS SHEET

Disease: Infection by *Shewanella* spp.
Pathogen: Different species of *Shewanella*
Susceptible hosts: Common carp, crucian carp, other carp species
Clinical signs: Skin ulcerations, haemorrhages and necrosis
Pathology: Oedema, and haemorrhages in the internal organs
Mortality: Usually low, can be high depending on other infections
Temperature: Disease outbreaks can be seen from 7°C water temperature in common carp
Prevention and treatment: Prevention is by optimisation of environmental conditions and a reduction in ectoparasites. Treatment is dependent on the result of the antibiogram

8.5.9 Vibrionaceae

Vibrionaceae are among the most important bacterial pathogens of marine fish and are ubiquitous in seawater, but can also be detected in brackish water and freshwater. The *Vibrio* content in fish is particularly high in the presence of large organic loads and in water with a high salinity. Some *Vibrio* species are harmless to fish, whilst others play a role as opportunistic infectious agents, and still others are primarily pathogenic to fish. In fish, *Vibrio alginolyticus*, *V. anguillarum*, *V. cholerae*, *V. fischeri*, *V. furnissii*, *V. harveyi*, *V. ichthyoenteri* (this includes *V. carcheriae* and *V. trachuri*), *V. logei*, *V. ordalli*, *V. pelagius*, *V. salmonicida*, *V. splendidus*, *V. tapetis*, *V. vulnificus* and *V. wodanis* can regularly be isolated. Diseases occur mainly at high water temperatures; and overstocking, organic water pollution and other stressors can contribute to disease outbreaks. Infected fish show haemorrhagic septicaemia, reddening and skin changes. Anaemia and necrosis of the internal organs can be found in the dissection. Depending on the fish species, different disease symptoms can occur. A diagnosis can be made if the pathogen is detected in the tissue of the affected fish and the clinical symptoms are present. Without the presence of symptoms of the disease, the detection of *Vibrio* spp., for example from the skin or from the intestines of fish, is meaningless. Vibriosis is a classic example of a stress-induced condition. Keeping conditions should be optimised and stress should be avoided. Due to increasing resistance, antibiotic treatment should only be carried out after an antibiogram. In carp species, especially *V. anguillarum, V. cholerae* and *V. vulnificus* occur.

Vibrio anguillarum

AETIOLOGY. *Vibrio anguillarum* is a motile, Gram-negative rod-shaped bacterium, which in the 19th and 20th centuries caused a disease known as the red plague, which led to considerable losses among eels. *V. anguillarum* is part of the physiological microbiome of marine ecosystems and seawater fish, but has also been detected in disease outbreaks in ornamental carp. The bacteria can survive for a very long time, over 4 years, in seawater.

PATHOGENESIS. Fish can become infected via the bacteria in the water or, in the case of previous damage, via the bacteria in the microbiome. Outbreaks of disease are promoted when copper or iron is present in the holding water. The disease can be peracute to chronic. A variety of virulence factors are described in *V. anguillarum*, including a range of secretory extracellular products that have haemolytic and proteolytic activates that are responsible for the induction of disease signs.

CLINICAL SIGNS OF THE DISEASE. Infections with *Vibrio anguillarum* lead to haemorrhagic septicaemia in fish. Typically, affected fish show discolouration necrotising lesions in the muscles, bleeding at the base

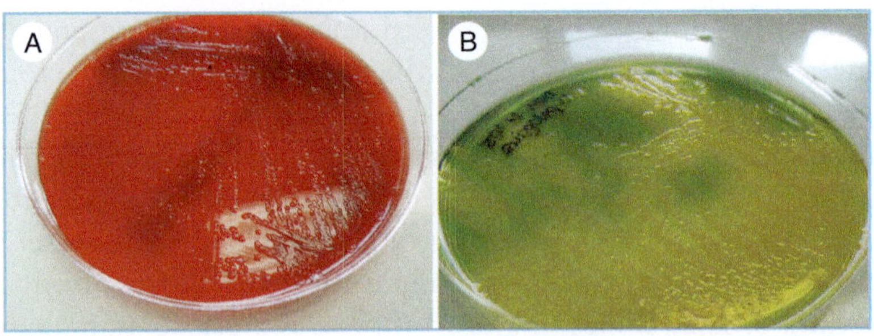

Figure 8.10 *Vibrio* bacteria colonies grown on (A) Columbia sheep blood agar (CSB agar) and (B) thiosulphate citrate bile salts sucrose agar (TCBS agar).

of the fins and in the mouth, and fin rot. The peracute form is manifested by anorexia, darkening and sudden death. The acute form leads to ulcerations and reddening of the skin, whilst the chronic form is associated with deep muscle lesions and eye deformations.

PATHOLOGY. In acute forms anaemia is present, and autopsy reveals petechial bleeding, splenomegaly and renal necrosis. In addition, necrotising enteritis may occur. Histopathologically, necrosis in internal organs can be seen.

DIAGNOSIS. *Vibrio anguillarum* grows on nutrient agar in off-white, round, glossy colonies (Fig. 8.10). Cultivation should take place at 15–37°C with an addition of 0.3–3.5% sodium chloride for up to 7 days. Species identification is possible by 16S rRNA sequencing.

TREATMENT. Formalin-inactivated vaccines are available for use in marine aquaculture. If carp species are affected, the environmental conditions must first be optimised. If antibiotic therapy is necessary, this should only be carried out after an antibiogram, due to the poor resistance situation. Since infected fish do not usually eat, treatment via the food is often not possible.

FACTS SHEET

Disease: 'Red plague' in eels, infections with *Vibrio anguillarum*
Pathogen: *Vibrio anguillarum*
Susceptible hosts: Common carp/koi carp (*Cyprinus carpio*), probably also other carp species
Clinical signs: Discoloration, necrotizing lesions, haemorrhages at the base of the fins and in the mouth, fin rot
Pathology: Haemorrhagic septicaemia, necrosis in internal organs
Mortality: Can be high
Temperature: 15–30°C
Prevention and treatment: Prevention by vaccination and by optimisation of environmental conditions. Treatment is dependent on the result of the antibiogram

Vibrio cholerae

AETIOLOGY. *Vibrio cholerae* is a rod-shaped, Gram-negative, motile bacterium that has occasionally been isolated, especially in freshwater fish, and is particularly associated with other bacterial infections. Goldfish are often affected and infections in common carp have also been described. In general, *V. cholerae* is a zoonotic pathogen that causes cholera in humans. When examining the isolated bacteria more closely, the variant *V. cholerae*-non-O1, which is non-pathogenic for humans, is isolated in almost all fish infections.

PATHOGENESIS. *V. cholerae* can survive in the aquatic environment, for example at 25°C for 32 days in freshwater and seawater conditions. Therefore, fish are likely to become infected through the water. Under laboratory conditions, the bacterium appears to be highly virulent for fish.

CLINICAL SIGNS OF THE DISEASE. Diseased fish show petechial haemorrhages and septiceamia, and infection has also been found in affected goldfish. Outbreaks of the disease are usually protracted.

PATHOLOGY. Alterations of the internal organs such as liver and kidney can be found in cases of a septicaemia.

DIAGNOSIS. *V. cholerae* grows on standard nutrient agar supplemented with 0–6% sodium chloride at pH values of 7–10. Species identification is possible by 16S rRNA sequencing.

TREATMENT. An antibiotic treatment is recommended after diagnosing *V. cholerae*. Additionally, the environmental conditions and especially the hygienic situation should be improved.

FACTS SHEET

Disease: Infections with *Vibrio cholerae*
Pathogen: *Vibrio cholerae*
Susceptible hosts: Common carp (*Cyprinus carpio*), goldfish (*Carassius auratus*), probably also other carp species
Clinical signs: Petechial haemorrhages
Pathology: Alterations of internal organs
Mortality: Unknown
Temperature: 15–30°C
Prevention and treatment: Prevention is by optimisation of environmental conditions. Treatment is dependent on the result of the antibiogram

Vibrio vulnificus

AETIOLOGY. *Vibrio vulnificus* is a motile, Gram-negative rod-shaped bacterium which is an important zoonotic agent and can cause severe wound infections in humans.

PATHOGENESIS. *V. vulnificus* is ubiquitous in the aquatic marine environment. Fish become infected by the uptake of bacteria from the surrounding water. Infection usually starts by colonisation of the skin or the gills.

CLINICAL SIGNS OF THE DISEASE. Great losses due to infections with *V. vulnificus* have been seen in aquaculture facilities in China.

PATHOLOGY. Histopathological examinations reveal that in grass carp an infiltration of inflammatory cells occurs in the submucosa of the intestine. Additionally, congestion in the liver and necrosis of the epithelial cells of the secondary lamellae have been recognised.

DIAGNOSIS. *V. vulnificus* grows on CSB agar with supplementation of 0.5–5% sodium chloride or on selective media for *Vibrio* spp., such as thiosulphate bile salts sucrose (TCBS) agar, in off-white to transparent grey, round, glossy colonies. Cultivation temperatures can range between 12°C and 37°C, but ideally at the keeping temperature of the affected fish. Species identification is possible by 16S rRNA sequencing.

TREATMENT. If carp species are infected with *V. vulnificus*, the environmental conditions should be optimised to improve the general health status of the fish. An antibiotic treatment can be carried out after an antibiogram.

FACTS SHEET

Disease: Infections with *Vibrio vulnificus*
Pathogen: *Vibrio vulnificus*
Susceptible hosts: Grass carp, probably also other carp species
Clinical signs: Death, gill alterations
Pathology: Intestinal alterations

Mortality: Can be high
Temperature: 12–30°C
Prevention and treatment: Prevention is by optimisation of environmental conditions. Treatment is dependent on the result of the antibiogram

8.5.10 Yersiniaceae

Formerly a genus belonging to the family Enterobacteriaceae, it has recently been reclassified and the new family Yersiniaceae established.

Yersinia ruckeri

AETIOLOGY. *Yersinia ruckeri* is primarily an infectious agent of salmonids. Since an infection with this bacterium rarely occurs in different carp species, the pathogen is only briefly considered here. *Y. ruckeri* is a Gram-negative, rod- to cocci-shaped bacterium. Most isolates of *Y. ruckeri* are motile by peritrichous flagella and are classified as biotype 1 isolates. Immotile isolates are also regularly occurring that are classified as biotype 2 isolates. The bacterium was first isolated in the USA in the 1950s and by the 1980s the number of diseases caused by an infection with *Y. ruckeri* increased in several European countries. *Y. ruckeri* can be isolated from water, aquatic invertebrates, various commercial and wild fish species and muskrats. Infections with *Y. ruckeri* are most prevalent in trout, especially in rainbow trout, where the disease caused by the bacterium is known as enteric redmouth disease (ERM). Nevertheless, the bacterium can also occur in other fish species, including different carp species. Additionally, the bacterium has been isolated from clinically healthy common carp and goldfish. The typical disease signs, such as haemorrhages in the mouth and the liver, known in rainbow trout might not be seen in carp and goldfish.

PATHOGENESIS. Fish usually become infected through the water. Since *Y. ruckeri* can also be isolated from bird droppings, these vertebrates may be a source of infection. *Y. ruckeri* binds avidly to rough surfaces in the water and is found associated with algae and pond sediment. The resulting biofilm in fish-keeping systems can lead to infection in the fish stock. The main route of entry of *Y. ruckeri* is through the gills, and thence to other organs via the bloodstream. In experimental bath infections, the gut appears to be a main target organ of the bacteria. Infection can be rapid and after 30 min *Y. ruckeri* can be detected in the gut. In contrast, the bacteria are found in the kidney only after 3 days, and in the liver, spleen, brain and heart only after 7 days post-infection.

CLINICAL SIGNS OF THE DISEASE. Typical signs of the disease in rainbow trout are septicaemia and haemorrhages on body surfaces, fins and internal organs. In many cases, exophthalmos can occur and conspicuous redness has been observed on the head, especially in the mouth area, but this does not necessarily occur in all animals. In chronic cases, the fish appear dark and lethargic, but can also act as asymptomatic carriers. Losses are low in chronic forms, but can increase dramatically in cases of suboptimal water quality or stress. In other fish species, the clinical signs of the disease can vary, and often unspecific signs are seen. In bighead carp (*Hypophthalmichthys nobilis*) and silver carp (*Hypophthalmichthys molitrix*) haemorrhages in the skin, especially on the abdomen, the base of the fins, around the mouth and the eyes, and at the opercula, are seen. In some fish, exophthalmos and pale gills can also be detected. Mortality rates up to 100% have been noted, and the infectious strain was motile and therefore believed to belong to biotype 1. Disease signs in the three Indian major carp, rohu, catla and mrigal, are also associated with haemorrhages on the skin and at the base of the fins, and darkening of the gills. Ulcerations of the skin and the fins have also been detected. Mortality levels of 20–30% can occur. The detected isolates were motile and therefore also seemed to belong to biotype 1. Only in experimental infections of Indian major carp were typical haemorrhages detected around the mouth. In aquaculture systems disease outbreaks in Indian major carp have been associated with high ammonia levels and high bacterial amounts in the water. Therefore, it can be assumed that suboptimal environmental conditions may trigger the infection. In common carp, descriptions of clinically relevant disease outbreaks in natural environments are lacking. In experimental infections of common carp fry by intraperitoneal application, disease signs occurred 2 days post-infection. The fry developed exophthalmos, darkening of the skin, petechiae and fusion of the gill arches, haemorrhages on the skin and the bases of the fins, as well as on the opercula and the lateral lines. Reddening around the mouth was not detected in the infected fish. The mortality rate can reach 85%.

PATHOLOGY. In bighead carp and silver carp, signs of inflammation and haemorrhages were seen in various organs such as kidney, liver and spleen. In some of the affected fish, swelling and accumulation of gas in the intestine also occurred. In the Indian major carp species, petechial bleeding has been seen in the liver, fat tissue and in the peritoneum of the abdominal cavity. In individual fish splenomegaly and necrosis in the head kidney are occasionally found. When common carp were experimentally infected most internal organs exhibited oedema, degeneration, necrosis and petechial bleedings. However, no changes were detected in the pancreatic tissue and in the spleen.

DIAGNOSIS. *Yersinia ruckeri* grows easily on the usual culture media, like tryptic soy agar (TSA) or Columbia sheep blood agar (CSB). The optimal temperature range is 20–28°C and colonies can be seen after 24 hour at an incubation temperature of 22°C. *Y. ruckeri* grows as round, elevated, shiny, whitish colonies without showing any haemolytic capacities. Biochemical tests such as the API 20E tests can be used for identification of the pathogen, but 16S rRNA sequencing is more reliable for species identification. Additional tests should be performed if the biotype of the isolate needs to be identified. This is especially important when a fish stock requires vaccination.

TREATMENT. Antibiotic treatment can be carried out after an antibiogram. In addition, the hygienic conditions should be checked. Vaccines are available for salmonids, but these are not usually used for carp species.

FACTS SHEET

Disease: Enteric redmouth disease (ERM)
Pathogen: *Yersinia ruckeri*
Susceptible hosts: Bighead carp, silver carp, rohu, catla, mrigal, probably also common carp
Clinical signs: Often unspecific in carp species, septicaemia and haemorrhages
Pathology: Haemorrhages in the internal organs
Mortality: Up to 100% under suboptimal environmental conditions
Temperature: 22–28°C. Disease signs can occur also at lower water temperatures
Prevention and treatment: Vaccination is available for salmonids. Prevention is by optimisation of environmental conditions. Treatment is dependent on the result of the antibiogram

8.6 Gram-Positive Disease-Causing Bacteria

Only a few bacteria that can be detected in diseased fish and found in the aquatic environment are Gram-positive. They are called this as they take up the crystal violet stain that is part of the Gram staining, a method for bacterial identification. Gram-positive bacteria stain purple in colour because, in contrast to Gram-negative bacteria, they have a thick peptidoglycan layer in their cell wall which retains the stain.

8.6.1 *Mycobacteriaceae*

Mycobacterial infections are probably the most common chronic infections in aquarium fish. The disease is a zoonosis.

Mycobacterium spp.

AETIOLOGY. Mycobacteria are long, acid-fast, alcohol-fast, aerobic and rod-shaped. They are Gram-positive, but are usually difficult to stain. More than 150 species of mycobacteria have been described and some of these species are important pathogens for humans. Tuberculosis is caused in humans by *Mycobacteria tuberculosis* that forms, together with other *Mycobacterium* species that induce signs of tuberculosis (*M. bovis*, *M. africanum*, *M. microti*, *M. canetti*, *M. pinnipedii*), the *Mycobacterium tuberculosis* complex. Mycobacteriosis in fish, also known as fish tuberculosis, is caused by non-tuberculoid *Mycobacterium* species, primarily *Mycobacterium marinum* and *Mycobacterium fortuitum.* Other mycobacteria occurring in fish that can also lead to fish tuberculosis are *Mycobacterium abscessus, M. anabati, M. chelonei, M. gordonae, M. montefiorense, M. neoaurum, M. piscium, M. platypoecilus, M. pseudoshottsii, M. ranae, M. scrofulaceum, M. siminae, M. smegmatis* and *M. ulceranus.* Infections with mycobacteria occur very regularly in many fish species, and are usually chronic and characterised by granulomas in multiple organs or tissues. In humans, the non-tuberculoid *Mycobacterium* species can induce local skin infections, such as granulomatous lesions. Therefore, fish tuberculosis is an important zoonosis. The skin alterations in humans are known as swimming-pool granuloma or fish fancier's finger. These skin changes heal very poorly and are very difficult to treat because the bacteria are resistant to most of the drugs that can be used. If there are skin injuries on the human hands and forearms, no work should be carried out on the aquarium or pond and there should be no direct contact with fish without adequate protective measures (gloves, thorough cleaning and disinfecting of the skin after work). Immunocompromised people, young children, the elderly and pregnant women should be careful when handling fish. Here, too, the simplest method of prevention is to wear gloves when working in a fish tank or pond and when handling the fish.

PATHOGENESIS. Mycobacteriosis in fish is primarily spread by shedding of bacteria from infected fish through the skin or from the gut. Other fish ingest these bacteria through the gastrointestinal tract. Transovarial transmission has also been described for certain ornamental fish. The bacteria can survive in the environment for 2 years and fish can be infected for 2–4 years before showing any signs of a disease. Therefore, the bacteria are easily introduced into fish-keeping systems with new fish that appear clinically healthy.

CLINICAL SIGNS OF THE DISEASE. In fish, emaciation, poor growth and a lack of breeding success are sometimes the only symptoms of an infection with mycobacteria. Skeletal deformations, non-healing skin ulcerations, fin erosions, exophthalmos, keratitis, abnormal swimming behaviour, and darkening and swelling of focal areas of the skin can also occur. As the infection is chronic, losses may not be high over a given time period, but regular losses can occur over longer periods of time.

PATHOLOGY. In autopsy, whitish nodules of about 1–4 mm in diameter are associated with several of the internal organs (Fig. 8.11).

DIAGNOSIS. The clinical picture already gives an indication of the disease. Granulomas can often be detected in the internal organs of infected fish. Some of these can be seen macroscopically as small nodules that can run through the entire organ. The granulomas are clearly recognisable in a fresh squeeze preparation of kidney, spleen, liver and also intestine. The granulomas can also be recognised in histological preparations. However, since other bacteria can also lead to granuloma formation, further clarification is necessary. This can first be done on an organ imprint or on histological preparation using the Ziehl–Neelsen stain or the Fite–Faraco stain. If acid-fast rods are detected, specialised laboratories are able to further classify the bacteria. Culturing on Lowenstein–Jensen medium or Middelbrook LH10 agar and biochemical testing are often required for an accurate diagnosis. The cultivation can take a long time and not all isolates can be cultivated.

TREATMENT. Mycobacterial infections are very common in aquarium fish, but can also occur in fish from aquaculture facilities. Despite this, only a few fish show symptoms of illness. Often, symptoms do not appear under optimal husbandry conditions. For this reason, the most important measure is to optimise the environment of the fish. The chemical and microbiological water parameters, the feeding and the set-up of the aquarium should be critically checked and improved if necessary. Other possible diseases such as parasitic infections and other bacterial infection should also be treated. Diseased fish should be removed whenever possible from the keeping system to prevent further spread of the pathogen. For fish that are showing signs of the disease, the prognosis is usually not good. Although combinations of different antibiotics have been described to treat the disease, the chances of success are very limited.

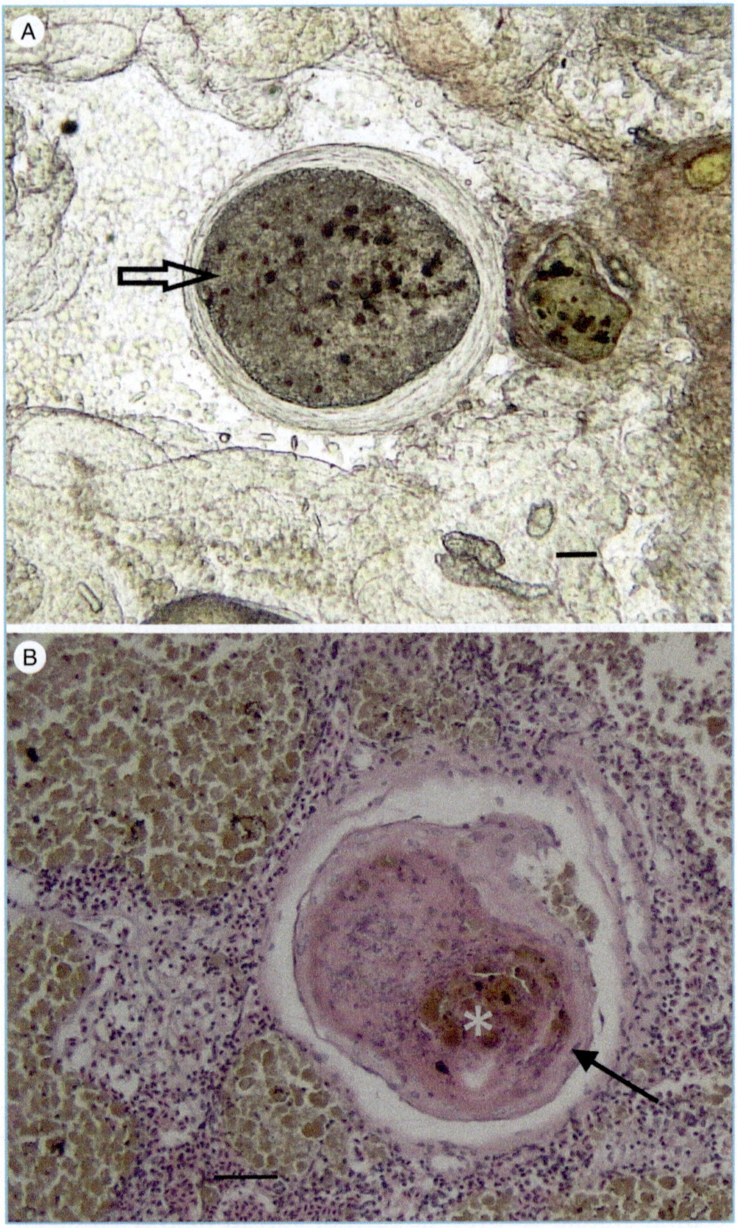

Figure 8.11 Microscopic image of carp spleen infected with mycobacteria. (A) Fresh mount showing granuloma (arrow). (B) Histological section of granuloma showing degenerated macrophages infected by bacteria (*) encapsulated by connective tissue (arrow). Indicative scale bars 100 μm.

FACTS SHEET

Disease: Mycobacteriosis
Pathogen: Non-tuberculoid species of Mycobacteria
Susceptible hosts: All carp species
Clinical signs: Skeletal deformations, non-healing skin ulcerations, fin erosions, exophthalmos, keratitis, abnormal swimming behaviour, and darkening and swelling of focal areas of the skin can occur
Pathology: Granulomas in multiple organs and tissues
Mortality: Can be high, but over a long period of time
Temperature: Wide temperature range
Prevention and treatment: Prevention is by optimisation of the environmental conditions to avoid disease outbreaks. Treatment is not possible.

8.6.2 Nocardiaceae

Nocardia spp.

AETIOLOGY. *Nocardia* spp. are Gram-positive, immobile, weakly acid-fast, aerobic, branched rods. The bacteria are facultative pathogens that are found in soil and on plants. *Nocardia asteroides* is the causative agent of false neon disease. The pathogen not only causes symptoms of a bacterial infection in neon tetras, but has also been detected in many other freshwater and saltwater fish.

PATHOGENESIS. Fish can become infected by ingesting contaminated feed. However, the bacteria only show a low level of infectivity. In the case of an infection with *Nocardia*, the fish will encapsulate the pathogens in connective tissue envelopes, which can be recognised as granulomas.

CLINICAL SIGNS OF THE DISEASE. *Nocardia* spp. cause symptoms like those of mycobacteria. The fish lose weight and often show exophthalmos, fading colours or distended abdomens.

PATHOLOGY. On dissection, whitish nodules (granuloma), caused by the encapsulation of the pathogens in the tissue, can be observed in the organs, the musculature and the subcutaneous tissue. Granulomas can also be detected microscopically in the internal organs.

DIAGNOSIS. The first signs of infection are usually based on clinical observations. A histological examination or an examination of a native squeeze specimen of the internal organs can confirm the presence of granulomas. Nevertheless, utilising such observations cannot rule out an infection with mycobacteria. To eliminate this possibility a Ziehl–Neelsen stain should be performed. If *Nocardia* spp. are involved, they are slightly acid-resistant, filamentous-branched and chain-like. Cultures must be carried out for clear differentiation; and molecular biological examinations are required for identification of the species.

TREATMENT. As with any bacterial infection, the first step is to optimise the housing conditions. Antibiotic treatment can be carried out after an antibiogram. In the early stages of the disease, a cure can be achieved. If the internal organs are already severely damaged, treatments are usually not successful.

FACTS SHEET

Disease: Nocardiosis ('False neon disease' in neon tetra)
Pathogen: *Nocardia* spp., especially *N. asteroides*
Susceptible hosts: Diverse fish species, including carp
Clinical signs: Loss of weight, exophthalmos, fading colours, distended bellies
Pathology: Granulomas in internal organs, musculature and subcutaneous tissue
Mortality: Can be high
Temperature: Wide temperature range
Prevention and treatment: Prevention is by optimisation of environmental conditions. Treatment is dependent on the result of the antibiogram

8.6.3 Streptococcaceae

Streptococcus spp.

AETIOLOGY. *Streptococcus* spp. are Gram-positive, immotile, coccoid bacteria. The pathogens are very widespread in the aquatic environment and occur as commensals and as saprophytes. Few species appear to be pathogenic to fish, although diseases can occur in both freshwater and saltwater fish. Most of the disease-causing isolates belong to the ß-haemolytic streptococci. *Streptococcus iniae*, for example, can cause severe disease signs in tilapia and several seawater fish species. *S. iniae* has also been detected in common carp.

Streptococcus agalactiae isolated from diseased tilapia has been shown to induce signs of disease in bighead carp.

PATHOGENESIS. Diseases occur mainly at high water temperatures and when the water is organically contaminated.

CLINICAL SIGNS OF THE DISEASE. Superficial haemorrhagic skin ulcerations are common and septicaemia can also occur. Infected fish can show lethargy, dorsal rigidity, erratic swimming behaviour and symptoms associated with dysfunction of the central nervous system.

PATHOLOGY. An infection with *S. iniae* can lead to meningoencephalitis and meningitis. Haemorrhages and inflammation in the internal organs can also occur.

DIAGNOSIS. Cultivation of *Streptococcus* spp. is possible on common culture media such as CSB agar at temperatures between 20°C and 30°C. The bacteria grow in white colonies and can be easily identified by 16S rRNA sequencing.

TREATMENT. In particular, the hygienic conditions in which the fish are kept must be optimised. If this measure alone does not lead to success and an improvement in fish health, treatment with antibiotics is possible after an antibiogram. Erythromycin and tetracycline have often been shown to be effective.

FACTS SHEET

Disease: Infection with *Streptococcus* spp.
Pathogen: *Streptococcus* spp., for example *S. agalactiae*, *S. dysgalactiae* and *S. iniae*
Susceptible hosts: Common carp (*Cyprinus carpio*), bighead carp (*Hypophthalmichthys nobilis*), silver carp (*Hypophthalmichthys molitrix*), probably other carp species are also affected
Clinical signs: Central nervous symptoms, haemorrhagic skin ulcerations
Pathology: Meningoencephalitis, meningitis, inflammation, and haemorrhages in the internal organs
Mortality: Can be high
Temperature: Higher water temperatures
Prevention and treatment: Prevention is by optimisation of environmental conditions. Antibiotic treatment according to the results of the antibiogram

8.7 Selected References for Further Reading

Adamek, M., Teitge, F., Jung-Schroers, V., Heling, M., Gela, D. *et al.* (2018) Flavobacteria as secondary pathogens in carp suffering from koi sleepy disease. *Journal of Fish Diseases* 41, 1631–1642.

Adhikary, A., Hazarika, G., Das, R., Sarma, K., Vimal, S. *et al.* (2022) Isolation, identification and characterisation of *Aeromonas veronii* from diseased Indian major carp, *Cirrhinus mrigala*. *Aquaculture International* 31, 997–1010.

Al-Haider, S.M., Al-Niaeem, K.S. and Resen, A.K. (2019) Prevalence and antibiotic resistance in *Aeromonas* species isolated from common carp (*Cyprinus carpio* L.) cultivated in floating cages at Al-Hilla River. *Basrah Journal of Agricultural Sciences* 32, 33–43.

Al-Haider, S.M., Al-Niaeem, K.S. and Resen, A.K. (2019) Isolation of *Citrobacter* species from common carp, *Cyprinus carpio* cultivated in floating cages at Al-Hilla river, Babylon province. *IOP Conference Series.: Earth and Environmental Science* 388(12036), 1–5.

Attia, M.M., Abdelsalam, M., Elgendy, M.Y. and Sherif, A.H. (2022) *Dactylogyrus extensus* and *Pseudomonas fluorescens* dual infection in farmed common carp (*Cyprinus carpio*). *Microbial Pathogenesis* 173(Pt A), 105867.

Austin, B. and Austin, D.A. (2016) *Bacterial Fish Pathogens – Disease of Farmed and Wild Fish*, 6th edn. Springer.

Bai, J., Huo, Y., Hu, X., Lü, A. and Sun, J. (2021) Characterization of pathogenic *Pseudomonas alcaligenes* isolated from Koi Carp in China. *Journal of Aquatic Animal Health* 33, 243–251.

Behera, B.K., Paria, P., Das, A., Bhowmick, S., Sahoo, A.K. *et al.* (2017) Molecular characterization and pathogenicity of a virulent *Acinetobacter baumannii* associated with mortality of farmed Indian Major Carp *Labeo rohita* (Hamilton 1822). *Aquaculture* 471, 157–162.

Behera, B.K., Bera, A., Paria, P., Das, A., Parida, P. *et al.* (2018) Identification and pathogenicity of *Plesiomonas shigelloides* in silver carp. *Aquaculture* 493, 314–318.

Berc, A., Petrinec, Z., Matasin, Z. and Kozarić, Z. (1999) *Yersinia ruckeri* septicaemia in experimentally infected carp (*Cyprinus carpio* L.) fingerlings. *Acta Veterinaria Hungarica* 47, 161–172.

Böttcher, K., Bulla, V. and Richter, R. (2021) Disease associated with *Yersinia ruckeri* infection in *Cyprinus carpio* in a common carp farm in Germany. *Bulletin of the European Association of Fish Pathologists* 41, 128–132.

Buller, N.B. (2022) *Bacteria from Fish and Other Aquatic Animals: A Practical Identification Manual*, 2nd edn. CABI Digital Library. CAB International, Wallingford, UK.

Chen, F., Sun, J., Han, Z., Yang, X., Xian, J.A. *et al.* (2019) Isolation, identification and characteristics of *Aeromonas veronii* from diseased crucian carp (*Carassius auratus gibelio*). *Frontiers in Microbiology* 26, article no. 2742.

Chen, H., Zhao, Y., Chen, K., Wei, Y., Luo, H. *et al.* (2022) Isolation, identification, and investigation of pathogenic bacteria from common carp (*Cyprinus carpio*) naturally infected with *Plesiomonas shigelloides*. *Frontiers in Immunology* 13, 872896.

Colquhoun, D. and Duodu, S. (2011) *Francisella* infections in farmed and wild aquatic organisms. *Veterinary Research* 42, 47.

Declercq, A.M., Chiers, K., Van den Broeck, W., Dewulf, J., Eeckhaut, V. *et al.* (2015) Interactions of highly and low virulent *Flavobacterium columnare* isolates with gill tissue in carp and rainbow trout. *Veterinary Research* 46, 25.

Delghandi, M.R., El-Matbouli, M. and Menanteau-Ledouble, S. (2020) Mycobacteriosis and infections with non-tuberculous Mycobacteria in aquatic organisms: A review. *Microorganisms* 8(9), 1368.

Dong, H., Nguyen, V., Kayansamruaj, P., Gangnonngiw, W., Senapin, S. *et al.* (2016) *Francisella noatunensis* subsp. *orientalis* infects striped catfish (*Pangasianodon hypophthalmus*) and common carp (*Cyprinus carpio*) but does not kill the hosts. *Aquaculture* 464, 190–195.

Favaro, L., Scanzio, T., Varello, K., Caffara, M., Righetti, M. *et al.* (2014) Mixed mycobacterial infection in an adult koi carp *Cyprinus carpio* L. *Journal of Fish Diseases* 37, 753–755.

Gauthier, D.T. and Rhodes, M.W. (2009) Mycobacteriosis in fishes: A review. *Veterinary Journal* 180(1), 33–47.

Gultepe, N. and Aydin, S. (2009) *Pseudomonas elongata* infection in scattered mirror carp (*Cyprinus carpio*): Bacteriology, gross pathology and treatment. *Journal of Animal and Veterinary Advances* 8, 835–838.

Guo, W., Han, R., Xu, W., Lu, Z., Li, Y. *et al.* (2023) The protective effect of inactivated *Flavobacterium columnare* vaccine in grass carp (*Ctenopharyngodon idellus*). *Frontiers in Cellular Infection Microbiology* 14, 1162975.

Han, Z., Sun, J., Jiang, B., Hu, X., Lv, A., Chen, L. and Guo, Y. (2021) Concurrent infections of *Aeromonas veronii* and *Vibrio cholerae* in koi carp (*Cyprinus carpio* var. koi). *Aquaculture* 535, 736395.

Harriff, M.J., Wu, M., Kent, M.L. and Bermudez, L.E. (2008) Species of environmental mycobacteria differ in their abilities to grow in human, mouse, and carp macrophages and with regard to the presence of mycobacterial virulence genes, as observed by DNA microarray hybridization. *Applied and Environmental Microbiology* 74, 275–285.

Hawke, J.P., Daniel, R., Strother, K., Sokolova, Y., Elliott, J. *et al.* (2021) *Streptococcus dysgalactiae*: A pathogen of feral populations of Silver Carp from a fish kill event. *Journal of Aquatic Animal Health* 33, 231–242.

Heckman, T.I., Yazdi, Z., Pomaranski, E.K., Sebastião, F.A., Mukkatira, K. *et al.* (2023) Atypical flavobacteria recovered from diseased fish in the Western United States. *Frontiers in Cellular Infection Microbiology* 13, 1149032.

Houghton, G., Wiegertjes, G.F., Groeneveld, R.A. and Van Muiswinkel, W.B. (1991) Differences in resistance of carp, *Cyprinus carpio* L., to atypical *Aeromonas salmonicida*. *Journal of Fish Diseases* 14, 333–341.

Jeremic, S. and Jakic-Dimic, D. (2003) Systemic *Citrobacter freundii* infection in carp. *Veterinarski Glasnik* 57, 3–9.

Jung-Schroers, V., Jung, A., Ryll, M., Bauer, J., Teitge, F. and Steinhagen, D. (2018) Methods for identification and differentiation of different *Shewanella* spp. isolates for diagnostic use. *Journal of Fish Diseases* 4, 689–714.

Jung-Schroers, V., Jung, A., Ryll, M., Bauer, J., Teitge, F. *et al.* (2019) Diagnostic methods for identifying different *Aeromonas* species and examining their pathogenicity factors, their correlation to cytotoxicity and adherence to fish mucus. *Journal of Fish Diseases* 42, 189–219.

Karunasagar, I., Karunasagar, I. and Pai, R. (1992) Systemic *Citrobacter freundii* infection in common carp, *Cyprinus carpio* L., fingerlings. *Journal of Fish Diseases* 15, 95–98.

Kim, D.J., Park, J.H., Seok, S.H., Cho, S.A., Baek, M.W. *et al.* (2005) Epitheliocystis in carp (*Cyprinus carpio*) in South Korea. *Journal of Veterinary Medical Science* 67, 119–120.

Kozińska, A., Figueras, M.J., Chacon, M.R. and Soler, L. (2002) Phenotypic characteristics and pathogenicity of *Aeromonas* genomospecies isolated from common carp (*Cyprinus carpio* L.). *Journal of Applied Microbiology* 93, 1034–1041.

Kumar, G., Mayrhofer, R., Soliman, H. and El-Matbouli, M. (2012) Novel Chlamydiales associated with epitheliocystis in grass carp (*Ctenopharyngodon idella*). *The Veterinary Record* 172, 47.

Lei, Q., Xiaojun, Z. and Keran, B. (2012) A new pathogen of gibel carp *Carassius auratus gibelio-Shewanella putrefaciens*. [J]. *Acta Microbiologica Sinica* 52, 558–565.

Lewisch, E., Menanteau-Ledouble, S., Tichy, A. and El-Matbouli, M. (2016) Susceptibility of common carp and sunfish to a strain of *Francisella noatunensis* subsp. *orientalis* in a challenge experiment. *Diseases of Aquatic Organisms* 121, 161–166.

Lian, Z., Bai, J., Hu, X. and Lu, A. (2020) Detection and characterization of *Aeromonas salmonicida* subsp. *salmonicida* infection in crucian carp *Carassius auratus*. *Veterinary Research Communications* 44, 61–72.

Liu, R., Lian, Z., Hu, X., Lü, A., Sun, J. *et al.* (2018) First report of *Vibrio vulnificus* infection in grass carp *Ctenopharyngodon idellus* in China. *Aquaculture* 499, 282–289.

Lu, Z., Gao, R., Duan, Y., Han, R., Guo, W. *et al.* (2021) Isolation and genetic characterization of *Flavobacterium columnare* from grass carp, *Ctenopharyngodon idellus*, in China. *Aquaculture* 541, 736762.

Luo, Z., Bai, X., Yunxia, X.U., Zhang, Z., Hao, S. *et al.* (2015) Recovery of *Pseudomonas aeruginosa* from diseased grass carp (*Ctenopharyngodon idella*) in China. *The Israeli Journal of Aquaculture – Bamidgeh* 67. DOI: 10.46989/001c.20687.

Machida, Y., Tang, B.C., Yamada, M., Sato, S., Nakajima, K. *et al.* (2021) Mycobacteriosis in cultured koi carp *Cyprinus carpio* caused by *Mycobacterium paragordonae* and two *Mycolicibacterium* spp. *Aquaculture* 539, 736656.

Majeed, S. and Gopinath, C. (2006) Cutaneous tuberculosis in the carp, *Cyprinus carpio* L. *Journal of Fish Diseases* 6, 313–316.

Markovic, M., Radojicic, M., Cosic, S. and Levnaic, D. (1996) Massive death of silver carp and bighead caused by *Pseudomonas fluorescens* bacteria. *Veterinarski Glasnik* 50, 761–765.

Mayrhofer, R., Menanteau-Ledouble, S., Pucher, J., Focken, U. and El-Matbouli, M. (2020) *Aeromonas* spp. suggested as the causative agents of red spot disease in northern Vietnamese grass carp *Ctenopharyngodon idella*. *Diseases of Aquatic Organisms* 139, 113–119.

Nayak, S.K., Shibasaki, Y. and Nakanishi, T. (2014) Immune responses to live and inactivated *Nocardia seriolae* and protective effect of recombinant interferon gamma (rIFN γ) against nocardiosis in ginbuna crucian carp, *Carassius auratus langsdorfii*. *Fish and Shellfish Immunology* 39, 354–364.

Pan, L., Yang, Y., Peng, Y., Li, D., Khan, T. *et al.* (2021) The novel pathogenic *Citrobacter freundii* (CFC202) isolated from diseased crucian carp (*Carassius auratus*) and its ghost vaccine as a new prophylactic strategy against infection. *Aquaculture* 533, 736190.

Pandey, V., Bhat, R.A.H., Chandra, S., Tandel, R.S., Dubey, M.K. *et al.* (2021) Clinical signs, lethal dose and histopathological lesions in grass carp, *Ctenopharyngodon idella* experimentally infected with *Edwardsiella tarda*. *Microbial Pathogenesis* 161(Pt. B), 105292.

Paperna, I. and Alves Dematos, A.P. (1984) The developmental cycle of epitheliocystis in carp, *Cyprinus carpio* L. *Journal of Fish Diseases* 7, 137–147.

Paździor, E., Pekala-Safinska, A. and Wasyl, D. (2019) Phenotypic diversity and potential virulence factors of the *Shewanella putrefaciens* group isolated from freshwater fish. *Journal of Veterinary Research* 63, 321–332.

Pekala-Safinska, A. (2018) Contemporary threats of bacterial infections in freshwater fish. *Journal of Veterinary Research* 62, 261–267.

Rajapakshe, A., Prasad, K.P. and Mukharjee, S. (2012) Gross clinical signs and haematological changes associated with artificial infection of *Edwardsiella tarda* in Koi Carp. *Journal of the National Aquatic Resources Research and Development Agency* 41, 1–10.

Rashid, M., Hossain, M.S. and Ali, M.F. (2013) Isolation and identification of *Aeromonas hydrophila* from silver carp and its culture environment from Mymensingh region. *Journal of the Bangladesh Agricultural University* 11, 373–376.

Rich, A.F. and Naguib, M. (2024) *Shewanella xiamenensis*-associated ulcerative dermatitis in koi carp (*Cyprinus rubrofuscus*). *Journal of Fish Diseases* 47(7), e13942.

Sellyei, B., Molnár, K. and Székely, C. (2017) Diverse chlamydia-like agents associated with epitheliocystis infection in two cyprinid fish species, the common carp (*Cyprinus carpio* L.) and the gibel carp (*Carassius auratus gibelio* L.). *Acta Veterinaria Hungarica* 65, 29–40.

Sharma, V., Katoch, R., Mahajan, A., Sharma, M., Chahota, R. *et al.* (2004) Pathogenic potential of *Vibrio anguillarum* isolate in common carps. *The Indian Veterinary Journal* 81, 616–619.

Sioutas, S., Hoffmann, R.W., Pfeil-Putzien, C. and Fischer-Scherl, T. (1991) Carp erythrodermatitis (CE) due to an *Aeromonas hydrophila* infection. Casuistic and experimental results. *Zentralblatt Fur Veterinarmedizin B* 38, 186–194.

Sood, N., Pradhan, P., Verma, D.K., Gupta, S., Chhimwal, R. *et al.* (2019) Epitheliocystis in rohu *Labeo rohita* (Hamilton, 1822) is caused by novel Chlamydiales. *Aquaculture* 505, 539–543.

Soriano-Vargas, E., Castro-Escarpulli, G., Aguilera-Arreola, M.G., Vega-Castillo, F. and Salgado, C. (2010) Isolation and identification of *Aeromonas bestiarum* in cultured common carp (*Cyprinus carpio* L.) from Santa Maria Chapa de Mota, Estado de Mexico, Mexico. *Veterinaria Mexico* 41, 111–115.

Sun, H.C. (2009) Isolation and identification of a pathogenic *Vibrio* sp. from silver carp (*Hypophthalmichthys molitrix*). *Freshwater Fisheries* 39(4).

Syrova, E., Kohoutova, L., Dolejska, M., Papezikova, I., Kutilova, I. *et al.* (2018) Antibiotic resistance and virulence factors in mesophilic *Aeromonas* spp. from Czech carp fisheries. *Journal of Applied Microbiology* 125, 1702–1713.

Teng, T., Liang, L.-G. and Xie, J. (2015) Isolation, identification, and detection, of the virulence factor of *Vibrio cholerae* in grass carp. *Israeli Journal of Aquaculture – Bamidgeh* 67. DOI: 10.46989/001c.20677.

Wahli, T. and Madsen, L. (2018) Flavobacteria, a never ending threat for fish: A review. *Current Clinical Microbiology Reports* 5, 26–37.

Wiklund, T. and Dalsgaard, I. (1998) Occurrence and significance of atypical *Aeromonas salmonicida* in non-salmonid and salmonid fish species: A review. *Diseases of Aquatic Organisms* 32, 49–69.

Xu, T. and Zhang, X-H. (2014) *Edwardsiella tarda*: An intriguing problem in aquaculture. *Aquaculture* 431, 129–135.

Xu, T., Rasmussen-Ivey, C.R., Moen, F.S., Fernández-Bravo, A., Lamy, B. *et al.* (2023) A global survey of hypervirulent *Aeromonas hydrophila* (vAh) identified vAh strains in the Lower Mekong River Basin and diverse opportunistic pathogens from farmed fish and other environmental sources. *Microbiology Spectrum* 11, e03705–22.

Yan, L., Jin, D., Yang, S., Li, X., Hu, S. *et al.* (2022) Pathogenicity of fish pathogen *Pseudomonas plecoglossicida* and preparation of its inactivated vaccine. *Microbial Pathogenesis* 166, 105488.

Zhang, D., Ke, X., Liu, L., Lu, M., Shi, C. and Liu, Z. (2018) *Streptococcus agalactiae* from tilapia (*Oreochromis* sp.) transmitted to a new host, bighead carp (*Aristichthys nobilis*), in China. *Aquaculture International* 26, 885–897.

Zheng, W., Cao, H. and Yang, X. (2012) Grass carp (*Ctenopharyngodon idellus*) infected with multiple strains of *Aeromonas hydrophila*. *African Journal of Microbiology Research* 6, 4512–4520.

Chapter 9

Myxozoan Diseases

9.1 Introduction

Myxozoa are a group of microscopic cnidarian organisms, which live as parasites primarily in aquatic invertebrates, mainly in bryozoans and annelids. In their life cycle, many myxozoans acquired fish as secondary intermediate hosts. This has apparently fostered their massive biodiversification by coevolution with the piscine hosts and by host-switching to other vertebrates, including amphibians, reptiles or mammals. Currently, myxozoa represent about one-fifth of known cnidarian biodiversity. A characteristic trait of myxozoan life cycles is the formation of small spores with polar capsules, a complex structure with an ejectable filament, which is similar to nematocysts, the stinging cells of free-living cnidaria (Fig. 9.1). In addition to morphological similarities between polar capsules and nematocysts, myxozoa possess minicollagen and nematogalectin genes similar to those that play an important role in the nematocyst structure of the cnidarian *Hydra*. This further supports a phylogenetic relationship between myxozoa and cnidaria.

The vast majority of myxozoa alternate in their life cycle between an annelid host and a vertebrate host. In freshwater, most of the myxozoa develop in an oligochaete, which releases actinospores. These stages anchor onto fish, release their sporoplasm into the host and develop into myxospores, which are released into the environment and eventually are ingested by the oligochaete. There they develop into actinospores to complete their life cycle. In most myxozoa, developmental stages are reduced to a few cells and lack specific cnidarian structures except for the characteristic polar capsules. The taxonomy of myxozoa is traditionally based on

 Carps: A Practical Approach to Disease Control and Health Management (eds David Hoole et al.)
DOI: 10.1079/9781836990024.0009

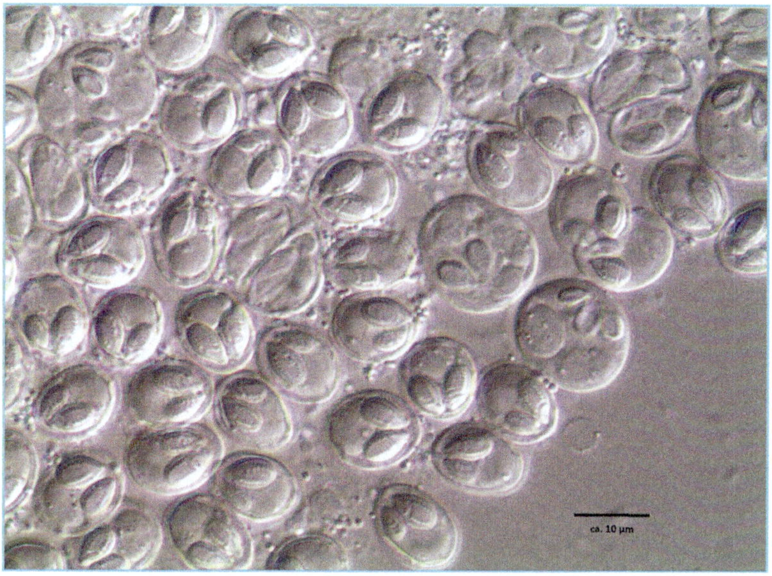

Figure 9.1 Myxospores with a *Myxobolus* type of morphology with a rounded to oval shape and two unequal apical polar capsules with a coiled polar filament. Fresh preparation, Nomarski optics. Indicative scale bar 10 µm. Courtesy of the Environment Agency, National Fisheries Laboratory, UK.

the morphology of the myxospore. Characters like the number and configuration of shell valves, the presence of projections, ribs, ridges or striations on the spore valves, number, position and relative size of the polar capsules and the number of turns of the polar filament are, among others, used to differentiate myxozoa at genus or species level. In addition, biological data such as the infected host and the site of the infection are used as taxonomic characters. However, recent molecular analyses do not support this classical spore-based classification, but reveal a paraphyletic and/or polyphyletic nature of many of the described genera. These analyses highlight four major clades among the myxozoa: a basal malacosporean clade with parasites of bryozoans, a clade with *Sphaerospora sensu stricto* species, a marine clade with polychaete-infecting myxosporeans and a predominantly freshwater oligochaete-infecting clade. Within the marine and freshwater clade, a clustering of species according to the organ system that they target in the vertebrate host can be observed: (i) coelozoic species from the biliary system or (ii) from the excretory/urinary tract; and (iii) a large group of histozoic species infecting skin, muscle, intestine and various other tissues.

In addition, phylogenetic analyses indicate that myxozoa diversified with the extensive radiation of their fish hosts, in particular percid and cyprinid fishes. Thus, a large number of myxosporean parasites infect cyprinid fish, and almost all tissues are affected. However, from a large number of species, the available information is restricted to the site of infection and spore morphology. Cyprinid hosts considered in this chapter are in particular infected by myxozoan parasites from the families Sphaerosporidae, Myxobilatidae and Myxobolidae. The latter family includes a large number of histozoic species, including *Myxobolus*, *Thelohanellus* and *Henneguya* species. Because knowledge on most species is cursory and, in particular, their pathogenic importance is not well documented, only a few species will be discussed here.

9.2 Gill Sphaerosporosis in Common Carp

9.2.1 Introduction

Sphaerosporids represent a particular clade of myxozoan parasites that form spherical or subspherical spores. They mostly live as coelozoic parasites in renal tubules and collecting ducts of the urinary tract of freshwater fish, while some species are histozoic. Most members of the sphaerosporid clade have extrasporogenic stages, which proliferate in blood and some other tissues, and exhibit specific movements. Gill sphaerosporosis in carp is caused by the histozoic *Sphaerospora molnari*, which develops myxospores in the epithelia of skin and gills of common carp, *Cyprinus carpio*.

9.2.2 Disease description

Sphaerospora molnari mainly infects carp fingerlings. Affected fish exhibit dystrophic changes and necrosis of gill tissue (Fig. 9.2A), but

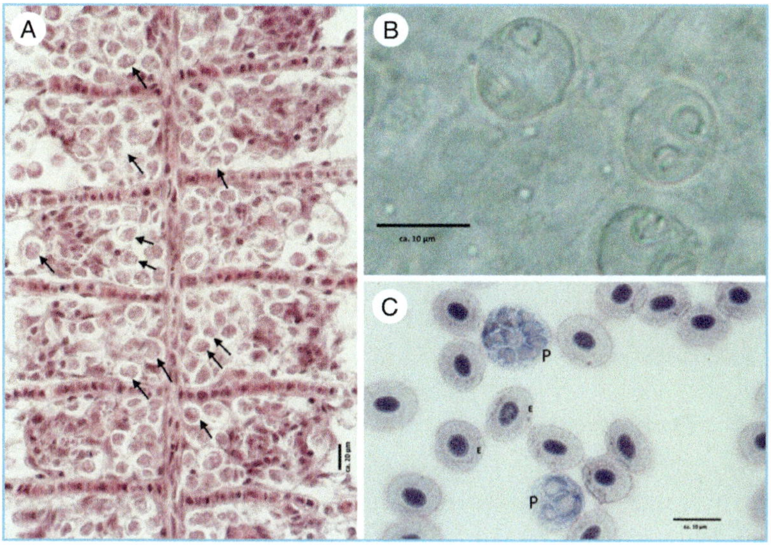

Figure 9.2 Gill sphaerosporosis. (A) Severe destruction of the gill tissue in common carp infected by *Sphaerospora molnari*. Numerous *Sphaerospora* myxospores (arrows) are developing in the interbranchial tissue. Histological section of a gill with HE staining. Indicative scale bar 20 µm. (B) *Sphaerospora molnari* myxospores with a subspherical morphology and two polar capsules opening at the anterior end of the spore. Fresh preparation. Indicative scale bar 10 µm. (C) Pre-sporogonic stages of *Sphaerospora* (P) in the blood of carp. Giemsa-stained blood smear with two pre-sporogonic stages at different stages of development and carp erythrocytes (E). Indicative scale bar 10 µm.

may also show necrosis of skin tissue in regions next to the gill cavity and surrounding the nasal pits. A heavily infected specimen may lose weight. The destruction of the tissue during the discharge of spores may cause secondary bacterial infections.

9.2.3 Pathogen

Sphaerospora molnari develops subspherical myxospores with two polar capsules opening at the anterior end of the spore and situated in a plane perpendicular to the straight sutural line of the spore (Fig. 9.2B). Spore development occurs in monosporic sporoblasts in epithelium cells of the gill or the skin. The spores are released into the water after necrotic destruction of infected cells. The life cycle most likely involves an invertebrate host, which has not yet been identified for any member of the *Sphaerospora sensu stricto* clade. The development of *S. molnari* in its vertebrate host includes the presence of pre-sporogonic stages in the blood and in the interstitial tissue of the kidney, but these stages have never been observed in the lumen of kidney tubules. Sphaerosporids infecting gills of *Carassius* species are morphologically, and most likely molecular biologically, different from *S. molnari*, and are ascribed to the species *S. carassii* and *S. chinensis*. The identification of these species needs to be validated by DNA sequencing.

9.2.4 Detection

Myxospores of *S. molnari* can be detected in fresh samples obtained by scraping tissue material from affected gill or skin areas. In addition, sporoblasts and myxospores are visible in methanol-fixed smear preparations of affected tissue after staining with Giemsa solution as well as in histological sections of affected gill or skin tissue. Pre-sporogonic stages in the blood can be detected in methanol-fixed and Giemsa-stained blood smears (Fig. 9.2C). Species identification requires morphometric measurements from fresh myxospores and sequencing of 18S rDNA nucleotide sequences.

9.2.5 Prevention and treatment

A treatment for *S. molnari* is not established.

FACTS SHEET

Disease: Gill sphaerosporosis

Pathogen: *Sphaerospora molnari* (common carp), *S. carassii* (*Carassius auratus*), *S. chinensis* (*C. gibelio*)

Susceptible species: Common carp, *Carassius* species

Mortality: Uncertain

Temperature: Uncertain

Prevention and treatment: Prevention is difficult. Only sanitary measures, such as liming the ponds, could be applied. No treatment protocol established

9.3 Swim Bladder Inflammation and Renal Sphaerosporosis in Carp

9.3.1 *Introduction*

The urinary tract of fish and amphibians is infected by several myxozoan species. In cyprinid fish in particular, *Sphaerospora* and *Hoferellus* species proliferate and sporulate in renal tubules and the collecting ducts of the urinary tract, which may induce clinical signs of a disease. In the context of this book, the two most important parasites in common carp, *Hoferellus* and *Sphaerospora*, are considered. The life cycle of sphaerosporid myxozoa involves an extrasporogonic proliferation in blood and other tissues, including swim bladder, of the infected fish host and the development of pansporoblasts and myxospores in the lumen of renal tubules. In *Hoferellus*, blood stages have also been detected in the bloodstream of carp, and the presence of elusive intracellular stages has been reported in epithelial cells of renal tubules of common carp.

9.3.2 *Disease description*

Swim bladder inflammation and renal sphaerosporosis may be the cause of a serious disease in carp fry, in particular in their 3rd to 4th month of life. Heavily infected fish appear black in colour, display locomotory disorders, often swim in circles and eventually die. In necropsy, a thickened swim bladder wall is seen, often with haemorrhages or yellow-brownish material adhering to the internal wall. The kidney is hypertrophied, tubules dilated, the epithelium then becomes atrophied and necrotic. These changes are associated with nephritis and loss of renal function.

9.3.3 *Pathogen*

In carp, swim bladder inflammation and renal sphaerosporosis are associated with an infection with *Sphaerospora dykovae*, previously identified as *S. renicola*. Myxospores of this species undergo some variation in shape during maturation. In the early stage of maturation, myxospores have large valve-cell nuclei at the posterior part of the spore, while fully mature spores are completely spherical. The polar capsules are of pyriform shape and relatively large in maturating spores, but of smaller size in mature spores. The spores develop in disporic pansporoblasts in the lumen of renal tubules, which causes dilatation of tubules, atrophied and necrotic tubule epithelium cells, and an associated nephritis. The developmental cycle of *S. dykovae* includes proliferation of extra-sporogenic stages in the bloodstream. These stages comprise primary cells, which contain a number of secondary and tertiary cells and are motile in fresh preparations. Heavily infected carp may be anaemic and may develop a leucocytosis. Extra-sporogenic blood stages invade the swim bladder, where they further proliferate and subsequently induce an inflammatory response of the connective tissue. This results in a thickening of the swim bladder wall, formation of oedema and haemorrhages (Figs 9.3 and 9.4). Necrotic material from the swim bladder can enter the peritoneal cavity and induce peritonitis. The prognosis for fish with swim bladder inflammation

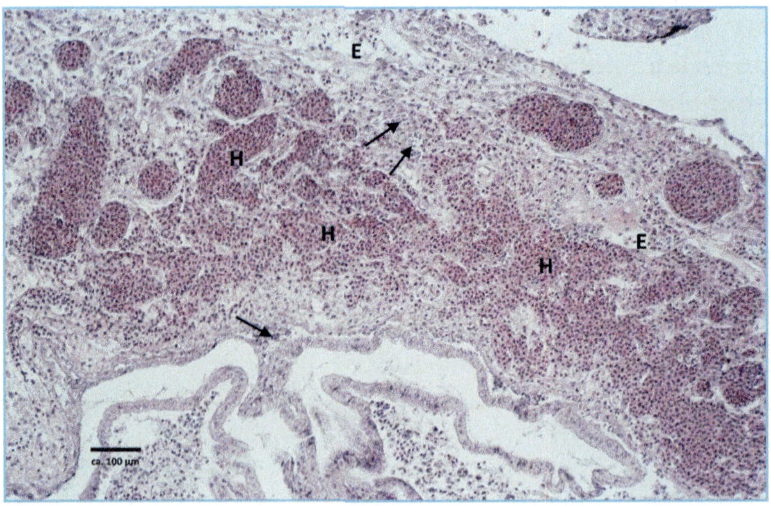

Figure 9.3 Histological section of severe swim bladder inflammation in juvenile common carp with haemorrhages (H), formation of oedema (E) and developing extra-sporogonic *Sphaerospora* stages (arrows.) HE staining. Indicative scale bar 100 µm.

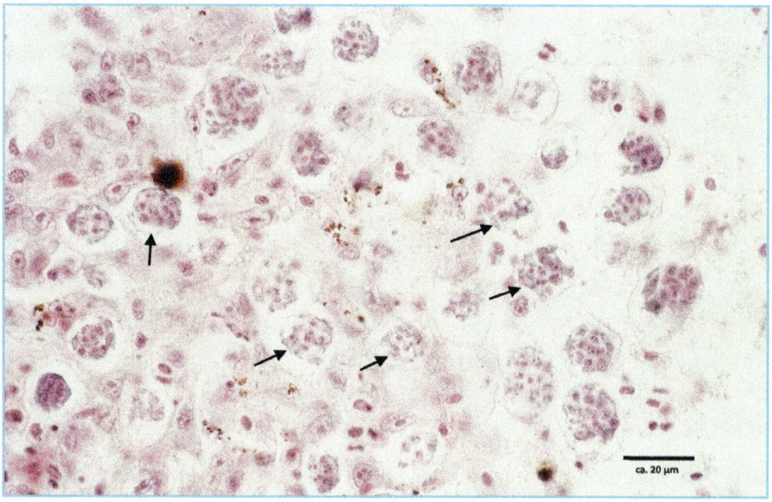

Figure 9.4 Extra-sporogonic *Sphaerospora* developmental stages (arrows) in the swim bladder of juvenile carp. Histological section with HE staining. Indicative scale bar 20 μm.

is poor and may increase mortality, particularly in the next winter following infection. In addition to *S. dykovae*, the kidney of carp may be co-infected with *Hoferellus cyprini*, which may contribute to renal pathology.

9.3.4 Detection

Myxospores of *S. dykovae* and *Hoferellus cyprini* can be detected in fresh tissue samples from the kidney of suspected specimens. In addition, sporoblasts and myxospores are visible in histological sections of affected renal tissue (Fig. 9.5). Pre-sporogonic stages in the swim bladder can be detected in histological sections of the tissue from affected fish and can also be visualised in methanol-fixed and Giemsa-stained blood smears (see Fig. 9.2C). Species identification requires morphometric measurements from fresh myxospores and sequencing of 18S rDNA or SSU rDNA nucleotide sequences. Species identification from blood stages also requires sequencing of the respective genes.

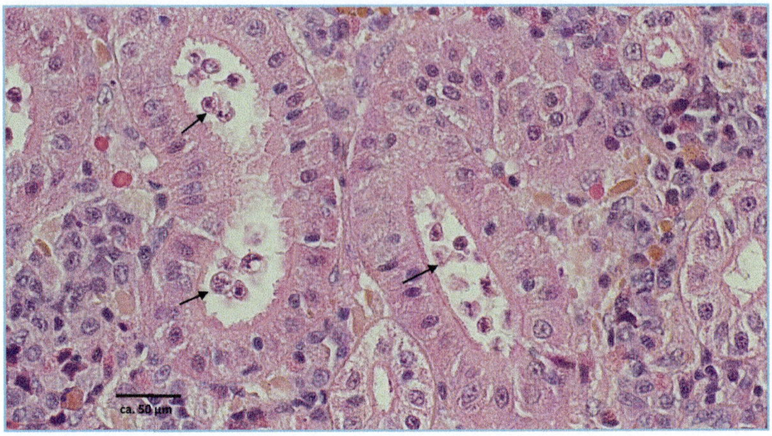

Figure 9.5 Developing myxospores of *Sphaerospora dykovae* in lumen of renal tubules (arrows). Histological section with HE staining. Indicative scale bar 20 µm. Courtesy of the Environment Agency, National Fisheries Laboratory, UK.

9.3.5 Prevention and treatment

A treatment for *S. dykovae* and *Hoferellus cyprini* is not established.

FACTS SHEET

Disease: Swim bladder inflammation and renal sphaerosporosis in carp
Pathogen: *Sphaerospora dykovae*, co-infection with *Hoferellus cyprini*.
Susceptible species: Common carp
Mortality: Uncertain; 5–15% in affected populations of young-of-the-year carp
Temperature: Uncertain
Prevention and treatment: Only sanitary measures, such as liming the ponds, could be applied. No treatment protocol established

9.4 Kidney Enlargement Disease in Goldfish

9.4.1 Introduction

The urinary tract of vertebrates is a common site for the development of myxospores from various myxozoan species. Among these, members of the genus *Hoferellus* are parasitic in the kidney of several cyprinid species, including several *Carassius* species.

9.4.2 Disease description

Heavily infected fish develop abdominal distension (ascites), display reduced swimming behaviour and may show accelerated breathing movements, which may be fatal. Necropsy reveals a marked hypertrophy of the kidney by a cystic papillomatous hyperplasia of the renal tubule epithelium. Many of these symptoms can be fatal to infected fish.

9.4.3 Pathogen

The disease is caused by an invasion of extra-sporogonic stages of *Hoferellus carassii* into epithelial cells of the renal tubules, which become metaplastic in response to the parasite. Infected cells, which may contain up to two parasites, elongate and protrude into the lumen of the renal tubule. Thereafter, the cells rupture and release parasite stages into the lumen, where they attach to the remaining intact epithelium and transform into spore-producing polysporic plasmodia. *Hoferellus carassii* spores, which are 'mitre-like', comprise many filaments at the posterior end of the myxospore and possess two drop-like polar capsules. Comprehensive data on development and life cycle are missing for *Hoferellus* species. For *H. cyprini*, two oligochaete species, *Branchiura sowerbyi* and *Nais* cf. *elinguis*, have been identified as invertebrate hosts, but this has not been confirmed for *H. carassii*. Pre-sporogonic blood stages similar to those seen in the life cycle of *Sphaerospora molnari* and *S. dykovae* have been recorded from the blood of goldfish infected with *H. carassii*.

9.4.4 Detection

Pre-sporogonic, sporogonic stages and myxospores of *Hoferellus carassii* can be detected in histological sections from the kidney of suspected fish. Myxospores are also visible during microscopic examination of fresh preparations of kidney samples. Species identification requires morphometric measurements from fresh myxospores and sequencing of 18S rDNA or SSU rDNA nucleotide sequences.

Species identification from blood stages also requires sequencing of the respective genes.

9.4.5 Prevention and treatment

Although a treatment for *H. carassii* has not been established, several husbandry practices may reduce infections. Ponds stocked with an affected fish population could be depopulated and carefully disinfected before re-stocking. The invertebrate host, probably an oligochaete, has not yet been determined and in earth ponds the elimination of these invertebrate hosts might be difficult. In addition, the source for infection may be outside the pond and the parasite may be introduced by the influx of actinospores via the water supply.

FACTS SHEET

Disease: Kidney enlargement disease in goldfish
Pathogen: *Hoferellus carassii.*
Susceptible species: Goldfish (*Carassius auratus*). Gibel carp (*Carassius gibelio*) may be infected, but do not develop a disease
Mortality: Uncertain
Temperature: Uncertain
Prevention and treatment: Only sanitary measures, such as liming the ponds, could be applied. No treatment protocol established

9.5 Intestinal Swellings in Scattered Mirror Carp

9.5.1 Introduction

Myxozoa from the *Myxobolus* clade are histozoic species that infect various tissues of freshwater fish. These parasites have a narrow host range and typically infect only one species or a limited number of closely related species. This clade contains numerous species that infect specific target tissues only. In many species from this clade, myxospores develop in cysts, which often are visible to the naked eye in the target tissue of the particular species. While fish from natural waters usually are hardly affected by these parasites, fish raised in intensive pond aquaculture may carry a high parasite burden and may be seriously affected, in particular, by numerous large cysts. This includes *Thelohanellus wuhanensis*, which infects the skin of juvenile gibel carp (*Carassius gibelio*), causing prominent swellings on the skin, and may seriously retard the growth of affected specimens. *Myxobolus honghuensis*, which causes nodular swellings in the pharynx of *Carassius gibelio*, may occlude the pharynx lumen during heavy infections and thus may cause mortality in pond-raised juveniles. Likewise, mass mortalities of juvenile mirror carp (a variety of common carp *Cyprinus carpio*) were observed in Chinese carp ponds in the context of an infection with the gut-dwelling species *Thelohanellus kitauei*, which also may provoke large cysts in the intestinal wall.

9.5.2 Disease description

Infected fish are unresponsive and lethargic and may have an enlarged abdomen. Necropsy may reveal an empty gut with a thinned intestinal wall where several white cysts are visible. In the body cavity, ascites fluid may be present. From the hyperplastic tissue of the intestinal tract, numerous myxospores with the morphology of *Thelohanellus* can be released.

9.5.3 Pathogen

Thelohanellus kitauei has pyriform myxospores with a blunt posterior and a tapered anterior end. These spores may be embedded in a balloon-like egg-shaped sack. A single polar capsule of pyriform shape is located near the anterior end of the spore and contains a coiled polar filament with 8–10 turns. The spores develop in cysts located in the intestinal propria and submucosa. Smaller cysts contain minute disporous plasmodia, while larger cysts may comprise numerous disporous plasmodia and mature spores. Other cysts contain long, slender plasmodia with a winding, worm-like appearance and several cross-sections of this plasmodium are usually visible in histological preparations of such cysts. This parasite may also develop giant polyp-like swellings, which contain numerous plasmodia with developing and mature spores. Cross-sections of these giant swellings may reveal numerous small cysts with a large number of plasmodia, with developing and numerous mature spores lying free within the cysts. These giant swellings are surrounded by abnormally developed intestinal mucosal tissue. Most likely, these cysts rupture and release the spores into the intestinal lumen, from where they are finally discharged to the environment with the faeces. Cysts of different size, in particular the giant cysts, may occlude in the intestinal lumen and lead to a loss of appetite and starvation of the host.

9.5.4 Detection

White nodular swellings containing *Thelohanellus kitauei* plasmodia and myxospores are visible in the intestine of suspected specimens. *Thelohanellus* myxospores can be detected in fresh preparations from these swellings during microscopic analysis. Pathological structure of the cysts, plasmodia, as well as developing and mature myxospores of *T. kitauei* can be observed in histological sections from altered tissue areas. Species identification requires morphometric measurements from fresh myxospores and sequencing of 18S rDNA or SSU rDNA nucleotide sequences.

9.5.5 Prevention and treatment

A treatment of *T. kitauei* is not established. Ponds stocked with an affected population of fish could be depopulated and carefully disinfected before re-stocking. The invertebrate host, probably an oligochaete, has not yet been determined and in earth ponds the elimination of these invertebrate hosts might be difficult. In addition, the source for infection may be outside the pond and the parasite may be introduced by the influx of actinospores via the water supply.

FACTS SHEET

Disease: Intestinal swellings in scattered mirror carp
Pathogen: *Thelohanellus kitauei.*
Susceptible species: Mirror carp (*Cyprinus carpio*)
Mortality: Moderate in affected stocks
Temperature: Undetermined; disease occurred at 25–29°C
Prevention and treatment: Only sanitary measures, such as liming the ponds, could be applied. No treatment protocol established

9.6 Nodular Swellings in the Pharynx of Gibel Carp

9.6.1 Introduction

This disease causes mortalities in juvenile gibel carp (*Carassius gibelio*) in aquaculture ponds in China, mainly during July to September. Since infected fish develop prominent nodular swellings in the pharynx, the disease is also assigned as 'throat cancer' of gibel carp.

9.6.2 Disease description

Affected specimens show signs of emaciation and exophthalmos and they behave lethargically with sluggish swimming movements. In necropsy, prominent nodular swellings and inflammation are seen in the pharynx of heavily affected specimens. When the nodular lesions rupture, liquefaction may occur. From a nodular lesion, numerous myxospores with a myxobolid morphology can be isolated.

9.6.3 Pathogen

Myxospores isolated from the pharyngeal swellings are of *Myxobolus*-type morphology with an elongated, pyriform shape and a blunted posterior and a pointed anterior side. In sutural view, the spore has a straight and narrow sutural ridge and a lentiform posterior end. In the pathogen identified as *M. honghuensis*, spores measure 15.1–19.5 × 9.0–11.0 μm in size and contain two polar capsules of equal size, which measure 7.9–8.1 × 3.0–4.4 μm, and comprise a coiled polar filament with 7–8 coils.

9.6.4 Detection

Nodular swellings occur in the pharyngeal region of affected specimens under the orbit and before epibranchial organs, although infections of gills have also been reported. Myxospores can be observed by microscopic examination of fresh preparations from ruptured swellings. In addition, the infection can be detected in histological sections of tissue samples collected from the affected region. Species identification requires morphometric measurements from fresh myxospores as well as sequencing of 18S rDNA or SSU rDNA nucleotide sequences.

9.6.5 Prevention and treatment

A treatment of *Myxobolus honghuensis* is not established.

FACTS SHEET

Disease: Nodular swellings in the pharynx of gibel carp
Pathogen: *Myxobolus honghuensis.*
Susceptible species: Gibel carp (*Carassius gibelio*)
Mortality: Up to 30% in affected stocks
Temperature: Undetermined
Prevention and treatment: Only sanitary measures, such as liming the ponds, could be applied. No treatment protocol established

9.7 Selected References for Further Reading

Alama-Bermejo, G., Jirku, M., Kodadkova, A., Peckova, H., Fiala, I. *et al.* (2016) Species complexes and phylogenetic lineages of *Hoferellus* (Myxozoa, Cnidaria) including revision of the genus: A problematic case for taxonomy. *Parasites & Vectors* 9, 13.

Chang, S., Neuhof, M., Rubinstein, N.D., Diamant, A., Philippe, H. *et al.* (2015) Genomic insights into the evolutionary origin of the myxozoa within Cnidaria. *Proceedings of the National Academy of Sciences of the USA* 112, 14912–14917.

Dykova, I. and Lom, J. (1982) *Sphaerospora renicola* n.sp., a myxosporean from carp kidney, and its pathogenicity. *Zeitschrift Für Parasitenkunde* 68, 259–268.

Egusa, S. and Nakajima, K. (1981) A new myxozoa *Thelohanellus kitauei*, the cause of intestinal giant cystic disease of carp. *Fish Pathology* 15, 213–218.

Feist, S.W. (2008) Myxozoan diseases. In: Eiras, J., Segner, H., Wahli, T. and Kapoor, B.G. (eds) *Fish Diseases.* Science Publishers, Enfield, New Jersey, pp. 613–682.

Fiala, I., Bartosova-Sojkova, P. and Whipps, C.M. (2015) Classification and phylogenetics of Myxozoa. In: Okamura, B., Gruhl, A. and Bartholomew, J.L. (eds) *Myxozoan Evolution, Ecology and Development.* Springer International, Cham, Switzerland, pp. 85–110.

Fontes, I., Hallett, S. and Mo, T.A. (2015) Comparative epidemiology of myxozoan diseases. In: Okamura, B., Gruhl, A. and Bartholomew, J.L. (eds) *Myxozoan Evolution, Ecology and Development.* Springer International, Cham, Switzerland, pp. 317–341.

Lisnerova, M., Fiala, I., Cantatore, D., Irigoita, M., Timi, J. *et al.* (2020) Mechanisms and drivers for the establishment of life cycle complexity in myxozoan parasites. *Biology* 9, 10.

Liu, Y., Lövy, A., Gu, Z. and Fiala, I. (2019) Phylogeny of Myxobolidae (Myxozoa) and the evolution of myxospore appendages in the *Myxobolus* clade. *International Journal for Parasitology* 49, 523–530.

Lom, J. and Dykova, I. (1992) *Protozoan Parasites of Fishes.* Elsevier, Amsterdam, The Netherlands.

Lom, J. and Dykova, I. (1995) Myxosporea (Phylum Myxozoa). In: Woo, P.T.K. (ed.) *Fish Diseases and Disorders: Protozoan and Metazoan Infections,* Vol. 1. CAB International, Wallingford, UK, pp. 149–148.

Okamura, B. and Gruhl, A. (2015) Myxozoan affinities and route to endoparasitism. In: Okamura, B., Gruhl, A. and Bartholomew, J.L. (eds) *Myxozoan Evolution, Ecology and Development.* Springer International, Cham, Switzerland, pp. 23–44.

Patra, S., Bartosova-Sojkova, P., Peckova, H., Fiala, I., Eszterbauer, E. *et al.* (2018) Biodiversity and host-parasite cophylogeny of *Sphaerospora* (sensu stricto) (Cnidaria: Myxozoa). *Parasites & Vectors* 11, 347.

Chapter 10

Protozoan Diseases

10.1 Introduction

The term Protozoa describes single-celled eukaryotes with a free-living or parasitic lifestyle that feed on organic matter originating from debris or from tissues. Such organisms with a unicellular organisation and heterotrophic metabolism are found in various systematic groups; therefore, the term Protozoa refers to a paraphyletic group of eukaryotic organisms and is used here to catch all eukaryotes that are not grouped within animals or fungi.

Most of the protozoan fish parasites have a direct biological cycle. Because of the high density of water, infective stages of fish parasites can float in the water column, which fosters their dispersal in fish populations. The high stocking density of fish in aquaculture facilities compared with natural waters further enhances the infection risk. Because of a high multiplication rate, an infestation with protozoan parasites may increase rapidly in densely stocked aquaculture facilities and may have devastating impacts on fish stocks. The specific conditions in intensified aquaculture, with a high amount of feed given to densely kept fish populations, may result in a reduced oxygen concentration in tank water, an accumulation of detritus and subsequently an increased growth of heterotrophic bacteria, which can constitute a food source for ciliates with a saprophytic lifestyle. These opportunistic ciliates can become pathogenic when water quality deteriorates. Changes in water quality may stress and impair the fish's immune responses against pathogens, which may further enhance the severity of the infection. In conclusion, every parasite infection may result in a disease situation in an affected

 Carps: A Practical Approach to Disease Control and Health Management (eds David Hoole *et al.*)
DOI: 10.1079/9781836990024.0010

aquaculture population, but for the treatment of these parasites, all of them require management and prophylactic measures in order to solve the problem.

Unicellular parasites of cyprinids are found in several phyla from the superkingdom Eukaryota. In this chapter, parasites will be discussed from the phylum Euglenozoa, class Kinetoplastida (*Ichthyobodo*, *Cryptobia*, *Trypanoplasma* and *Trypanosoma*), the phylum Amoebozoa (*Neglaeria*, *Vahlkampfia*), the superphylum Alveolata, phylum Ciliophora (*Ichthyophthirius*, *Apiosoma*, *Epistylis*, *Chilodonella*, *Tetrahymena*, trichodinids), the infraphylum Apicomplexa (Coccidia), and the infraphylum Dinozoa (*Piscinoodinium*). These parasites will be discussed in a sequence that reflects their target tissues.

10.2 Parasites and Epibionts of Skin and Gills

10.2.1 *Ichthyobodo* (syn. *Costia*)

Introduction

Ichthyobodo is an important parasite of gills and skin of a large variety of fish species from marine and freshwater habitats. In particular, in aquaculture facilities, infestations with *Ichthyobodo* can be fatal and can cause serious losses. Vulnerability to the infection is particularly prominent in juvenile fish and fish in stress situations, such as crowded conditions, too low water temperature, or inappropriate water parameters.

Disease description

Clinical signs are non-specific and depend on infection intensity. In light infections, fish may rub themselves against tank walls or submersed objects, jump, and may present accelerated breathing movements. With increasing infections, fish become listless and a milky-bluish excessive mucous layer covers the skin and the eyes, making them appear cloudy. In addition, in particular in the ventral part of the body, congested blood vessels can be seen (Fig. 10.1A). In histology, the skin of heavily infected fish presents vacuolation and oedema, followed by degeneration and detachment of the epidermis. Infected gills have a greyish appearance and experience an occlusion of the intralamellar space by hyperplasia and proliferation of the branchial epithelium.

Pathogen

Ichthyobodo has bi-flagellated cells of about $10 \times 5\,\mu m$, which alternate between a free-swimming moving stage and an attached feeding stage. The free-swimming, non-feeding stage has a flat, slightly asymmetric body. In one plane, the body is of oval shape (Fig. 10.1B), in the other plane it is flat with a convex dorsal side and a concave ventral side. Two unequal flagella originate from a flagellar pocket at the ventral side of the body. The cell also possesses a small contractile vacuole, a central nucleus and a kinetoplast, a structure that contains a dense mass of mitochondrial DNA and is a characteristic structure of organisms from the class Kinetoplastida. The cells swim in a spiral way, are sometimes hesitant and search for an attachment site. On a suitable host, the cells attach to epithelial cells (Fig. 10.1C), their body changes to a pyriform shape and develops an attachment plate, which sends a finger-like protrusion into the epithelial cell. In the feeding form, the flagella are retained and point away from the attachment site. Multiplication of *Ichthyobodo* occurs by binary fission and transmission is direct from fish to fish. It is not clear whether *Ichthyobodo* forms cysts and their presence needs to be confirmed.

Because *Ichthyobodo* cells from different hosts and from different habitats have a similar morphological appearance, just one species, *I. necator*, has been described. However, genetic analyses of isolates from several hosts and several locations reveal the presence of a species complex with a high genetic diversity. At least two species were identified in samples from cyprinids, and both species were recovered from goldfish and carp. This indicates that one *Ichthyobodo* species can infect several fish species, and the infection can be transferred to fish from several cyprinid species.

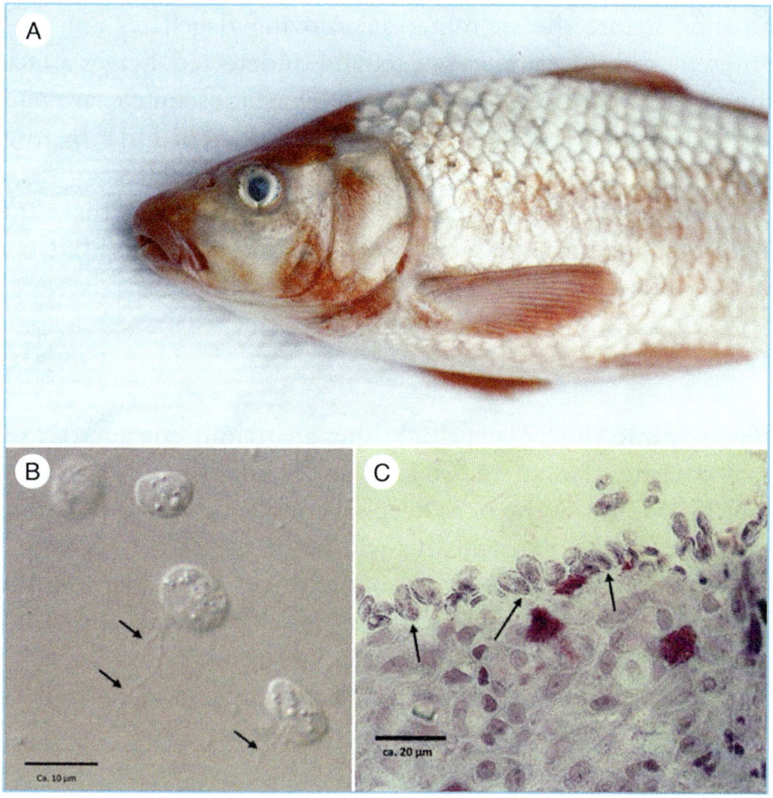

Figure 10.1 *Ichthyobodo* infection. (A) A coloured koi with a severe *Ichthyobodo* infestation. On the skin a milky excessive mucous layer can be seen, the eyes appear cloudy, and numerous congested blood vessels are visible. (B) Bi-flagellated *Ichthyobodo* cells of oval shape. Flagella are indicated by arrows. Fresh preparation, differential interference contrast (DIC) image. Indicative scale bar 10 µm. (C) Histological section of *Ichthyobodo* cells (arrows) attached to the skin of an infected carp. HE staining. Indicative scale bar 20 µm. (B) courtesy of the Environment Agency, National Fisheries Laboratory, UK.

Detection

The parasite can be detected in fresh samples obtained by scraping off mucus and cells from skin and gills of a possibly infected fish during observation with a microscope at medium to high magnification. The detection of the parasites is easier from alive or recently dead fish, when moving flagellates can be observed. Using fish that have been dead for an extended period, or in samples that were collected some time before the examination, moving flagellates are usually not present and infection might remain undetected. Stages attached to the tissue can be detected in histology. Microscopic examination allows diagnosis at genus level, which is sufficient for a treatment plan.

For species identification or diagnosis from samples that cannot be subjected to immediate microscopic examination, PCR-based detection methods are available.

Prevention and treatment

Ichthyobodo infections of cyprinid fishes are mainly considered as secondary parasites that affect fish kept under unfavourable conditions. Therefore, an improvement in keeping conditions has to be included in every treatment plan and this will be decisive for its sustainable success. Outbreaks in populations of ornamental fish, like goldfish or koi, can be treated by baths with chemicals, such as formalin or malachite green. However, these substances cannot be used for fish for human consumption because of legal restrictions in many countries.

FACTS SHEET

Disease: Ichthyobodosis
Pathogen: *Ichthyobodo* spp.
Susceptible hosts: Common carp, goldfish and other cyprinids
Mortality: Moderate to high in susceptible and stressed populations
Temperature: 2–30°C
Prevention and treatment: Prevention is by adopting normal sanitary practices. A range of treatments such as formalin, malachite green and sodium chloride are available for ornamental carp. In fish for human consumption only sodium chloride could be used. It is recommended that the national legislation on treatment of fish is consulted prior to any action being taken

10.2.2 Cryptobia branchialis

Introduction

Several kinetoplastid flagellates are reported as ectozoic species attached to the gills of various fish species and in some reports they are considered as harmful for their hosts. In particular, *Cryptobia branchialis* has been related to mortality of carp, goldfish and other fish species. *C. branchialis*, which feeds on bacteria and detritus, has been reported from the gills of juvenile fish and fish fry kept in organically polluted water.

Disease description

Reports on the presentation of the disease are controversial. Some publications report an excess production of mucus, hypertrophy of the respiratory branchial epithelium and swollen secondary gill lamellae, associated with increased mortality. These alterations were observed, in particular, in fry and juvenile fish, while older specimens remained unaffected. Other publications could not relate the presence of *Cryptobia* epibionts to a particular pathology and assumed that additional factors were responsible for the observed mortality. The flagellate spreads in a population by direct transmission from fish to fish.

Pathogen

The bi-flagellated cryptobiid flagellates are an important component of the heterotrophic micro-plankton mainly of freshwater habitats. *Cryptobia branchialis* has a tear-shaped cell of about 10–12 × 3–5 μm and two flagella, which arise from a flagellar pocket. One short flagellum extends freely from the cell, while a longer flagellum extends posteriorly and with its free end attaches the flagellate to cells of the gill epithelium. The cell harbours a contractile vacuole next to the flagellar pocket and digestive vacuoles in the posterior part of the cell. Multiplication occurs by binary fission. Because the description of the organism relies on light microscopic and electron microscopic observations, the epibionts/ectoparasites on the gills of different fish species from different habitats may indeed originate from different parasite species.

Detection

The parasite can be detected in fresh samples scraped off the gills of suspected fish during observations with a light microscope at medium to high magnification. This has to be done immediately after collection in order to observe the movement of the flagellates. In addition, the flagellates can be detected in histological sections of the gills.

Prevention and treatment

Ectoparasitic *Cryptobia* are reported from the gills of fish kept in water containing a high load of organic detritus and heterotrophic bacteria. Therefore, improvement of the keeping conditions is the most effective treatment plan.

FACTS SHEET

Disease: Cryptobiosis
Pathogen: Epibiont cryptobia, like *C. branchialis*
Susceptible hosts: Fry and juveniles of common carp, goldfish and other cyprinids
Mortality: Uncertain, low
Temperature: Up to about 25°C
Prevention and treatment: Improving water quality, in particular reducing the organic load in water. In the ornamental carp, the parasite burden could be reduced by malachite green or formalin. However, the parasite infestation may re-occur. It is recommended that the national legislation on treatment of fish is consulted prior to any action being taken

10.2.3 Ichthyophthirius multifiliis

Introduction

The ciliate *Ichthyophthirius multifiliis* is responsible for the white spot disease or 'ich' of fish, which causes serious losses in aquaculture populations of virtually all freshwater fish species from all climatic zones. Its global distribution and its wide host range make *I. multifiliis* one of the most dangerous parasites of finfish in freshwater aquaculture. The species can be considered as a primary pathogen, because it cannot survive in aquatic habitats without an infected host and outbreaks of the disease often occur when the parasite is introduced into a naïve population or when the immune response of fish is suppressed during stress conditions.

Disease description

Infected fish that develop small white spots on the skin and fins (Fig. 10.2A) rub themselves against tank walls or submersed objects, jump, and may present accelerated breathing movements. With increasing parasite infestation, the fish become listless, parts of the mucous layer detach from the skin, and the number of spots increases. Heavily affected specimens are usually isolated from the shoal, and stay in calm and well-aerated areas of the pond or tank. These fish often have a dark coloration, excessive mucus production, cloudy appearance of the eyes and numerous white spots on skin and fins. In histology, parasite trophonts are found in the epithelial layer of skin and gills. Due to the parasite's movements, the epithelial layer thickens and covers the trophonts, while superficial blood vessels in the skin are congested. In heavily infected fish, the skin epithelium is eroded, gills are pale, with a marked hyperplasia of the epithelium, and the trophonts tend to migrate towards larger blood vessels. Because of the erosion of the skin, the osmoregulation of the fish is impaired and secondary infections with bacterial pathogens flourish. In many cases, the infection is associated with serious fish losses when left untreated.

Pathogen

Ichthyophthirius multifiliis, a holotrichous ciliate with cilia distributed over the entire cell (Fig. 10.2B), belongs to the class Oligohymenophorea, order Hymenostomatida. It has a life cycle with pelagic, highly motile theronts, which invade epithelia of the skin and gills of fish. In these locations the theront transforms to a feeding stage, the trophont, which burrows in the epithelial layer. At this stage the parasite has a characteristic horseshoe-shaped macronucleus and constantly rotates with a slow movement. It feeds on disrupted cells and grows rapidly from an initial size of 30–40 μm to about 1000 μm, when it is visible as a white spot. Mature trophonts actively escape from the epithelium into the water where they secrete a proteinaceous cyst, which allows them to attach to any submerged surface. The encysted tomont undergoes a series of palintomic

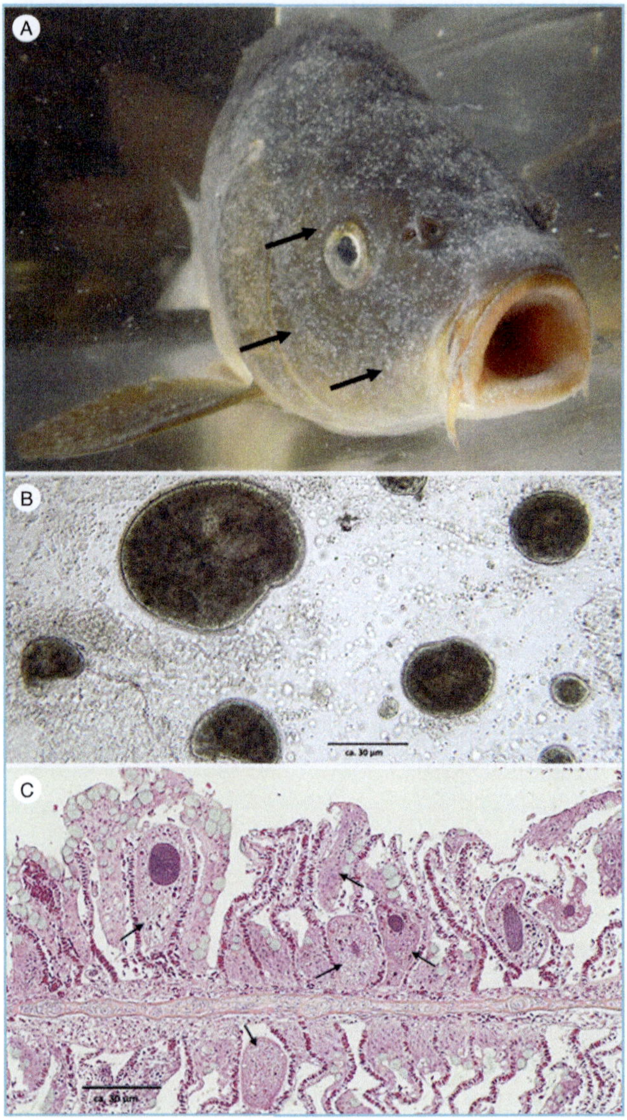

Figure 10.2 *Ichthyophthirius multifiliis*. (A) Common carp with a severe *Ichthyophthirius multifiliis* infestation. On skin and fin, numerous white spots (arrows), which all resemble a parasite, can be seen. (B) Holotrichous ciliated *Ichthyophthirius* trophonts of different size. Indicative scale bar 30 μm. (C) Histological section of a carp gill infested with several *Ichthyophthirius* trophonts (arrows). The infections destroyed the branchial architecture. HE staining. Indicative scale bar 30 μm. (C) courtesy of the Environment Agency, National Fisheries Laboratory, UK.

divisions, in which no growth of the daughter cells is observed between subsequent divisions. This process results in 200–800 tomites, which differentiate into spindle-shaped theronts that leave the cyst, swim through the water and search for a susceptible fish to complete their life cycle. Theronts that cannot reach a susceptible fish within a few days will die. At 20°C for instance, the infectivity of theronts decreases significantly after 24 hour. The time needed for completion of the entire life cycle is temperature dependent and ranges from 3 to 5 days at about 25°C, 6–10 days at 20°C, 14–20 days at 15°C and up to 5 weeks at 10°C. However, some parasites may show a delayed development and therefore infections may occur even at a later time point. Dormant cysts have not been reported. Trophonts that leave the host in a premature stage fail to develop into tomonts and die.

Because of the short life cycle and the high number of theronts released from a tomont, the parasite burden may rapidly increase in fish in aquaculture facilities. In cyprinids, this is observed particularly in summer, associated with reduced oxygen levels, or when fish are kept at high stocking density, for instance in wholesale or retail stores.

Several strains of *Ichthyophthirius multifiliis* have been recognised by means of antisera against parasite immobilisation antigens (i-antigens), which constitute abundant surface membrane proteins of the parasite. These i-antigens may differ between isolates, but the genetic basis for this remains unknown. Genetic differences between strains can be recognised by comparing nucleotide sequences encoding mitochondrial genes like mitochondrial NADH dehydrogenase I and cox-1. It is not clear whether different isolates have different virulences.

Detection

The pathogen can be detected in fresh samples scraped off the skin and gills of suspected infected fish and by these samples being observed with a light microscope using a low- to medium-power lens. In order to assess the severity of the infection, it is advisable to examine samples from skin as well as from gills. The ciliate can also be detected in histological sections of skin and gills (Fig. 10.2C). In addition, a PCR-based detection method is available and could be used, for instance, for screening water samples for the presence of theronts.

Prevention and treatment

An infection is usually introduced into a fish population by infected fish, contaminated equipment, or via aerosols generated from infected water. In addition, an outbreak may occur after events stressing the fish, like transportation, sorting or crowding. An effective treatment of ich is still missing. A wide array of chemicals has been suggested, but none of these substances has a universal effectiveness. Outbreaks in populations of ornamental fish can be treated by repeated baths with malachite green, formalin, or combinations of both with other reactive dyes. However, trophonts, and in particular encysted tomonts, are not affected by these chemicals and can be a source for a re-emergence of the infection. Because of legal restrictions, the use of the above-mentioned chemicals is not feasible for fish used for human consumption in many countries. Here, the main treatment strategy is to reduce the transmission success of the parasite by reducing the stocking density of the fish and increasing the water exchange rate of the tank or pond. Fish that recover from infection develop immobilising antibodies binding to i-antigens, surface membrane proteins of the ciliate, and these fish are protected against re-infection. Despite several attempts to develop a vaccine on the basis of these antigens or on other possible antigens, a vaccine providing effective, long-lasting protection is not currently available.

FACTS SHEET

Disease: White spot disease of finfish, ich
Pathogen: *Ichthyophthirius multifiliis*
Susceptible hosts: All cyprinids
Mortality: Moderate to high
Temperature: 5–27°C
Prevention and treatment: Avoid introduction, reduction of transmission success by lower stocking density, increased water exchange rate, ornamental stocks can be treated with a bath with malachite green, formalin or other chemicals. It is recommended that the national legislation on treatment of fish is consulted prior to any action being taken

10.2.4 *Chilodonella*

Introduction

In contrast to the obligate fish parasites *Ichthyophthirius multifiliis*, several ciliates with saprophytic or grazing feeding habits may be harmful to fish, in particular in high-density aquaculture. Ciliates of the genus *Chilodonella* (Phyollopharyngea: Chilodonellidae) are predominantly free-living and graze on bacteria, diatoms, filamentous algae and cyanobacteria in biofilms using their cytostome, a specialised mouth structure. While grazing on biofilm substrates of fish gills and skin, the cytostome of *Chilodonella* may directly penetrate fish epithelial cells and incorporate cellular content. In particular, *Chilodonella hexasticha* and *C. piscicola* are described as fish parasites. Losses due to *Chilodonella* infestation have been reported from aquaculture populations of several freshwater fish species, including cyprinids.

Disease description

Clinical signs associated with a *Chilodonella* infection are not specific and include a grey appearance of the skin due to a hypersecretion of mucus. Reduced swimming activity often near the water surface or near pond edges occurs, and sometimes increased gill ventilation. In severe cases, emaciation, sunken eyes and scale loss can be observed. In gills, a *Chilodonella* infestation causes gill epithelial hypertrophy, gill lamellar fusion and subsequent necrosis.

Pathogen

Chilodonella has a flattened cell body; the ventral ciliature is reduced to two ciliary rows close to the body margin (Fig. 10.3A, B). The right ventral ciliary band is arched and longer than the straight row on the right side of the cell. The cytostome, the mouth structure of the cell, can attach to surfaces (Fig. 10.3C). Using specialised cilia associated with this mouth structure, a vacuum is produced which aids feeding on bacteria and mucus from skin and gill of infested fish. *Chilodonella*

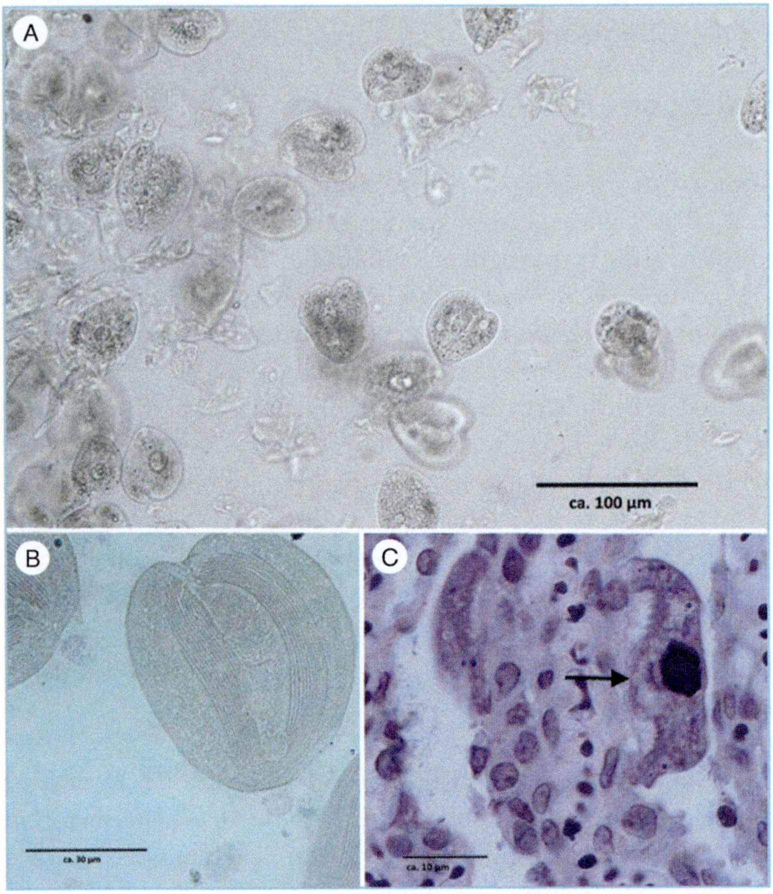

Figure 10.3 *Chilodonella* infestation. (A) Fresh preparation of *Chilodonella piscicola* cells showing a heart-shaped body. Indicative scale bar 1000 µm. (B) *Chilodonella piscicola* cells showing silver-impregnated ciliature. Indicative scale bar 30 µm. (C) Histological section of *Chilodonella* (arrow) attached to the gill surface. Giemsa stain. Indicative scale bar 10 µm. (A) courtesy of the Environment Agency, National Fisheries Laboratory, UK.

has a nuclear dualism and possesses an oval macronucleus. Whilst most *Chilodonella* species are free-living, several species have been isolated from gills and skin of bony fish, and in particular *C. hexasticha* and *C. piscicola* have been associated with mortality events in aquaculture populations of fish. *Chilodonella* spp., like many ciliates, prefer relatively low levels of oxygen in water, and in such conditions epizootics develop rapidly. It is assumed that bacteria that multiply under these conditions are the preferred food items for ciliates from this genus and thus foster their multiplication. High stocking densities of fish, fluctuation in temperature and high organic load are further factors that might be associated with a *Chilodonella* infection. The temperature range at which *Chilodonella* infections occur ranges from 4°C to 20°C, but infections are also found at 26–31°C. The transmission of the parasite is direct from fish to fish, and multiplication occurs by binary fission and by a sexual process known as conjugation, which allows a recombination of genetic material

Detection

The parasite can be detected morphologically in fresh scrape samples from skin and gills of suspected infected fish (Fig. 10.3A). Morphological differentiation of the parasite species mainly relies on characteristics such as cell shape and oral ciliature. Generally, *C. hexasticha* has a round body shape, and with a body size of 30–66 × 20–50 μm it is slightly smaller than the heart-shaped *C. piscicola* with a body size of 30–80 × 20–60 μm. In addition, methods for a PCR-based species detection and differentiation based on a comparison of the nucleotide sequence of the mitochondrial SSU rDNA gene are available.

Prevention and treatment

Chilodonella is a ciliate that is considered to be an opportunistic pathogen for fish species in aquaculture. Disease outbreaks mainly occur when fish are kept in crowded conditions, with reduced oxygen levels in the water, increased organic load and unstable water temperature. Outbreaks can be prevented by reducing stocking density, feeding at an adequate feeding rate, monitoring of water quality and taking measures to ensure good water quality. Several chemicals, mainly formalin, hydrogen peroxide or reactive dyes like malachite green or potassium permanganate, are used to control disease outbreaks in ornamental stocks. These substances cannot be used for the treatment of fish for human consumption, because of legal restrictions in many countries.

FACTS SHEET

Disease: Chilodonellosis
Pathogen: *Chilodonella* species like *Chilodonella hexasticha* or *C. piscicola*
Susceptible hosts: Cyprinids, in particular common carp, goldfish
Mortality: Low to moderate
Temperature: 4–30°C
Prevention and treatment: Ensure good water quality, sufficient oxygen level in water, reduced organic load, regular health monitoring, and regular water exchange. Acute outbreaks in ornamental stocks can be treated by baths with hydrogen peroxide, formalin or other chemicals. It is recommended that the national legislation on treatment of fish is consulted prior to any action being taken

10.2.5 Tetrahymena

Introduction

Ciliates from the genus *Tetrahymena* live among decomposing organic material, where they feed on organic matter and bacteria. In aquaculture facilities, tetrahymenids can become invasive and pathogenic for fish when the water is warm, has a high ammonia level and is rich in organic matter. In such conditions *Tetrahymena* spp. can penetrate the skin and the muscle of the fish, and in severe cases enter the bloodstream and reach capillaries in liver, kidney, eyes and brain.

Disease description

Clinical signs of *Tetrahymena* epizootics include mucous hypersecretion, raised scales and whitish areas of sloughing skin. In severe cases, skin ulcers can occur, usually associated with *Tetrahymena* in skin epithelium, the subdermal fat tissue and the muscle. In goldfish, *Tetrahymena* can also be observed in the gills.

Pathogen

Tetrahymena is a holotrichous ciliate (Class Oligohymenophorea, Subclass Hymenostomata), with a pear-shaped, radial symmetric body of about 100 × 20 μm in size. The buccal cavity is a deep sac-like indentation with a specialised ciliature. About four species from the genus *Tetrahymena* may invade fish, including *T. corlissi* and *T. pyriformis. T. corlissi* is considered the most common and most pathogenic species, which may also infect koi and goldfish.

Detection

An infection with *Tetrahymena* can be detected by light microscopic observations of fresh scrape samples from affected skin or in histological sections of skin samples from suspected infected fish. The morphological identification of species relies on the analysis of the ciliature in silver-impregnated specimens.

Prevention and treatment

Tetrahymena infections of cyprinid fish mainly occur in populations of weakened fish kept under unfavourable conditions in water with low oxygen content and rich in organic matter. Improvement of the keeping conditions is of paramount importance for a successful treatment of the infection. Ornamental fish populations can be treated with parasiticidal chemicals such as formalin. However, this will not be effective in fish populations with deep muscular infections.

FACTS SHEET

Disease: Tetrahymenosis
Pathogen: *Tetrahymena* species, like *T. corlissi, T. pyriformis*
Susceptible hosts: Ornamental common carp, goldfish
Temperature: ~25°C
Prevention: Improvement of keeping conditions by reduced stocking density, customised feeding, improved oxygen supply and water exchange

10.2.6 Sessile ciliates: Apisoma, Ambiphrya, Epistylis and trichodinids

Introduction

The skin and gills of freshwater fish are populated by various sessile ciliates, in particular species from the subclass Peritrichia within the class Oligohymenophorea, which also comprises the subclass Hymenostomata with *Ichthyophthirius* and *Tetrahymena.* Peritrich ciliates are characterised by prominent peristomal cilia, which encircle the oral cavity of the cell counter-clockwise before entering the oral cavity. The somatic ciliature is reduced. In some species, an equatorial locomotorial circle is present, or a circle with locomotorial cilia at the adoral side of the cell. Ciliates from the suborder Sessilida attach permanently to the substrate or the body surface of aquatic organisms, and may even secrete a prominent stalk. Ciliates from the suborder Mobilia attach temporarily using an adhesive disc at the adoral side of the cell, which is supported by a complex skeletal apparatus. Sessile ciliates are present on fish from natural habitats and, in particular, the infection with trichodinids is linked to high levels of organic matter in water. Sessile ciliates are usually considered as epibionts, but with increasing infestation they can behave as commensals or parasites, and be involved in disease outbreaks.

Disease description

Sessile ciliates on fish skin or gills are considered as epibionts, which use the fish as a moving substrate that carries them to their food items: organic debris particles and planktonic bacteria. Their attachment does not affect skin or gill epithelium cells or induce any pathological response. Sessile ciliates are sometimes considered as responsible for losses in aquaculture populations of fish, mainly when the fish are kept in organically polluted water or when these fish also suffer from additional bacterial infection. In this case, the sessile ciliates usually proliferate because of the increased abundance of food from organic matter or increased bacterial load in the water. Likewise, trichodinid ciliates live as epibionts on skin and gills of fish, glide over these surfaces and temporarily attach with their adhesive disc. They also feed on waterborne organic particles, bacteria and detritus particles from the fish surface. The irritation caused by the attachment via the adhesive disc is negligible. However, in fish affected by other factors, or in vulnerable life stages, such as fry or juveniles, trichodinids can proliferate and may harm the epithelium at the attachment site. In such circumstances trichodinids can also feed on disrupted cells and invade skin or gill tissue, causing damage. In these fish, affected skin areas respond with excessive mucus production, and the fish appear listless, swim near the water surface and stop feeding.

Pathogen

Sessile ciliates possess a peristomal ciliature and attach to fish skin or gills by means of a specialised region of the cell, the scopula, in the case of *Ambiphrya* and *Apiosoma*, or by means of a secreted stalk as in *Epistylis*. While *Ambiphrya* has a permanent circular fringe of cilia used in attachment, these cilia are absent in attached *Apiosoma* cells. *Apiosoma* cells are elongated, conical with a wider oral side and possess a compact macronucleus in the adoral half of the body (Fig. 10.4A, B, C). *Epistylis* secretes a non-contractile stalk (Fig. 10.5), which carries several zooids. Trichodinids have a more or less flat body with an oral ring of cilia and three adoral rings of cilia, which have locomotory or tactile functions (Fig. 10.6A, B). A prominent feature of trichodinids is the proteinaceous cytoskeleton of the adhesive disc at the adoral side of the body. This cytoskeleton consists of a complex ring of several elements with flat projections. To these skeleton elements myofibrillar bundles are attached, which allow these elements to be deformed and aid the trichodinids to attach. These skeletal elements are visible in live specimens (Fig. 10.6B) and can be impregnated with a silver stain in fixed specimens (Fig. 10.6C). The macronucleus of trichodinids is elongated, often horseshoe-like. Numerous trichodinid species have been reported from skin and gills of fish.

The transmission of *Ambiphrya, Apiosoma* and *Epistylis* occurs from fish to fish by migratory stages with a locomotory ring of ciliates, while trichodinids migrate from fish to fish. Multiplication occurs by binary fission.

Detection

Sessile ciliates are best observed in fresh samples scraped from skin (Fig. 10.6A) and gills of susceptible fish and examined with a light microscope or in histological preparations of these tissues (Fig. 10.7).

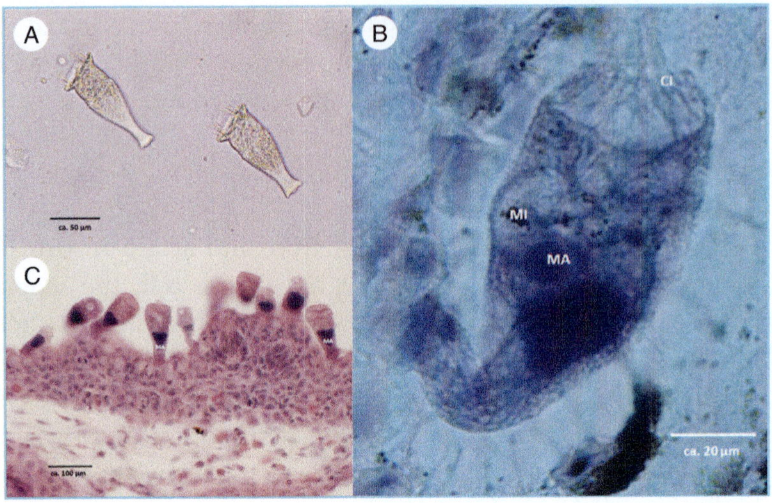

Figure 10.4 *Apiosoma* infestation. (A) Fresh preparation of *Apiosoma* with an elongated conical cell shape possessing a wider oral end with a visible peristomal ciliature. Indicative scale bar 50 µm. (B) *Apiosoma* cell attached to skin cells. Note the presence of the macronucleus (MA), micronucleus (MI) and peristomal ciliature (CI) in the parasite. Osmium-fixed and haematoxylin-stained preparation. Indicative scale bar 20 µm. (C) Histological section of *Apiosoma* attached to the oesophagus epithelium of a juvenile fish. HE staining. Indicative scale bar 100 µm. (A) courtesy of the Environment Agency Fish Disease Laboratory, UK.

For identification, the observation of a live specimen is important; the morphological identification of trichodinids is based on the structure of the skeletal elements observed in silver-impregnated specimens (see Fig. 10.6C).

Prevention and treatment

Sessile ciliates are considered as harmless ectocommensals rather than parasites. They thrive in water with high concentrations of organic matter and planktonic bacteria, which serve as food for these ciliates. While sessile peritrichs are considered to be harmless, trichodinids may become harmful in massively infected fish, particularly when challenged by additional stressors. Successful treatment protocols need to rely on the improvement of keeping conditions. This

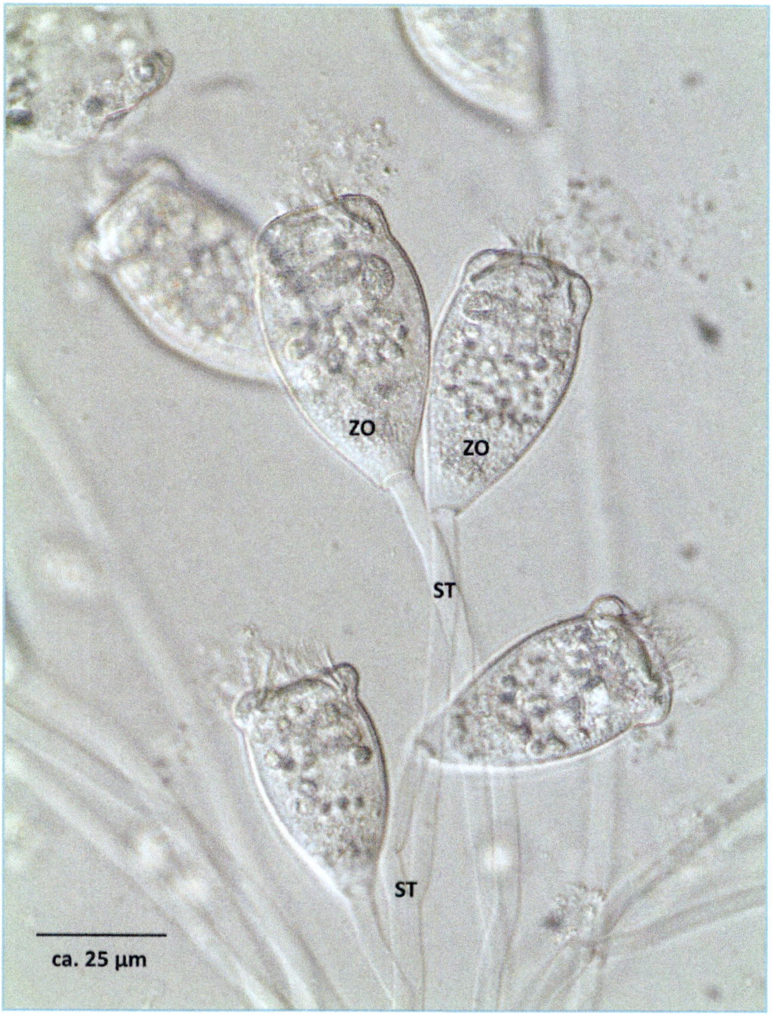

Figure 10.5 *Epistylis*. Fresh preparation of *Epistylis* zooids (ZO) containing several food vacuoles and the peristomal ciliature visible, attached by a non-contractile stalk (ST). Indicative scale bar 25 µm. Courtesy of the Environment Agency Fish Disease Laboratory, UK.

involves, in particular, customised feeding, increased water exchange and improved removal of organic matter. Ornamental stocks may be treated with sodium chloride or parasiticidal chemicals such as formalin.

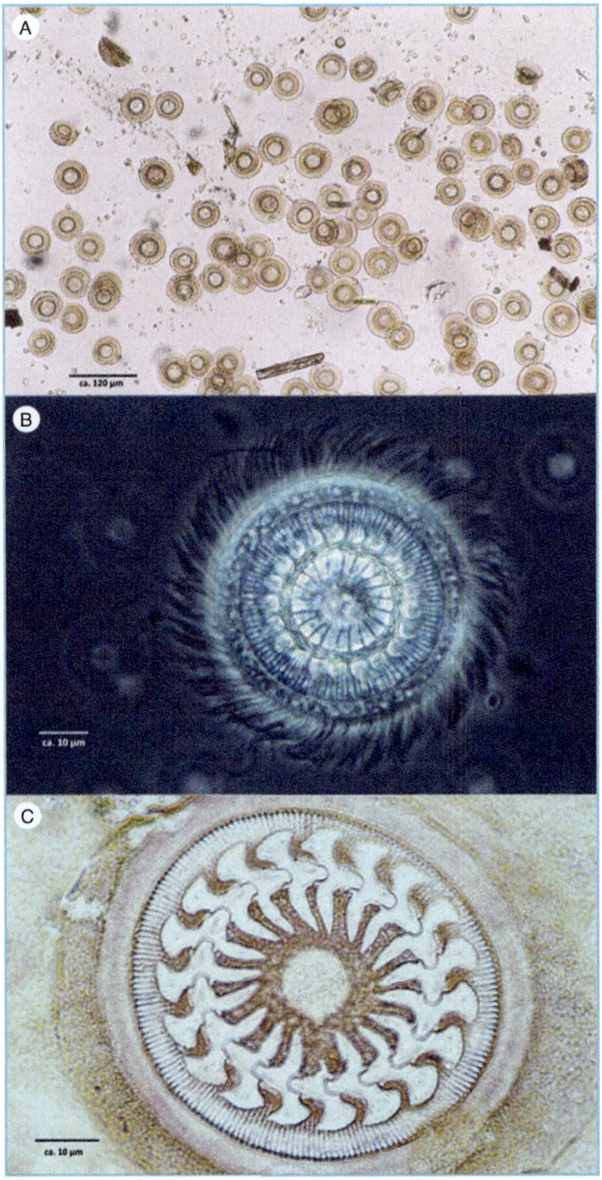

Figure 10.6. Trichodinid infestation. (A) Trichodinids in a fresh sample scraped from the skin of a fish. Indicative scale bar 120 µm. (B) Fresh preparation of a *Trichodina* cell showing the oral rings of ciliates and the cytoskeleton elements. Indicative scale bar 10 µm. (C) Formalin-fixed and silver-impregnated *Trichodina* cell showing cytoskeleton elements. Indicative scale bar 10 µm. (A) courtesy of the Environment Agency Fish Disease Laboratory, UK.

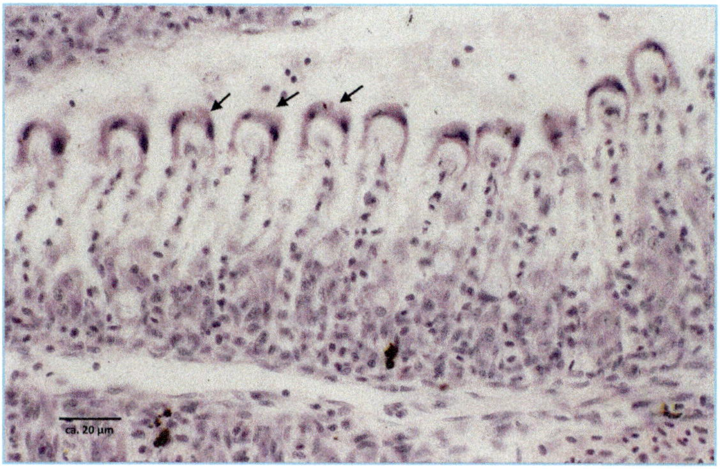

Figure 10.7 Histological section of trichodinids (arrows) attached to the tip of secondary lamellae of gills. HE staining. Indicative scale bar 200 µm.

FACTS SHEET

Disease: Sessile ciliates
Pathogens: *Ambiphrya*, *Apiosoma*, *Epistylis*, trichodinids
Susceptible hosts: All cyprinids
Mortality: Low
Temperature: 2–30°C
Prevention and treatment: Improvement of keeping conditions by reduced stocking density, customised feeding, improved oxygen supply and water exchange

10.2.7 Piscinoodinium

Introduction

Piscinoodinium (Alveolata, Dinozoa, Dinoflagellata) are parasitic dinoflagellates, a group of organisms that have an important role in the primary production of organic matter in freshwater and marine habitats. About 60% of the known dinoflagellate species are photosynthetic, while some species have a mixotrophic feeding habit and incorporate particulate organic matter by phagocytosis when light intensity is insufficient for photosynthesis. Several groups of dinoflagellates are adapted to a parasitic way of life and live as ectoparasites or endoparasites on or in various aquatic organisms. Freshwater fish are mainly infested by species from the genus *Piscinoodinium*, which can cause serious fish losses.

Disease description

Infested fish have a skin with a velvety, matt appearance, covered by high amount of mucus. Heavily infected fish scratch their skin on submerged objects and show signs of impaired respiration, including flushing of the gills and gasping. In these fish, the infection may be a serious cause of mortality.

Pathogen

Dinoflagellate cells possess two unequal flagella, one transversal in an equatorial cell groove (cingulum), the other longitudinal in a longitudinal groove, the sulcum. The cell boundary is formed by a three-membrane complex and, in several species, thecal cellulose plates reinforce this complex.

The life cycle of *Piscinoodinium* comprises a free-swimming migratory stage, which attaches to skin and gill epithelium cells of susceptible fish, where it transforms to a pyriform-to-ovoid feeding stage, the trophont. The trophont forms an attachment disc with numerous rod-like organelles, rhizocysts, which penetrate into epithelial cells and thus attach the parasite firmly to the host cell. The adhesive disc most likely serves in the uptake of nutrients by pinocytosis from disintegrated cells. The trophont has a brownish-yellow colour and contains a chloroplast and several starch granules, although phagocytic vacuoles have not been observed. The cell wall is covered by a theca that does not form thecal plates. After a period of feeding, the trophont drops off from the attachment site, falls to the bottom of the waterbody and transforms into a rounded tomont without secreting an envelope. Within the tomont, a series of cell divisions result in the formation of approximately 130–250 tomites, which subsequently transform into gymnospores. These gymnospores swim in a spiral movement through the water to find a suitable host, where they attach by the sulcal flagellum. At 25°C, the trophont grows for about 6 days on an infected fish. The development of gymnospores in the tomont takes about 6 hour in the light and 8 hour in the dark, and gymnospores live for approximately 15–24 hour. During this time-span, they have to find a suitable host where they can attach and transform into a trophont. The time for completing this development is expanded at lower temperatures.

Detection

Trophonts and sometimes tomonts of the parasite can be observed in fresh samples of mucus from skin and gills when examined by light microscopy. In addition, the infection can be detected in histological preparations of affected tissue. The nucleotide sequence of the gene encoding the small subunit ribosomal RNA has been sequenced and could also be used for molecular diagnosis.

Prevention and treatment

The treatment of *Piscinoodinium* is difficult because the tomont stage is not affected by commonly used parasiticidal chemicals. Infections on ornamental fish can be treated by a prolonged bath with chemicals such as methylene blue, acriflavine, potassium permanganate and sodium chloride. These chemicals cannot be used for fish for human consumption in many countries, because of legal restrictions. In addition, these treatments may harm the fish, making application difficult. Therefore, it is advisable to perform a careful parasitological examination of all stocks before they are introduced into aquaculture facilities in order to avoid introduction of the parasite into populations of susceptible fish.

FACTS SHEET

Disease: Oodinosis (velvet disease)
Pathogen: *Piscinoodinium*
Susceptible hosts: Grass carp, bighead carp, goldfish
Mortality: Low to moderate
Temperature: 20–30°C
Prevention and treatment: Parasitological examination of fish stocks before their introduction into the aquaculture facility. Prolonged bath of parasiticidal chemicals for

treating the infection in populations of ornamental fish. It is recommended that the national legislation on treatment on fish is consulted prior to any action being taken

10.3 Parasites of the Intestinal Tract

10.3.1 Intestinal Coccidia

Introduction

Apicomplexan organisms are parasites of various tissues in both cold- and warm-blooded vertebrates and several species cause serious diseases in humans or livestock animals. All apicomplexans are intracellular parasites and are embedded in a parasitophorous vacuole. In fish, apicomplexans primarily parasitise the intestinal tract, but also occur in the swim bladder, liver, kidney and blood cells. While reports on the presence of apicomplexan haemoparasites in fish are scarce and the impact of these parasites on fish populations is largely unknown, gut-dwelling coccidia in particular are frequently found in fish. Because juvenile and adult fish develop a strong immunity against these parasites after an initial infection, disease symptoms and losses are mainly observed in fry raised in ponds, although the impact of the infection on fish health in aquaculture populations remains unclear.

Disease description

Gut-dwelling coccidian parasites infect intestinal epithelial cells. In heavily infected specimens, the intestinal epithelium is destroyed (Fig. 10.8A, B); the fish stop feeding, experience a loss of electrolytes via the intestine, experience an increase of secondary bacterial infections and may eventually die. In particular, fish fry or juveniles with a heavy infection of enteritic coccidians swim listlessly, may cast white pseudofaeces and die.

Pathogen

A large number of coccidian species are described from the intestinal tract of different fish hosts, including cyprinids. In the old literature, these parasite species are characterised by morphological traits of the infective stages, the oocysts and the sporocyst. Modern phylogenetic studies have revealed an astonishing diversity of coccidian parasites of fish. In general, gut-dwelling coccidian species (Alveolata: Apicomplexa) are intracellular parasites of enterocytes. The host acquires the infection by ingestion of an infectious stage, usually encysted in an oocyst, which usually contains four sporocysts (Fig. 10.8C). In the intestinal tract, motile sporozoites excyst from

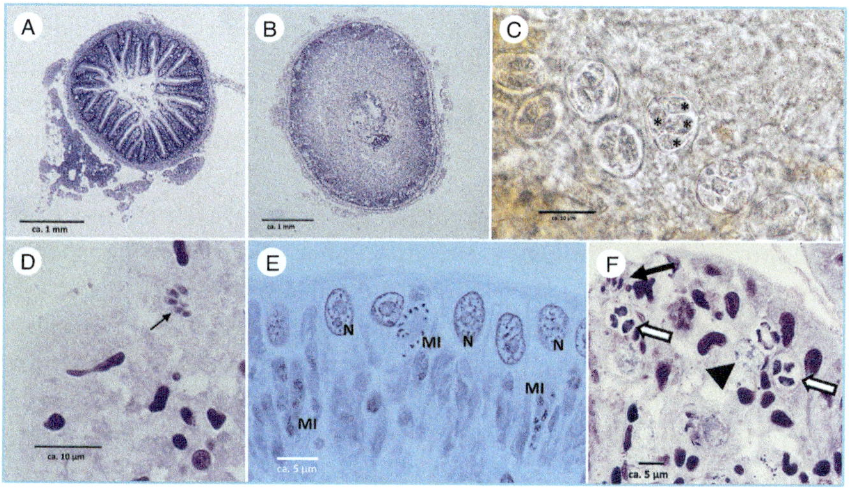

Figure 10.8 Enteritic coccidiosis. (A) Histological preparation of the cross-section of the anterior gut segment of healthy common carp showing high and slender mucosal folds. Giemsa staining. Indicative scale bar 1 mm. (B) In comparison to (A), histological preparation of the cross-section of the anterior segment of a common carp with severe intestinal coccidiosis. Note that the mucosal architecture is completely destroyed. Indicative scale bar 1 mm. (C) Fresh preparation of oocysts of the *Goussia* type with four sporocysts (stars). Indicative scale bar 10 µm. (D) Histological section of carp intestinal mucosa with intracellular merozoites (arrow). Giemsa staining. Indicative scale bar 10 µm. (E) Histological section of carp intestinal mucosa showing a *Goussia* infection with intracellular merogonic developmental stages (ME), a macrogamont (MI) and nuclei of enterocytes (N). Giemsa staining. Indicative scale bar 5 µm. (F) Histological section of carp intestine infected by *Goussia* showing intracellular merozoites (black arrow), macrogamont (black arrowhead) and oocysts (white arrow), and nucleus of enterocytes (N). Giemsa staining. Indicative scale bar 5 µm. (C) courtesy of the Environment Agency Fish Disease Laboratory, UK.

the sporocyst and invade into enterocytes, where they undergo an asexual merogonic multiplication (Fig. 10.8D). As a result of this multiplication, several merozoites are liberated from the infected enterocyte and infect additional enterocytes, where they undergo an additional merogonic multiplication cycle. The merozoites of the second merogony develop in enterocytes into microgametes (Fig. 10.8E) or macrogametes. During a subsequent sexual reproduction, the macrogamete is fertilised by a microgamete and subsequently sporulates to an oocyst comprising sporocysts (Fig. 10.8F). These encysted stages are excreted into the environment via the faeces and can be transmitted to a susceptible host during feeding. In some species, a paratenic host such as tubificid worms or other invertebrate vectors may be involved in the life cycle. Most coccidian parasites of the fish's gut develop in the apical part of the enterocyte; however, there are a few species, such as *Goussia janae*, which develop intracellularly, but in a characteristic epicellular position. The sporocyst morphology varies among fish coccidia. Many species form sporocysts with a simple longitudinal suture (*Goussia* species), which is regarded as an ancestral feature, but in some species (*Eimeria* species) univalved sporocysts containing a plug-like Stieda body are present. In the intestine of cyprinid fishes, *Goussia carpelli*, *G. sinensis* and *G. cheni* are found infecting enterocytes in a dispersed, enteritic manner, while *Eimeria subepithelialis* forms nodule-like aggregates of macrogamonts and sporogonic stages (Fig. 10.9). These aggregates can be seen as white nodules in the posterior part of the intestine in early spring and may even obstruct the intestinal passage.

Detection

Because enteritic coccidia of fish sporulate in the intestinal epithelium, oocysts and sporocysts can be found during microscopic observation of intestinal scrapings (Fig. 10.8C). After an initial infection, oocysts may remain in intestinal tissue in low numbers embedded in 'yellow bodies' for a prolonged period of time. These dispersed yellow bodies containing oocysts can also be seen in light microscopic observation of intestinal scrapings. Infections with nodular coccidia are seasonal and can be detected in intestinal scrapings collected at an

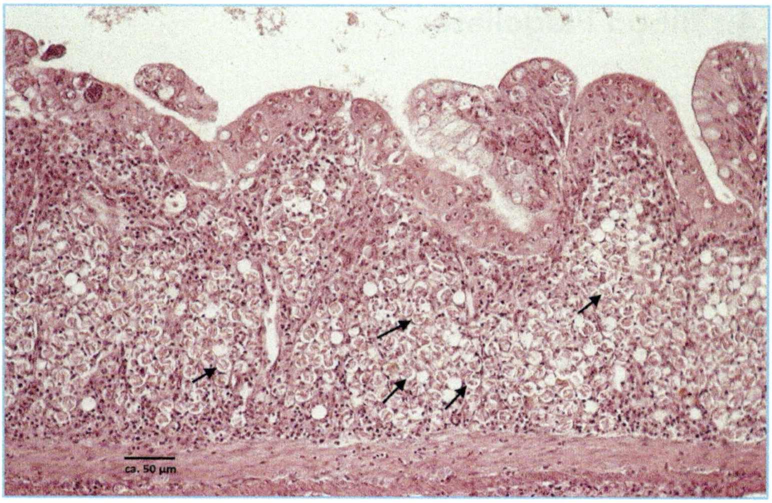

Figure 10.9 Histological section of carp intestine with a massive aggregation of *Eimeria subepithelialis* oocysts (arrows). HE staining. Indicative scale bar 50 µm.

appropriate time, such as when infections are high. Developmental stages of coccidia can also be detected in histological preparations of the gut (Figs. 10.8D–F and 10.9). Flotation methods to separate and concentrate oocysts from the faeces of susceptible fish have not been established for routine diagnosis.

Prevention and treatment

Prevention requires normal sanitary measures. Except from some experimental trials with coccidiostatic substances such as amprolium or toltrazuril, treatment strategies for coccidia are not established.

FACTS SHEET

Disease: Intestinal coccidiosis
Pathogen: *Goussia* and *Eimeria* species
Susceptible hosts: Common carp, grass carp, bighead carp, silver carp, goldfish, *Carassius* species, catla
Mortality: Uncertain
Temperature: Up to 25°C
Prevention and treatment: Prevention is difficult. Only sanitary measures, such as liming the ponds, could be applied. No treatment protocol established

10.4 Blood Flagellates

10.4.1 Introduction

Kinetoplastid flagellates are found as blood parasites in cold-blooded and warm-blood vertebrates, including humans. Several species are well adapted to their hosts, develop a low parasitaemia and are not associated with disease symptoms, but some species are serious pathogens of humans or livestock animals. In fish blood, species from two different kinetoplastid developmental lines are found: cryptobiid (*Trypanoplasma*) species from the Parabodonida group; and trypanosomes from the Trypanosomatida. While infections with trypanosomes and trypanoplasms are seldom associated with disease outbreaks, the infections may cause disease and mortality in susceptible genetic host lines, and during unfavourable fish-keeping conditions.

10.4.2 Disease description

Clinical signs are absent in low parasitaemia, but heavily infected specimens are listless, gather in calm areas of the pond or tank, or are often located near the water inlet or aeration points. In infected fish the skin colour turns dark, and sunken eyes and pale gills occur. Heavily infected juveniles may eventually die.

10.4.3 Pathogen

In fish blood, bi-flagellated trypanoplasms and mono-flagellated trypanosomes can be found. *Trypanoplasma* (Fig. 10.10) have one long recurrent flagellum, which originates from an anterior flagellar pocket and is attached to the body surface, forming an undulating membrane. A short anterior flagellum also originates from the flagellar pocket, as does a preoral ridge, a cellular structure reinforced by microtubules. Next to the flagellar pocket a large, elongated kinetoplast is located, a structure that contains densely packed mitochondrial DNA. In contrast in trypanosomes, this DNA is not organised in mini- and maxicircles, and the nucleus

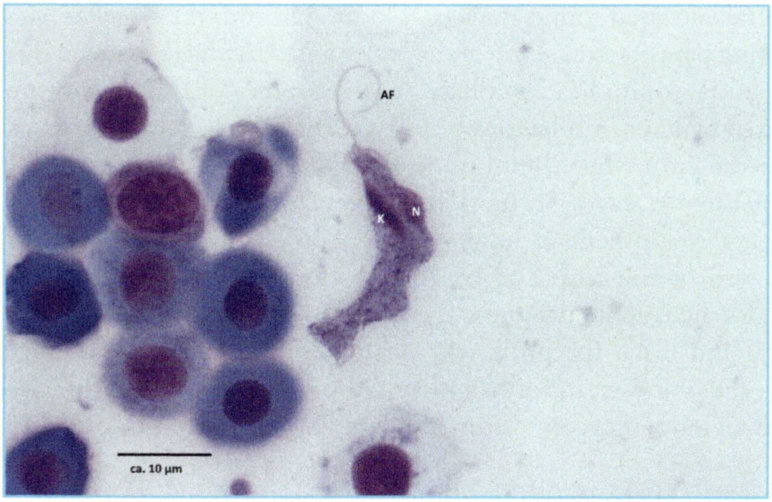

Figure 10.10 Blood smear from common carp showing *Trypanoplasma borreli* with an anterior flagellum (AF) and a recurrent flagellum, which is attached to the body surface. In the cell, a nucleus (N) and a kinetoplast (K) are visible. Osmium-fixed and stained with May–Grünwald–Giemsa. Indicative scale bar 10 µm.

is rounded and lies next to the kinetoplast. The morphology varies during parasitaemia. Shortly after infection, a rapid multiplication of trypanoplasms occurs by binary fission, and short and slender cells are then present in the bloodstream. In later infections, broad and larger flagellate forms dominate in the blood. Trypanoplasms are transmitted by fish-biting leeches, which ingest bloodstream forms together with the bloodmeal into their crop. Here a rapid multiplication of the trypanoplasms takes place, although all parasite stages occurring in the leech are infective for fish and can start an infection when the leech regurgitates them into the host's bloodstream during the feeding process. From the blood of cyprinids, several species, including *Trypanoplasma borreli*, *T. carassii* and *T. agitans*, have been described, but the identity of these species needs to be confirmed.

Trypanosoma species possess one flagellum, which originates from a flagellar pocket near the posterior end, passes anteriorly along the body to which it is attached and then extends from the anterior end as a free flagellum. The parasite swims with the free end of the flagellum pointing in the direction of travel. Parasites often show a wriggling movement when stationary. A small kinetoplast comprising a centrally located round nucleus is located next to the flagellar pocket. During parasitaemia, long and slender parasitic stages are seen during the initial multiplication phase. Trypanosomes of fish are also transmitted by haematophagous biting leeches, which ingest the parasites into the crop where rapid multiplication and a transformation into epimastigote stages occur. The result of this multiplication process is the development of long slender parasite stages, which attach to the wall of the crop and transform to extremely long and slender metacyclic trypomastigote stages. These move within the crop of the leech and reach the proboscis sheath, from where they are injected into the bloodstream of fish during the next bloodmeal of the leech. In the blood of cyprinids mainly *Trypanosoma carassii* is found.

10.4.4 Detection

The parasite can be detected in fresh blood samples or in blood smears stained with Romanowsky or Giemsa stain. The flagellates can be concentrated by means of the haematocrit centrifuge technique in which a haematocrit capillary is filled with peripheral blood from a suspected infected fish, spun in a haematocrit centrifuge and then the area above the buffy coat of leucocytes is checked microscopically for the presence of blood flagellates.

10.4.5 Prevention and treatment

Because the transmission relies on the presence of fish-biting leeches, the parasite can be excluded when leeches are not present. A treatment of infected fish is not established.

FACTS SHEET

Disease: Trypanosomiasis
Pathogen: *Tryanoplasma* and *Trypanosoma* spp.
Susceptible hosts: Common carp, goldfish, grass carp, bighead carp, silver carp
Mortality: Moderate in susceptible species or genetic strains
Temperature: Up to 25°C
Prevention and treatment: Removal of fish-biting leeches. No treatment protocol of infected fish established

10.5 Systemic Amoebic Diseases

10.5.1 Introduction

Free-living amoebae may settle on skin or gills of fish and feed on mucus, adherent bacteria and sloughed-off cells. In marine salmonid aquaculture, amoebic gill disease (AGD), caused by *Neoparamoeba perurans*, may be responsible for serious losses. In freshwater, gill infections with amoebae have been reported occasionally from salmonids, as well as from catfish and cyprinids. In addition, free-living amoebae may enter the digestive tract with feed, where they start proliferating and may invade tissues of the host. In particular, in goldfish a granulomatous disease in kidney and liver has been reported from which amoebae could be isolated. Overall, the circumstances under which such amoebae develop a pathogenic potential for fish remain unclear, but some stage of immunosuppression in affected fish is suspected.

10.5.2 Disease description

Affected specimens may present dyspnoea, skin lesions, ascites and raised scales. In specimens with systemic infection, granulomas with concentrically arranged layers of infected cells at the periphery and necrotic central areas in liver and kidney (Fig. 10.11) occur.

10.5.3 Pathogen

In cells of the peripheral tissue layers, amoebic organisms can be identified in histological and electron microscopic sections (Fig. 10.12). In electron microscopic images, the development of pseudopodia and phagocytic vacuoles have been observed.

10.5.4 Detection

The parasite can be detected in histological and electron microscopical samples from peripheral layers of granulomas. For a species identification, the isolation of amoebae from infected tissues on agar plate cultures is required. Mostly, organisms from the genera

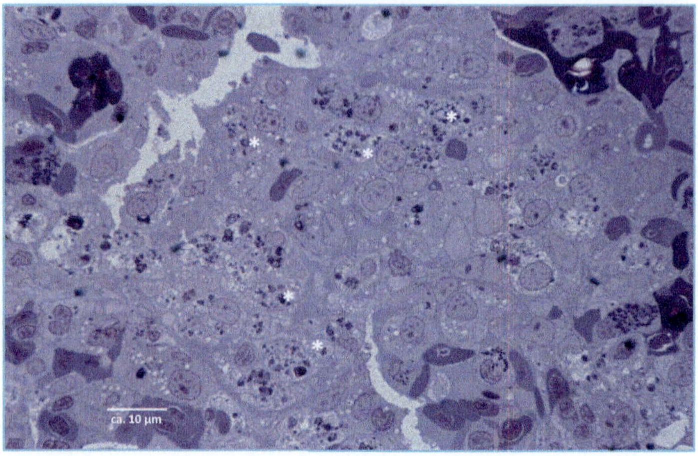

Figure 10.11 Histological section of goldfish liver cells infected with amoebae (stars). Toluidine blue staining. Indicative scale bar 10 µm.

Acanthamoeba, *Vahlkampfia* or *Naegleria* can be isolated. Because specimens from these free-living genera have the potential to infect animals, such amoebae are assigned as 'amphizoic'.

10.5.5 Prevention and treatment

Systemic infections of fish with amoebae were mainly observed in water with high organic load and abundant heterotrophic bacteria, which serve as food for the free-living amoebae. Together with low oxygen content in the water, these stressful conditions may impair the immunocompetence of fish and promote infection of the fish. For a successful treatment, water quality needs to be improved. A chemical treatment of infected fish is not available.

FACTS SHEET

Disease: Systemic amoebosis
Pathogen: Amphizoic amoebae, like *Acanthamoeba*, *Vahlkampfia* or *Naegleria*
Susceptible hosts: *Carassius* species, in particular goldfish
Mortality: Moderate in affected stocks
Temperature: Up to 25°C
Prevention and treatment: Prevention is by improvement of water quality, increased frequency of water exchange. No treatment protocols are established.

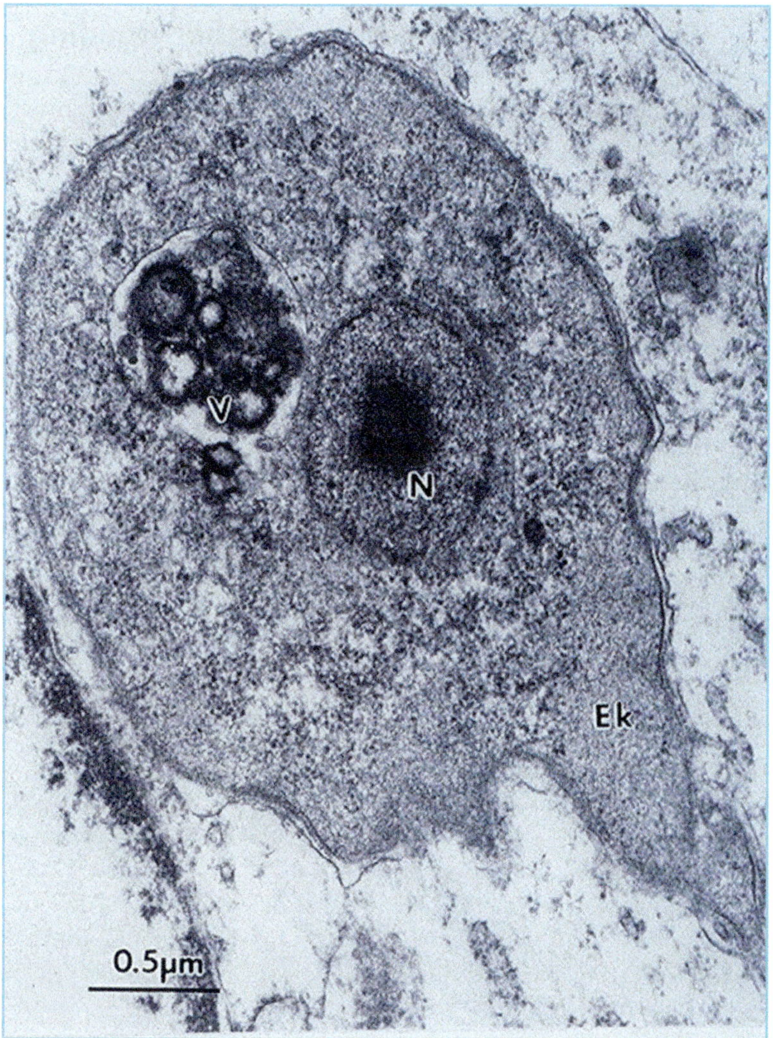

Figure 10.12 Electron microscopic image of goldfish liver cell infected with intracellular amoebae. Note presence in the parasite of a nucleus (N), food vacuole (V), ectoplasma (Ek) and the formation of pseudopodia. Indicative scale bar 0.5 μm.

10.6 Selected References for Further Reading

Alvarez-Pellitero, P. (2008) Diseases caused by flagellates. In: Eiras, J.C., Segner, H., Wahli, T. and Kapoor, B.G. (eds) *Fish Diseases*. Science Publishers, Enfield, New Jersey, pp. 421–515.

Bastos Gomes, G., Jerry, D.R., Miller, T.L. and Hutson, K.S. (2017) Current status of parasitic ciliates *Chilodonella* spp. (Phyllopharyngea, Chilodonellae) in freshwater fish aquaculture. *Journal of Fish Diseases* 40, 703–715.

Callahan, H.A., Litaker, R.W. and Noga, E.J. (2005) Genetic relationships among members of the *Ichthyobodo necator* complex: Implications for the management of aquaculture stocks. *Journal of Fish Diseases* 28, 111–118.

Colorni, A. (2008) Diseases cause by ciliophora. In: Eiras, J.C., Segner, H., Wahli, T. and Kapoor, B.G. (eds) *Fish Diseases*. Science Publishers, Enfield, New Jersey, pp. 569–612.

Davies, A.J. and Ball, S.J. (1993) The biology of fish coccidia. *Advances in Parasitology* 32, 294–366.

Dickerson, H.W. and Dawe, D.L. (1995) *Ichthyophthirius multifiliis* and *Cryptokaryon irritans* (Phylum Ciliophora). In: Woo, P.T.K. (ed.) *Fish Diseases and Disorders: Protozoan and Metazoan Infections*, Vol. 1. CAB International, Wallingford, UK, pp. 181–227.

Dykova, I. (2008) Amoeboid protists as parasites of fish. In: Eiras, J.C., Segner, H., Wahli, T. and Kapoor, B.G. (eds) *Fish Diseases*. Science Publishers, Enfield, New Jersey, pp. 397–420.

Gomez, F. and Skovgaard, A. (2015) The molecular phylogeny of the type-species of *Oodinium* Chatton, 1912 (Dinoflagellata: Oodiniaceae), a highly divergent parasitic dinoflagellate with non-dinokaryptic characters. *Systematic Parasitology* 90, 125–135.

Haider, G. (1964) *Monographie der Familie Urceolariidae (Ciliata, Peritricha, Mobilia)*. VEB Gustav Fischer.

Isaksen, T.E., Karlsbakk, E., Repstad, O. and Nylund, A. (2012) Molecular tools for the detection and identification of *Ichthyobodo* spp. (Kinetoplastida), important fish parasites. *Parasitology International* 61, 675–683.

Jahangiri, L., Shinn, A., Pratoomyot, J. and Bastos Gomes, G. (2020) Unveiling associations between ciliate parasites and bacterial microbiomes under warm water fish farm conditions – a review. *Reviews in Aquaculture* 13, 1097–1118.

Karvonen, A., Rintamäki, P., Jokela, J. and Valtonen, E.T. (2010) Increasing water temperature and disease risks in aquatic systems: Climate change increases the risk of some, but not all, diseases. *International Journal for Parasitology* 40, 1483–1488.

Lom, J. and Dykova, I. (1992) *Protozoan Parasites of Fishes.* Elsevier, Amsterdam, The Netherlands.

Lom, J. (1995) Trichodinidae and other ciliates (Phylum Ciliophora). In: Woo, P.T.K. (ed.) *Fish Diseases and Disorders: Protozoan and Metazoan Infections*, Vol. 1. CAB International, Wallingford, UK, pp. 229–262.

MacColl, E., Therkelsen, M.D., Sherpa, T., Ellerbrock, H., Johnston, L.A. *et al.* (2015) Molecular genetic diversity and characterization of conjugation genes in the fish parasite *Ichthyophthirius multifiliis. Molecular Phylogenetics and Evolution* 86, 1–7.

Molnar, K. (1995) Phylum Apicomplexa. In: Woo, P.T.K. (ed.) *Fish Diseases and Disorders: Protozoan and Metazoan Infections*, Vol. 1. CAB International, Wallingford, UK, pp. 263–287.

Noga, E.J. and Levy, M.G. (1995) Dinoflagellida (Phylum Sarcomastigophora). In: Woo, P.T.K. (ed.) *Fish Diseases and Disorders: Protozoan and Metazoan Infections*, Vol. 1. CAB International, Wallingford, UK, pp. 1–25.

Shaharom-Harrison, F.M., Anderson, I.G., Siti, A.Z., Shazili, N.A.M., Ang, A.J. *et al.* (1990) Epizootics of Malaysian cultured freshwater pond fishes by *Piscinoodinium pillulare* (Schäperclaus 1954) Lom 1981. *Aquaculture* 86, 127–138.

Simpson, A.G.B., Gill, E.E., Callahan, H.A., Litaker, R.W. and Roger, A.J. (2004) Early evolution within Kinetoplastids (Euglenozoa), and the late emergence of trypanosomatids. *Protist* 155, 407–422.

Simpson, A.G.B., Stevens, J.R. and Lukes, J. (2006) The evolution and diversity of Kinetoplastid flagellates. *Trends in Parasitology* 22, 168–174.

Steinhagen, D. and Davies, A. (2008) Diseases caused by Apicomplexa. In: Eiras, J.C., Segner, H., Wahli, T. and Kapoor, B.G. (eds) *Fish Diseases.* Science Publishers, Enfield, New Jersey, pp. 517–567.

Todal, J.A., Karlsbakk, E., Isaksen, T.E., Plarre, H., Urawa, S. *et al.* (2004) *Ichthyobodo necator* (Kinetoplastida) – complex of sibling species. *Diseases of Aquatic Organisms* 58, 9–16.

Woo, P.T.K. and Ponyton, S.L. (1995) Diplomonadida, Kinetoplastida and Amoebida (Phylum Sarcomastigophora). In: Woo, P.T.K. (ed.) *Fish Diseases and Disorders: Protozoan and Metazoan Infections*, Vol. 1. CAB International, Wallingford, UK, pp. 27–96.

Chapter 11

The Flukes

11.1 Introduction

The group of parasites commonly termed 'the flukes' represents a diverse assembly of flatworms that can cause wide-ranging impacts on carp culture throughout the world. They are primarily divided into two classes, based on structure and life cycle strategies. The monogenean flukes are mainly ectoparasites and primarily have direct life cycles with intermediate stages occurring within the waterbody. Direct infection usually takes place from fish to fish. In contrast, the digenean flukes, sometimes called trematodes, are endoparasites and include complex life cycles involving at least two hosts. The fish can serve as the definitive host in which the adult parasite resides, or as the intermediate host, which serves as a host for the life cycle stage: the metacercarial cyst. The inclusion of at least one intermediate host in the digenean life cycle adds another level of complexity to the control of these parasites.

11.2 Monogeneans

In fish, monogeneans inhabit a range of 'external' organs and in their adult stages primarily infect the skin, gills, pharynx and buccal cavity. They attach to their host via a large posterior attachment organ, the opisthohaptor, which comprises two large central hooks surrounded at the extremities of this 'haptor' by 12–16 marginal hooks. This organ, which is useful in the identification of the species of monogenetic fluke, is also associated with the pathology caused by the parasite. Two major monogenean families that are commonly

 Carps: A Practical Approach to Disease Control and Health Management (eds David Hoole *et al.*)
DOI: 10.1079/9781836990024.0011

found associated with carp culture are the dactylogyrids and gyrodactylids. A third group, the Polyopisthocotylida, represented by *Diplozoon paradoxum*, which is an unusual creature comprising two fused individuals, does occur on the gills of cyprinid fish, e.g. common bream (*Abramis brama*), but has occasionally been located on carp.

The dactylogyrids, which are primarily located on the gills, and the gyrodactylids, which can be found mainly on the skin and fins of carp, have two different reproductive strategies, which has implications on their transmission routes and control strategies. The dactylogyrids are somewhat typical monogeneans as they lay eggs, which release a free-living ciliated oncomiracidium stage into the waterbody. Whilst this free-living stage is susceptible to a range of parameters, the egg, in comparison, is resistant and attaches to a range of substrates, including pond substrata, vegetation, surfaces of tanks and pond nets. It is the oncomiracidium that completes the life cycle by attaching to a fish. In comparison, the gyrodactyid flukes give birth to live young, which directly infect fish. This efficient and yet unusual reproductive strategy has meant that the flukes are sometimes called the 'generation worm', which gives them a high reproductive potential and means that they can rapidly infest carp ponds once established.

11.2.1 Dactylogyrids

Epidemiology and host range

These gill parasites can increase rapidly under carp farm conditions primarily due to the direct life cycle involving the egg and liberated free-living stage. The oncomiracidium has a short lifespan of a few hours and hatching is controlled by temperature, changes in light intensity and diurnal rhythms. This has implications for the periods of high infection, which usually occur at higher temperatures appropriate to the resident temperature regime. The dactylogyrids have an extensive host range, but tend to predominate in the cyprinids, and particularly in confined conditions, which aids infection and transmission. The dactylogyrids have a world distribution within carp production units, and the species of parasite involved can vary depending on fish species infected and geographical location.

For example, *Dactylogyrus vastator*, which is endemic in Southeast Asia, now has a wider range of distribution within Europe, North America and some countries in the Middle East, and is very similar to another dactylogyrid, *D. extensus*, found in the same geographical range and host species. Goldfish (*Carassius auratus*) are known to be parasitised by at least seven species of dactylogyrids, with *D. vastator* and *D. intermedius* being the dominant species located on the gills, although *D. extensus* has been located on ornamental carp from Pakistan. *D. extensus* has also been isolated from the gills of common carp (*Cyprinus carpio*) in South America, highlighting its global range. In addition, *D. formosus*, which is also frequently found on goldfish, is widely distributed across Eurasia. *D. extensus*, *D. anchoratus*, *D. arcuatus* and *D. cyclocirrus* have been located on the gills of *C. carpio* obtained from North Africa, and in India *D. extensus* and *D. lamellatus* have been located on the gills of common carp and grass carp (*Ctenopharyngodon idella*), respectively, and *D. glossogobii* has been found in gills of rohu (*Labeo rohita*). Of the ten *Dactylogyrus* spp. known from the Far East, nine species (*D. achmerowi*, *D. anchoratus*, *D. extensus*, *D. falciformis*, *D. minutus*, *D. molnari*, *D. mrazeki*, *D. sahuensis*, *D. vastator*) have been observed in common carp in Hungary. This somewhat limited reference to the range of theses monogeneans is an acknowledgement of the adaptability of the dactylogyrids to infect carp production around the world.

Pathology

The location of the dactylogyrids on the gills of carp means that they can induce severe damage on the infected fish (Figs. 11.1 and 11.2) and thus can compromise the whole carp population, affecting both general health status and economic profitability of the production unit. Infection can cause hyperplasia of the gill epithelium and can lead to deformation of the gill lamellae. In addition, in some instances of heavy infection, the parasite can induce excessive mucus production. The exact location on the gills can vary and some

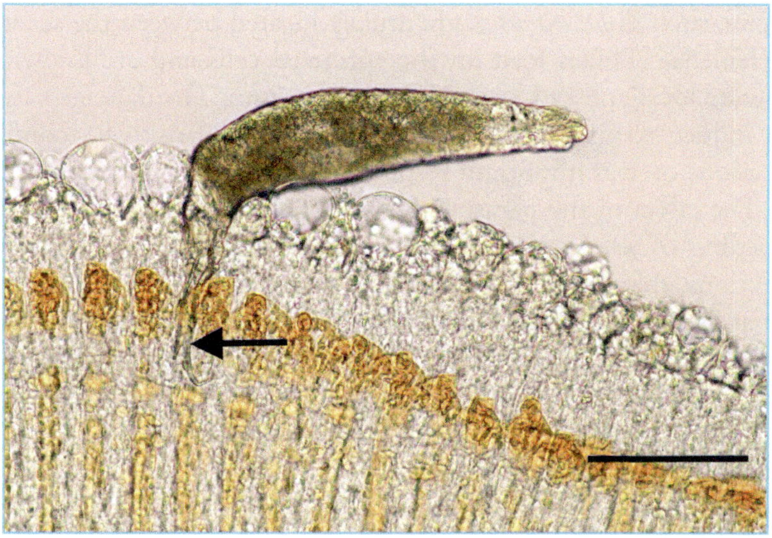

Figure 11.1 Whole wet preparation of *Dactylogyrus* sp. attached via hamuli (hooks) to the gill of carp (arrow). Indicative scale bar 0.5 mm. Courtesy of the Environment Agency, National Fisheries Laboratory, UK.

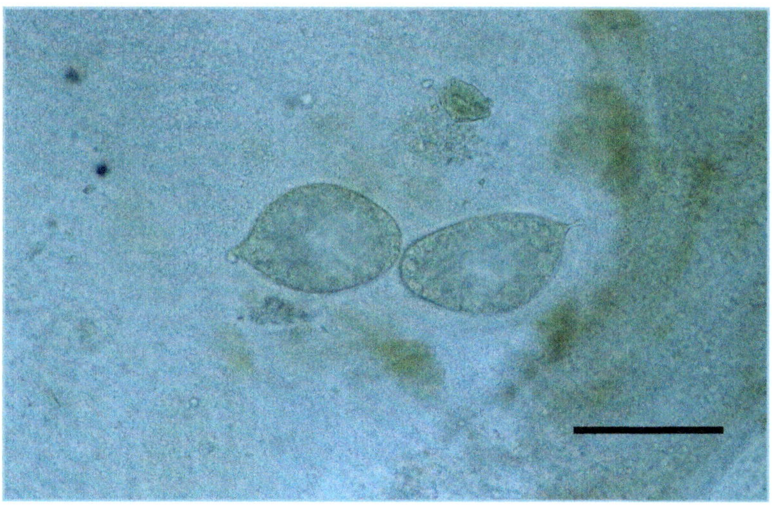

Figure 11.2 Whole wet preparation of two eggs from *Dactylogyrus* sp. Indicative scale bar 50 μm.

species, such as *D. extensus*, are usually located between the secondary lamellae. Flukes feed on the epithelial cells and are known to generate local and systemic immune responses. The damage caused can induce respiratory distress and increase the risk of secondary infections on this important respiratory organ.

The effect of the parasites on gill structure and the subsequent impeding of respiratory capabilities, particularly in high-intensity infections, may result in behavioural changes in infected fish. This is manifested in the congregation of fish in areas of higher oxygen content, such as on the surface of the pond or adjacent to water inlets.

Detection

Clinical outbreaks, particularly those associated with high level of infection, make diagnosis simple as it is usually based on wet mounts of the isolated parasite or infected gills. The parasite generally has a characteristic stretching and recoiling motion, and the species characteristics of the form of the haptor and associated hook/sucker pattern aid in classification by morphometric analysis. To aid identification, flukes can be fixed in formalin or alcohol. The wide diversification of the dactylogyrids has recently led to the application of molecular technologies using parasite 18S ribosomal DNA subunit and the internal transcribed spacer regions, 18S and 28S rDNA subunit regions and mitogenome analysis. In addition, environmental DNA has been utilised to characterise *Dactylogyrus* infections in ornamental fish.

11.2.2 Gyrodactylids

Epidemiology and host range

The viviparous nature of the gyrodactylids makes them unique amongst monogeneans and the fact that the young are fully formed at birth means that they can spread rapidly on individual fish and throughout a carp production unit, including within the ornamental sector (Fig. 11.3). The gyrodactylids have a complex speciation,

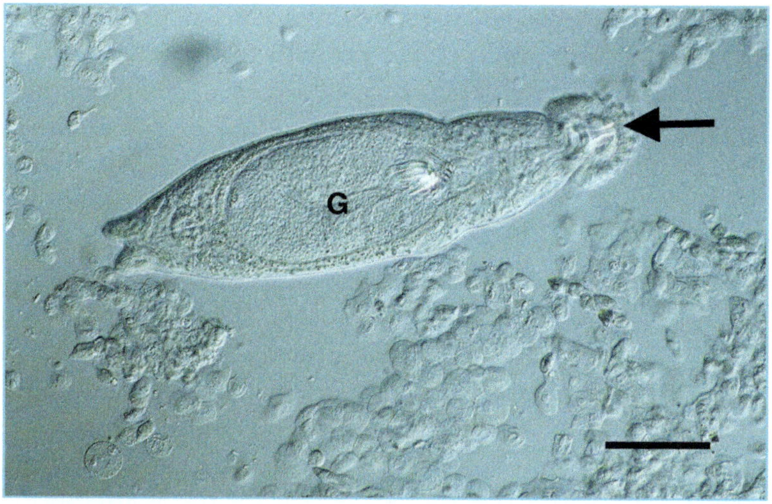

Figure 11.3 Whole wet preparation of *Gyrodactylus* isolated from gill of a carp. Note posterior haptor used to attach the parasite to the gill (arrow) and the characteristic next generation of parasite (G) within the parent parasite. Indicative scale bar 0.1 mm.

particularly when associated with the carp group. *Gyrodactylus katharineri* occurs in carp in Europe, Asia and North America and is thought to be synonymous with *G. cyprini* and *G. mizellei. Gyrodactylus sprostonae* has been located on common carp and Prussian carp (*Carassius gibelio*) in the Southern Caspian Basin and *G. shulmani*, *G. stankovici* and *G. kherulensis* have been found on common carp in Germany. In some instances, gyrodactylids have been reported from gills of common carp, for example *Gyrodactylus dzhalilovi*, *G. magnus*, *G taimeni* and *G. matovi* on common carp in Iraq. In the Indian subcontinent, *G. elegans* has been located on the surface of major mrigal carp (*Cirrhinus mrigala*).

This diverse group of monogeneans is also an issue within the ornamental trade, with *G. cyprini* being observed on the fins of koi carp, and *G. kobayashi* and *G. chiensis* being prominent in the goldfish (*Carassius auratus*). Such infections are observed during examinations of imports usually from the Far East or South-east Asia. In contrast, *Gyrodactylus gurleyi* and *Gyrodactylus nordmann* have been noted parasitising goldfish in North America. The global spread of the gyrodactylids is partly due to their reproductive potential, which is heightened in fish production units with high stocking densities. Transmission is affected by a range of environmental factors, e.g. temperature, and parasite intensities are greater in spring, although minimum infestations have been observed in the rainy season in Asia.

Pathology

The pathological effects are usually greater in small fish with high parasite intensities. The irritation caused by the parasite can lead to fish 'flashing', where rapid movements occur, and sometimes scraping against obstacles in the water body, including natural vegetation and netting. This gives the impression that the fish is attempting to remove the monogenean from the host surface. Infections in young common carp can cause severe problems as the fish can turn dark blue in colour, become emaciated and may die. Most damage is caused by the feeding activity of the parasite or by its attachment on the host surface. Whilst feeding can cause foci of small haemorrhages and epithelial hyperplasia, this type of pathology can be more extensive when associated with the action of the large attachment haptor. Such sites of attachment can lead to secondary infections and the generation of excessive mucus. It is perhaps not surprising that gyrodactylids in cultured, wild and ornamental carp have been associated with an immune response, which has indicated that systemic cytokine responses are important in controlling the monogenean infection.

Detection

Diagnosis of gyrodactylids is usually an easy process and the parasites can be located using light microscopy. Samples can be obtained from the surface of the skin or fins using skin scrapes and placing samples collected in freshwater on a glass slide. Care must be taken with carrying out this process on live fish, and anaesthesia, for example with MS222, can be utilised to reduce stress on the host. When the host is dead then perhaps more robust collection processes can be applied and the usual excessive mucus production occurring means that shaking the isolated gills, or indeed whole fish, is required to remove the mucus present. Parasites removed from either live or dead hosts can be fixed in 10% formalin and whole mounts made using standard procedures. Identification of species can sometimes be achieved based on haptor structure and its skeletal elements, as well as reproductive sclerites. Some recent developments have adopted semi-automated identification techniques with digital images. Sometimes this can be improved with staining of the skeletal elements and sclerites using a combination of 40 mM chromotrope 2R containing 3 mM phosphotungstic acid and 0.05% acetic acid. Such techniques require extensive staining, which can take days. More recently molecular techniques have been utilised, for example the identification of *G. kobayashii* from goldfish using phylogenetic analysis of a reference sequence spanning 5.8S and ITS-2.

11.2.3 Prevention strategies and treatments of dactylogyrids and gyrodactylids

As in all cases of infection, prevention is a better option than cure, hence any imported fish should be screened for infection of dactylogyrids and gyrodactylids prior to introduction into the carp production unit. In addition, nursery ponds should be fed by water that is isolated from the water source used for rearing ponds, and separate implements should be used for manipulating the fish in the two localities. Monogeneans can be removed from ponds by draining the water, allowing to dry and treating with lime. Monogeneans usually thrive at high host stocking densities; hence, a reduction in stocking densities may reduce mortalities. A range of chemical treatments have been proposed for the treatment of monogenean parasites, although when considering this control strategy, the cost-effectiveness and environmental impact must be considered, and compliance with local legislative requirements is required. Indeed, all applications should first be subjected to a trial to ascertain the correct dose and effects on the host. Saltwater baths are perhaps the most common treatment for external parasites. Typically for monogeneans of freshwater fish, a saltwater bath consisting of 40–50 ppt for 5 min is recommended. Whilst skin parasites, such as gyrodactylids, can succumb rapidly to salt treatment, gill parasites such as dactylogyrids tend to be more resistant due to their gill location. Other applications include treatments with formalin and potassium permanganate, in addition to benzimidazoles such as mebendazole, insecticides such as trichlorphon, and anthelmintics such as praziquantel. Whilst the application of such treatments requires very careful consideration and expertise, they have proved effective in some cases, for example water bath treatment with mebendazole (1 mg/L) for 24 hour resulted in a significant reduction of the number of *Dactylogyrus minutus* on carp fry. In addition, there have been several attempts to isolate active ingredients of naturally occurring plants that may be active against monogenean parasites. For example, *D. minutus* on common carp were killed after approximately 61 min and 8 min exposure to aqueous rosemary extract solution at 100 g and 200 g per litre of water, respectively. The active ingredient was thought to be 1,8-cineole, which may have some effect on the parasite when applied orally in feed at 60–100 ml aqueous extract per 100 g feed.

FACTS SHEET

Disease: Dactylogyrids and gyrodactylids

Pathogen: Many species, such as *Dactylogyrus vastator, D. intermedius, D. extensus, D. anchoratus, D. arcuatus* and *D. cyclocirrus*. The gyrodactylids also have many species, including *Gyrodactylus katharineri, G. cyprini, G mizellei, G. sprostonae*

Susceptible hosts: Many cyprinid species, including *Cyprinus carpio, Carassius auratus, Ctenopharyngodon idella, Labeo rohita, Cirrhinus mrigala*

Mortality: Infection causes hyperplasia, haemorrhages and increased mucus production. Mortalities can be high in heavy infections

Temperature: Parasite intensities can be greater with increasing temperature

Prevention and treatment: Infection can be reduced or prevented by good husbandry practices and avoiding introduction of contaminated fish stock or water. Treatments are available but have varying success and range from inexpensive salt treatment to more strategic application of anthelminthic

11.3 Digeneans (the Trematodes)

The class of parasitic flatworms has a wide range of hosts, of which many include a fish at some stage in the digenean life cycle. The complex life cycle comprises a definitive host, which harbours the adult stage of the parasite, and at least one intermediate host, which is infected by the sporocyst, redial and/or metacercarial stages. Intermediate hosts include several invertebrates, e.g. gastropods, but several vertebrates, especially fish, can serve this important role. The presence of free-living stages, such as the oncomiracidium and cercarial stages, does make control of this class of parasitic worms challenging, but a knowledge of transmission routes facilitates control strategies that target either the definitive host or intermediate host(s). Within enclosed conditions that are sometimes utilised for carp production and maintenance, for example in the ornamental trade, transmission can be limited due to the requirement for a natural host, for example a bird or gastropod. This means that in enclosed conditions, such as aquaria, fish cannot harbour flukes such as the eye fluke (*Diplostomum*), blood fluke (*Sanguinicola*) or black spot (*Posthodiplostomum*) unless they have acquired the infection from outdoor facilities prior to entering indoor enclosures.

Most of the digeneans infecting fish belong to the Prosostomata, which in the adult stage are characterised by two suckers: a ventral sucker used for attachment and an oral sucker located at the anterior end of the parasite and associated with the mouth. Flukes are usually hermaphrodite with a blind-ending gut, male and female reproductive structures and vitellaria. The form of the body is used for identification and diagnosis. Carp can therefore serve as definitive and intermediate hosts for digenean infections.

11.3.1 Sanguinicola (the blood flukes)

Epidemiology and host range

Sanguinicoliasis is a serious disease of cultured carp that can cause high mortalities in fry. The most prevalent species worldwide is *Sanguinicola inermis*, which can occur within a range of cyprinid fish, including wild and farmed carp groups, as well as ornamental fish such as koi. In addition, *Sanguinicola armata* has been located in grass carp fingerlings. The life cycle of *S. inermis* comprises the carp definitive host (Fig. 11.4) and a gastropod intermediate host,

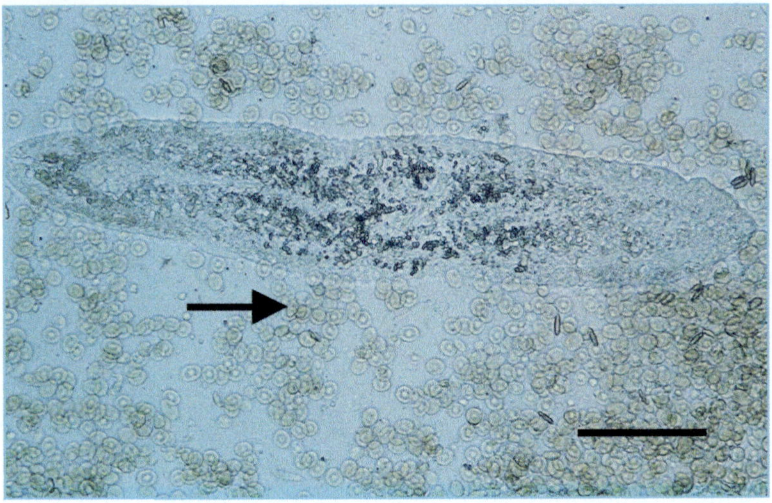

Figure 11.4 Whole wet preparation of *Sanguinicola inermis* isolated from the heart of a carp. Image also contains host red blood cells (arrow). Indicative scale bar 0.25 mm.

usually of the genus *Lymnaea*. The fish is infected by penetration of the skin or fins by the cercarial stage liberated from the gastropod intermediate host. Although a large number of the flukes remain in the penetration site, several migrate to the heart and gill vessels. Production of triangular-shaped eggs occurs and these eggs become located in a range of organs, e.g. liver, kidneys and gills (Fig. 11.5). It is those located in the latter organ that continue the life cycle by the release of the free-living miracidia, which penetrate the intermediate host. The whole life cycle and parasite development are temperature dependent, which usually means that infection in fish is closely linked with the development of the gastropod intermediate host. In temperate regions, therefore, the shedding of cercariae and the infection of fish usually occur in elevated temperature levels, such as in spring.

Pathology

Sanguinicoliasis can have a range of pathological effects on the carp host, primarily because the fish harbours the penetrating cercarial

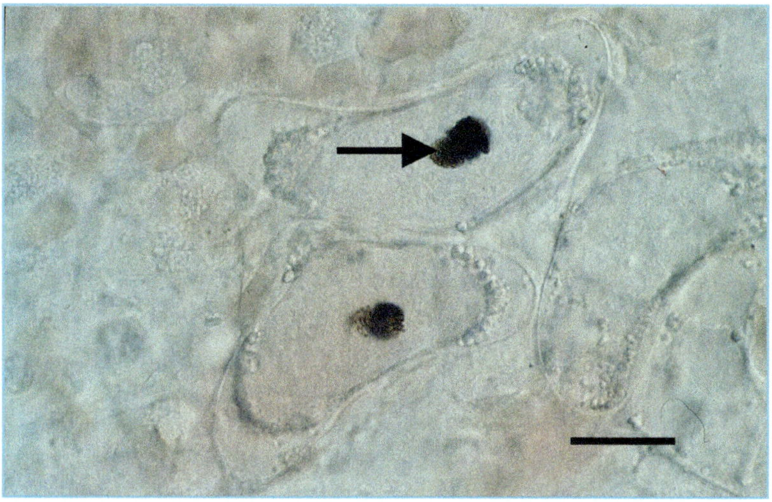

Figure 11.5 Whole wet preparation of *Sanguinicola* eggs within the kidney tissue of a carp. Note the typical triangular shape of the egg and the presence of enclosed miracidium with dark eye spot (arrow). Indicative scale bar 10 μm.

stage, the adult fluke and the eggs. The penetrating cercariae and/or entrapped eggs can cause severe pathological responses and even death. The eggs can form foci of inflammation and in the gills cause rupture of capillaries, epithelial hyperplasia and haemorrhages. Such damage of the gill tissue can cause respiratory distress. There is substantial evidence that there is an intense cellular and humoral immune response against infection, which can lead to parasite death.

Fish that are initially infected with *S. inermis* exhibit lethargy and reduced food intake, which can lead to poor growth and condition. As infection levels increase, carp may exhibit spiral swimming movements, become dark in colour and develop exophthalmia. Effects on the gills result in bulging opercula and fish tend to congregate at the surface of water and in areas of higher oxygen content.

Detection

Whilst behavioural changes in infected fish may be the first indication of parasitisation, they cannot be relied on for diagnosis.

Diagnosis of sanguinicoliasis requires light microscopic examination of dissected fish. The heart is removed and placed in appropriate solution, e.g. teleost ringer, and teased apart. Any flukes present are released and are noticeable by their 'snake-like' swimming movements, which cease once they attach to the container. Depending on temperature, adult flukes can be found in the heart and gill vessels at 90 days post-infection, but decline in numbers approximately 120 days post-infection. The eggs are more long lived, and whilst they can be seen in wet mounts of gill tissue at 40–1000 × magnification, clearer identification can be obtained by processing infected organs for light microscopic analysis, following histological preparation, sectioning and staining in standard stains such as haematoxylin and eosin. Whilst there have been some limited molecular studies on the parasites, such techniques are not currently available for diagnosis.

FACTS SHEET

Disease: Sanguinicoliasis
Pathogen: *Sanguinicola inermis* and *S. armata*
Susceptible hosts: *Cyprinus carpio, Ctenopharyngodon idella*
Mortality: Can occur in heavy infections
Temperature: Infection more prominent in high temperatures, due to presence of infected intermediate hosts
Prevention and treatment: Good husbandry practices including reduction of intermediate hosts. Application of anthelminthic can be used but may not be cost effective

11.3.2 The eye flukes: Diplostomum spp. and Tylodelphys spp.

Epidemiology and host range

As well as serving as definitive host for several trematodes, carp can also harbour a range of flukes as intermediate host. Perhaps the most commonly found and researched are the flukes that infect the eyes, the so-called 'eye fluke' (Fig. 11.6), of which there are two main genera, *Diplostomum* and *Tylodelphys*. The fish is infected by a penetrating cercarial stage liberated from an infected mollusc, usually of the

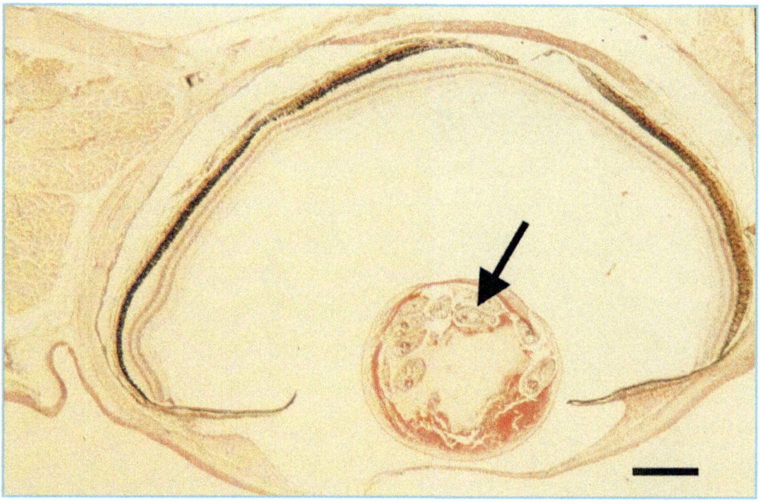

Figure 11.6 Histological section of the eye of a carp. Note the presence of *Diplostomum* infection within the lens (arrow). Indicative scale bar 1 mm.

genus *Lymnaea*, although *Radix* spp. and *Galba* spp. have also been noted as first intermediate hosts. Within the fish the cercarial stage transforms into the metacercarial stage, which remains unencysted, and in *Diplostomum* spp. is termed the 'diplostomule' (Fig. 11.7). Eventually the metacercaria resides in the eye after migration from the site of penetration.

Speciation within the 'eye flukes' has raised much controversy over many years, although it is recognised that the two genera mentioned occur in carp. Often postmortems just recognise the presence of eye fluke, noting *Diplostomum* sp. rather than ascertaining the exact species involved. It would appear that the most commonly recorded parasite, as well as that on which most research has been carried out, is *Diplostomum spathaceum*, which resides in a range of host fish, including carp. This has led to the disease being called diplostomiasis, diplostomastosis, parasitic cancer or, simply, eye fluke disease. *Tylodelphys* spp. are the causative parasites of the disease termed tylodelphosis.

D. spathaceum has been found in a range of cyprinids, including common carp, the Prussian carp, crucian carp (*Carassius carassius*) and grass carp and in ornamental cyprinids, i.e. goldfish. In

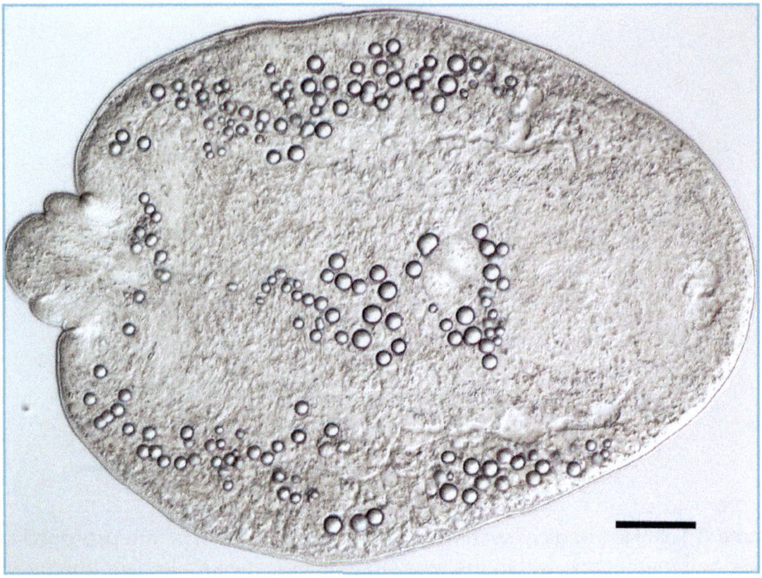

Figure 11.7 Whole wet preparation of the diplostomule stage of *Diplostomum* isolated from the lens of a carp. Indicative scale bar 50 µm. Courtesy of the Environment Agency, National Fisheries Laboratory, UK.

contrast, experimental exposures in laboratory conditions have indicated that crucian carp were resistant to infection, which may indicate that, in some conditions, this host may be physiologically unsuitable. In addition, other species of *Diplostomum* have been recorded in carp species, such as *D. rutili* in Prussian carp, *D. helveticum* in crucian carp and *D. volvens* in common carp. *Tylodelphys clavata* has also been observed in goldfish, common carp and grass carp.

The prevalence and intensity of eye fluke infections in fish is governed by a range of environmental conditions that impact on the life cycle. One of the prime drivers is the abundance and infection of the molluscan first intermediate host, which tend to increase in warm temperatures such as in spring. In temperate climates the optimum temperature for fish infection is approximately 17°C and the cercarial stage usually stays infective for a short period of time, e.g. 5 hour.

Pathology

The complex interaction between the eye fluke and the fish intermediate host means that pathology can be associated with the penetrating cercariae, the migrating metacercarial stage or the un-encysted metacercaria located in the eye. In carp fingerlings, infection of at least three cercariae per fish have been noted as having lethal results. *Diplostomum* spp. have been located within a range of sites within the nervous system of the fish (e.g. eye, brain, spinal cord), whilst *Tylodelphys* spp. tend to be primarily located in the vitreous humor of the eye. It is the pathology associated with the eye infections that appears to cause the majority of issues. Parasites can be associated with lens dislocation, cataracts and an exophthalmic response, all of which can cause blindness. This can lead to reduced food intake and increased probability of predation by, for example, the bird definitive host. An intense immune response comprising both humoral and cellular components has been noted in some fish, e.g. rainbow trout, prior to the parasite infecting the somewhat immunologically privileged eye.

Detection

General diagnosis of eye fluke infection is fairly simple at the macro level. As infection can be associated with cataract formation, external signs can be a cloudiness of the eye lens. Note, however, that some nutritional disorders can induce a similar pathological response and it is essential that the presence of the parasite be ascertained. This is achieved by removing the complete eye and placing it in a suitable solution, e.g. teleost ringer. The eye orbit can be dissected using a sharp scalpel and the vitreous humor examined at low magnification. To observe the fish lens, it is suggested that it be removed from the eye. The unencysted metacercaria can be observed within the lens and usually has a flexing motion. Both eyes should be examined and the number of parasites located can vary, with large numbers (e.g. 550) having been recorded. To view the pathology, the whole eye can be processed using conventional histological techniques and sections obtained stained with haematoxylin and eosin.

The absence of any clearly defined reproductive organs in metacercariae has proven a challenge for diagnosis to species level and is probably the reason why several authorities diagnose to genus level. Recently there has been extensive use of molecular biological techniques to aid in diagnosis to species level; for example, the complete sequences of the nicotinamide adenine dinucleotide dehydrogenase subunit 3 (nad3) mitochondrial gene, distance-based analysis of cytochrome c oxidase one barcodes and, in some specimens, internal transcribed spacer (ITS-1, 5.8S, ITS-2) sequences, and molecular sequencing of the partial mitochondrial *cox*1 and ribosomal 28S rDNA genes, as well as of ribosomal ITS1-5.8S-ITS2 region.

FACTS SHEET

Disease: Diplostomiasis (diplostomastosis) and tylodelphosis
Pathogen: Primary pathogen *Diplostomum spathaceum*, but others exist, such as *D. rutili*, *D. helveticum*, *D. volvens*. Also *Tylodelphys clavata*
Susceptible hosts: A range of cyprinid hosts, including *Cyprinus carpio*, *Carassius gibelio*, *Carassius carassius*, *Ctenopharyngodon idella*
Mortality: Rare and primarily associated with heavy infections. Chronic debilitating effects
Temperature: Important parameter as affects intermediate host population. Optimum temperature 17°C in temperate regions
Prevention and treatment: Good husbandry practices including reduction of intermediate hosts. Application of anthelminthic can be used but may not be cost-effective. Various mechanical techniques to disrupt life cycle

11.3.3 The encysted flukes

Epidemiology and host range

Considered by some as a minor group of trematode infections of fish are the metacercariae, which become encysted on penetration of the fish host, primarily due to the host reaction against the invading parasite. This response can be stimulated by a range of trematode infections, but in the cyprinids, including common carp and grass carp, *Posthodiplostomum cuticola* (syn. *Neasus cuticola*) (Fig. 11.8), *Centrocestus formosanus* (Fig. 11.9), *Petasiger* spp. and *Apharyngostrigea cornu* predominate.

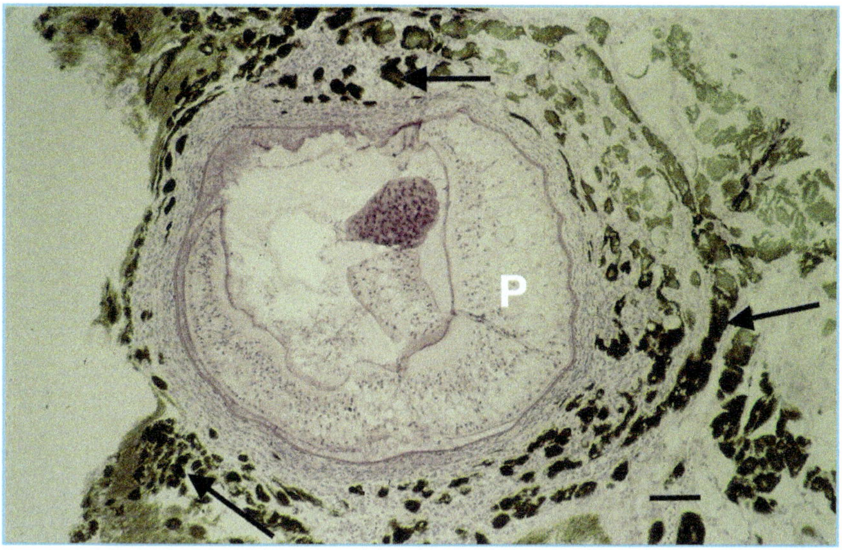

Figure 11.8 Histological section through the skin of a carp showing infection with *Posthodiplostomum* (P). Note the presence of extensive melanisation (arrows) of infection site giving the typical 'black spot'. Indicative scale bar 100 μm.

Posthodiplostomum cuticola is perhaps the most widely distributed and has been located in carps (e.g. *Cyprinus carpio*, *Carassius gibelio*) and in ornamental fish such as goldfish. In the Indian subcontinent it has also been found in rohu (*Labeo rohita*). Parasite cysts are located on the general body surface and fins where infection stimulates the formation of cysts characterized by the accumulation of melanin. This gives rise to the common name 'blackspot' for the infection. Fish become infected by the penetration of cercariae released from a molluscan first intermediate host, for example *Planorbis* spp. In contrast, *Apharyngostrigea cornu* encysts within the internal organs of the fish host; and in common carp, crucian carp and grass carp these cysts are primarily located on the peritoneum and mesenteries. Although infections can occur at any time of the year, they usually occur with rising water temperatures and predominate in situations where infection of the fish can be maximised by the cercariae, such as warm springs and average pond water levels.

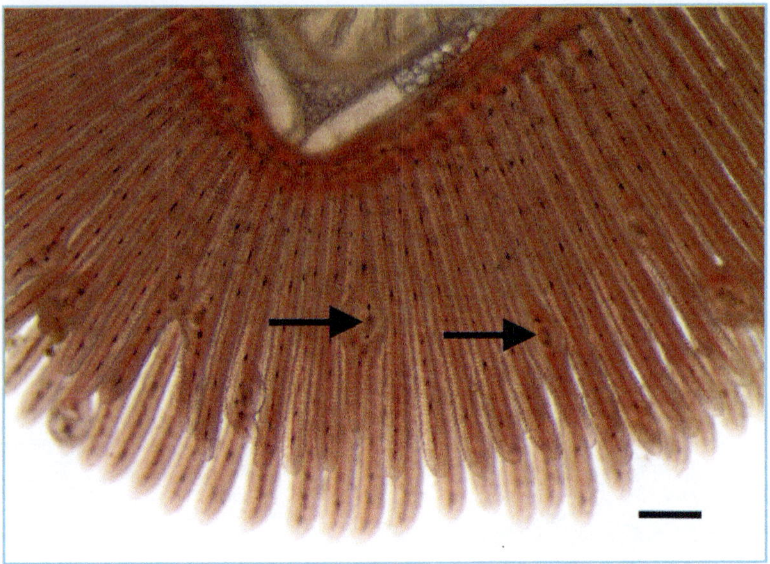

Figure 11.9 *Centrocestus formosanus* cysts within the gills of grass carp (arrow). Indicative scale bar 1 mm. Courtesy of the Environment Agency, National Fisheries Laboratory, UK.

Pathology

The development of encysted flukes can be associated with several pathological responses. This is perhaps most noticeable in the case of 'black spot' (Fig. 11.8), which is temperature dependent and usually takes approximately 33 days at a temperature of 22°C to develop. These cysts, which can be up to 1 mm in diameter, can occur anywhere on the fish's surface, but are particularly hazardous on the gills, where respiratory efficiency can be affected. Cysts formed on internal organs by infection with *A. cornu* can be particularly problematic. They are variable in size (e.g. 390 × 920 µm), can coalesce to form large lumps and can cause severe haemorrhaging and even death.

Detection

Diagnosis of both cyst infections is usually confirmed by macroscopical examination. Black spot is usually very easy to locate and, like the cysts of *A. cornu*, the presence of the parasite can be confirmed by routine histological processing and examination. Recently molecular characterisation and the application of MALDI–TOF have been used, but such studies are in their infancy as a diagnostic technique.

FACTS SHEET

Disease: Encysted flukes
Pathogen: *Posthodiplostomum cuticola* and *Apharyngostrigea cornu* predominate
Susceptible hosts: Range of cyprinids such as *Carassius gibelio*, *Carassius auratus*, *Labeo rohita*
Mortality: Severe haemorrhaging may cause death
Temperature: Disease tends to be more prevalent at higher temperatures
Prevention and treatment: Good husbandry practices, including reduction of intermediate hosts. Application of anthelminthic can be used but may not be cost-effective

11.3.4 Prevention strategies and treatments of trematodes

As trematodes have a complex life cycle requiring several hosts, the main route of infection occurs by the movement of fish, snails or water. Whilst the transmission via water and snails can be managed by good hygiene practices, for example the use of disinfectants to treat waders and nets, the shipment of fish can prove to be more problematic. The most effective control is to obtain fish from disease-free stock, although anthelmintics, such as praziquantel, can be applied. For example, praziquantel (Droncit) fed at a dose of 330 mg/kg body mass killed metacercariae of *Diplostomum spathaceum* infecting the eye lens of grass carp, and parasite death was achieved when infected grass carp and silver carp (*Hypophthalmichthys molitrix*) were exposed to a solution containing 1 mg praziquantel/litre for at least 90 hour.

The presence of other hosts besides the carp in the life cycle does open up the opportunity to break the life cycle by targeting the bird or gastropod hosts. Whilst most authorities agree it is difficult to completely eliminate birds from carp ponds, several techniques have been attempted such as bird-scaring devices, covering ponds with netting or shooting birds. A snail intermediate host in the case of sanguinicoliasis also means that a range of molluscicides, such as copper sulphate, has been applied with some success. However, whilst this could be considered in some closed ornamental tank systems, it is not recommended in larger extensive open systems where severe environmental impacts can arise and toxicity to the host can occur.

In addition, a range of physical applications has been suggested as a possible mechanism to reduce the levels of cercarial infections in water. These include the use of electrical current where plates set in water inflows 2.8 cm apart with a potential drop of 220–310 V are effective in killing cercariae. The use of ultrasonics at 3 s duration at frequencies of 50–100 kHz have been noted to destroy eggs, miracidia and cercariae. However, such physical applications are fraught with danger to the farm worker and pond life.

11.4 Selected References for Further Reading

General fluke parasitology

Bullard, S.A. and Overstreet, R.M. (2008) Digeneans as enemies of fishes. In: Eiras, J.C., Segner, H., Wahli, T. and Kapoor, B.G. (eds) *Fish Diseases*, Vol. 2. Science Publishers, Enfield, New Hampshire.

Hoole, D., Bucke, D., Burgess, P. and Wellby, I. (2001) *Diseases of Carp and Other Cyprinid Fish*. Blackwell Science, Oxford, UK.

Karvonen, A., Bagge, A.M. and Valtonen, E.T. (2005) Parasite assemblages of crucian carp (*Carassius carassius*) – is depauperate composition explained by lack of parasite exchange extreme environmental conditions or host unsuitability? *Parasitology* 131, 273–278.

Schäperclaus, W. (1991) *Fish Diseases*, 5th edn, Vol. 2. Amerind Publishing Co./Oxonian Press, New Delhi, India.

Whittington, I.D. and Chisholm, L. (2008) Diseases caused by Monogena. In: Eiras, J.C., Segner, H., Wahli, T. and Kapoo, B.G. (eds) *Fish Diseases*, Vol. 2. Science Publishers, Enfield, New Hampshire.

Monogenea

Ahmadi, A., Borji, H., Naghibi, A., Nasiri, M.R. and Sharifiyazdi, H. (2017) Morphologic and molecular (28S rDNA) characterisation of

Dactylogyrus spp. in *Cyprinus carpio* and *Ctenopharyngodon idella* in Mashhad, Iran. *Canadian Journal of Veterinary Research* 81, 280–284.

Chaudhary, A., Chiary, H.R. and Singh, H.S. (2017) First molecular confirmation of the *Dactylogyrus anchoratus* and *D. vastator* (Monogenea, Dactylogyidae) from *Carassius auratus* in Western India. *Bioinvasions Records* 6, 79–85.

Ling, F., Tu, X., Huang, A. and Wang, G. (2016) Morphometric and molecular characterisation of *Dactylogyrus vastator* and *D. intermedius* in goldfish (*Carassius auratus*). *Parasitology Research* 115, 1755–1765.

Lu, C., Ling, F., Ji, J., Kang, Y.-J. and Wang, G.-X. (2013) Expression of immune-related genes in goldfish gills induced by *Dactylogyrus intermedius* infections. *Fish and Shellfish Immunology* 34, 372–377.

Tu, X., Ling, F., Huang, A. and Wang, G. (2015) An infection of *Gyrodactylus kobayashi* Hukuda, 1940 (Monogenea) associated with the mortality of goldfish (*Carassius auratus*) from central China. *Parasitology Research* 114, 737–745.

Tu, X., Liu, L., Qi, X., Chen, W., Wang, G. and Ling, F. (2016) Characterization of Toll-like receptor gene expression in goldfish (*Carassius auratus*) during *Dactylogyrus intermedius* infection. *Developmental and Comparative Immunology* 63, 78–83.

Tu, X., Liu, L., Qi, X., Huang, A., Ling, F. *et al.* (2019) Cytokine gene expression profiles in goldfish (*Carassius auratus*) during *Gyrodactylus kobayashii* infection. *Fish and Shellfish Immunology* 86, 116–124.

Zhang, D., Zou, H., Wu, S.G., Li, M., Jakovlic, I. *et al.* (2018) Sequencing, characterization and phylogenomics of the complete mitochondrial genome of *Dactylogyrus lamellatus* (Monogena: Dactylogyridae). *Journal of Helminthology* 92, 455–466.

Trematodes

Chappell, L.H., Hardie, L.J. and Secombes, C.J. (1994) Diplostomiasis: The disease and host-parasite interactions. In: Pike, A.E. and Lewis, J.W. (eds) *Parasitic Diseases of Fish*. Samara Publishing, Cardigan, UK, pp. 59–86.

Enabulele, E.E., Awharitoma, A.O., Lawton, S.P. and Kirk, R.S. (2018) First molecular identification of an agent of diplostomiasis, *Diplostomum pseudospathaceu* (Niewiadomska 1984) in the United Kingdom and its genetic relationship with populations in Europe. *Acta Parasitologica* 63, 444–453.

Huguenin, A., Depaquit, J., Villena, I. and Ferte, H. (2019) MALDI–TOF mass spectrometry: A new tool for rapid identification of cercariae (Trematoda, Digenea). *Parasite* 26, 11.

Iqbal, Z., Shukerova, S.A. and Minhas, I.K. (2014) Occurrence of black spot disease in *Labeo rohita* (Hamilton) fry in carp fish hatchery Lahore, Pakistan. *Canadian Journal of Pure and Applied Sciences* 8, 2727–2731.

Kirk, R.S. and Lewis, J.W. (1994) Sanginicolasis in cyprinid fish in the UK. In: Pike, A.E. and Lewis, J.W. (eds) *Parasitic Diseases of Fish.* Samara Publishing, Cardigan, UK, pp. 101–118.

Locke, S.A., Al-Nasiri, F.S., Caffara, M., Drago, F., Kalbe, M. *et al.* (2015) Diversity, specificity and speciation in larval Diplostomidae (Platyhelminthes: Digenea) in the eyes of freshwater fish, as revealed by DNA barcodes. *International Journal for Parasitology* 45, 841–855.

Chapter 12

The Tapeworms (Cestoda)

12.1 Introduction

The usually large size of the tapeworms that infect fish invariably results in carp farmers, aquarists and anglers fearing the worst for the fish stocks. Whilst this is not always the case, there are several cestodes that infect carp whose presence can have some significant effects on individual fish, the fish population and economic profitability of the production unit. There are approximately 800 tapeworm species occurring in fish, classified into 12 orders. Several of these orders occur in marine fish, although it is the Pseudophyllidea, Proteocephalidea and Caryophyllidea that predominate in freshwater fish. Whilst fish can serve as definitive and intermediate host in cestode life cycles, carp/cestode life cycles usually comprise two or three hosts, with the adult tapeworm being located in the gut of the fish definitive host, and juvenile stages occurring in the one or two intermediate hosts. This complex life cycle strategy does have implications for control mechanisms. It means that intermediate hosts can be targeted, but the presence of infected intermediate hosts in a pond environment may mean that eradication of the life cycle in a fish farm can be difficult and any introduced non-infected fish can be rapidly incorporated into the parasite life cycle.

The adult stage of the tapeworm comprises a fairly standard body plan consisting of a scolex, or 'head', and a long tape structure. The scolex consists of several types of adhesive structures such as suckers, tentacles and hooks by which the cestode attaches to the gut wall of the definitive host. The tape comprises a series of segments, the proglottids, which contain the male and female reproductive organs, tapeworms being hermaphrodite. Tapeworms lack an intestine, and

 Carps: A Practical Approach to Disease Control and Health Management (eds David Hoole *et al.*)
DOI: 10.1079/9781836990024.0012

nutrients and other requirements are taken in over the surface, the tegument, either by diffusion or by a series of active processes. Some tapeworms, such as the caryophyllideans that occur in cyprinid fish, lack a segmented body, and are therefore easily distinguished by their monozoic form. It is the body form of the adult worm that has been used to distinguish different orders, genera and species, although more recently genetic characterisation has been adopted.

Fish can serve as both definitive and intermediate host, in the latter of which the juvenile stages of the parasite, the metacestode, occur. It is when the carp acts as the definitive host that severe problems can occur, highlighted by two intestinal tapeworms: the Asian carp tapeworm (now known as *Schyzocotyle acheilognathi*) and the *Khawia* spp. tapeworms that cause the disease khawiosis.

12.2 The Asian Carp Tapeworm

12.2.1 Pathogen

This parasite is indigenous to the Far East, and is thought to originate from China in the southern area and the Amur River. The worm is notorious for its rapid colonisation capacity and now occurs in many countries around the world. The expansion of the geographical range is thought to be expedited by the commercial exploitation and movement of fish, particularly ornamental and farmed carp. This fantastic coloniser has also expanded its host range and can be found in over 40 species of cyprinid fish and in other orders such as the Cyprinidontiformes, Atheriniformes, Osmeriformes, Perciformes, Acipenseriformes and Siluriformes. In the USA it has been recognised as a threat to several endangered fish species. The rapid expansion of the Asian carp tapeworm and its presence in a range of fish species make it a serious threat to carp production worldwide.

The Asian carp tapeworm is now known as *Schyzocotyle acheilognathi*, although in the past the worm has been referred to as *Bothriocephalus acheilognathi*, *B. gowkonongensis*, *B. opsariichthydis* and *B. phoxini*.

The worm can grow up to 320 mm in length and is approximately 4 mm in width. The scolex has a characteristic viper-shaped form, comprising two long grooves, the bothria (Fig. 12.1).

The segments mature towards the posterior aspect of the tape and contain eggs, approximately 0.034 mm × 0.048 mm, and appear in various shades of amber colour. Eggs are released in the faeces of the fish and after a period of development, which is determined by environmental temperature, the free-living coracidium is released via the operculum. A copepod serves as the intermediate host and the species that predominates in the life cycle varies between water-bodies and geographical location. Species such as *Cyclops abyssorum*, *Thermocyclops taihokuensis*, *Ectocyclops phateratus* and *Mesocyclops leuckarti* have been recorded as being infected. A procercoid stage

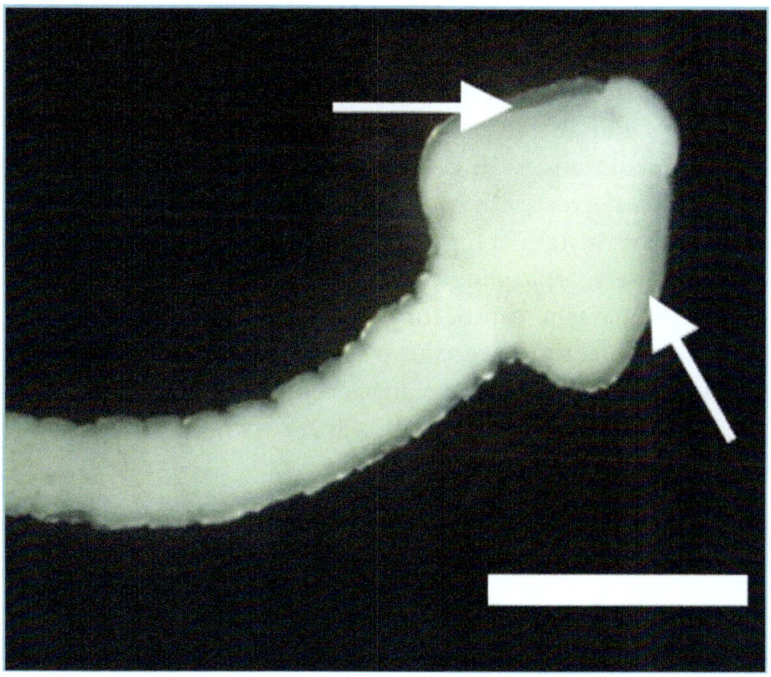

Figure 12.1 Whole preparation of *Schyzocotyle acheilognathi* showing characteristic 'viper-shaped' scolex with bothria as two grooves (arrows) at extremities of the scolex. Indicative scale bar 1 mm. Courtesy of the Environment Agency, National Fisheries Laboratory, UK.

develops in the haemocoel of the crustacean, and fish become infected by eating the infected intermediate host.

12.2.2 Epidemiology and host range

The adult parasite has been located in a range of carp species, for example common carp (*Cyprinus carpio*), grass or Chinese carp (*Ctenopharyngodon idella*), crucian carp (*Carassius carassius*) and ornamental fish, such as goldfish (*Carassius auratus*) and koi carp. The ubiquitous nature of this parasite in wild, cultured and ornamental fish means that it has been documented in almost every continent, except Antarctica.

The transmission of infection within a waterbody is affected by a range of environmental factors, such as water quality and temperature. Since the eggs of the parasite are sensitive to desiccation and can only survive for approximately 2 hour when frozen, in temperate climates the parasite is thought to overwinter in the fish host. In such areas of the world (e.g. Eastern Europe) with rising temperatures and an increase in numbers of the intermediate host, incidence and intensity in 0 + carp increases towards midsummer, followed by a decrease towards autumn. However, in areas of the world that have sustained high temperatures, egg production occurs throughout the year, leading to a less seasonal pattern of infection. Although the parasite can occur in all ages of carp, reports tend to indicate that infection is greatest in 0 + and 1 + fish, where it can cause severe mortality and substantial economic loss. The tapeworm is not gender specific, although in a few surveys females have been noted to be more parasitised.

12.2.3 Pathology

Symptoms of infection can include emaciation, sluggish movements and the tendency for fish to congregate at the surface of the water. Although adult parasites can be found throughout the gut of the host fish, they tend to be primarily located at the anterior part of the intestine, posterior to the opening of the bile duct (Fig. 12.2). At large intensities and in small fish, infection can cause blockage of the

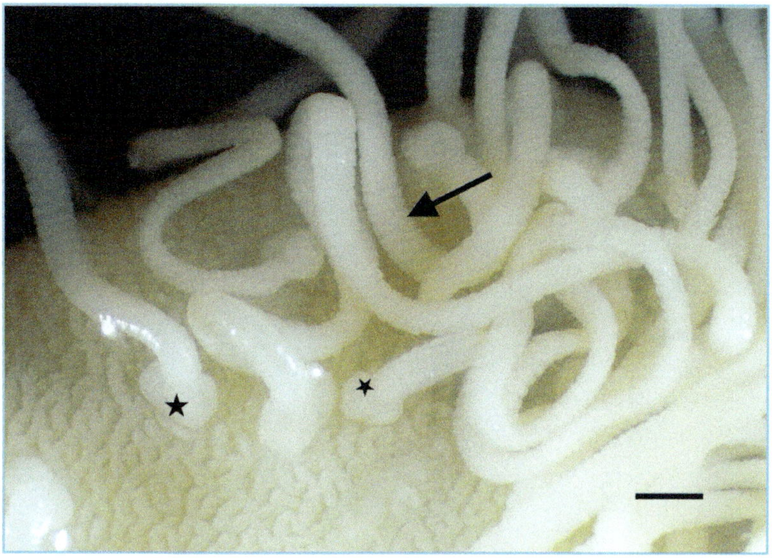

Figure 12.2 Several *Schyzocotyle acheilognathi* attached to the gut wall of carp via the characteristic 'viper-shaped' scolex (stars). Note the presence of a segmented body structure (arrow). Indicative scale bar 1 mm. Courtesy of the Environment Agency, National Fisheries Laboratory, UK.

gut and, on some occasions, rupture of the intestinal wall. Infection can thus lead to an enlargement of the body ventral to the dorsal musculature and posterior to the gills. The parasite remains within the gut by attaching to the gut wall by means of its bothria.

These attachment organs engulf the intestinal folds and induce local inflammatory responses leading to separation of the epithelial cells, loss of the gut epithelium and occasionally necrosis (Fig. 12.3). Such an intense local response is also linked with the migration and accumulation of host leucocytes at the site of infection, and the induction of a range of systemic humoral and cellular immune responses. Whilst the immune response can be quite intense, the presence of large intensities, which are thought to occur over an extensive period of exposure, have led to suggestions that, on some occasions, the immune response may not be protective, although there is also evidence that the genotypes of individual fish may be important in contributing to the intensity of the response to infection. This systemic activity is not limited to the alteration of

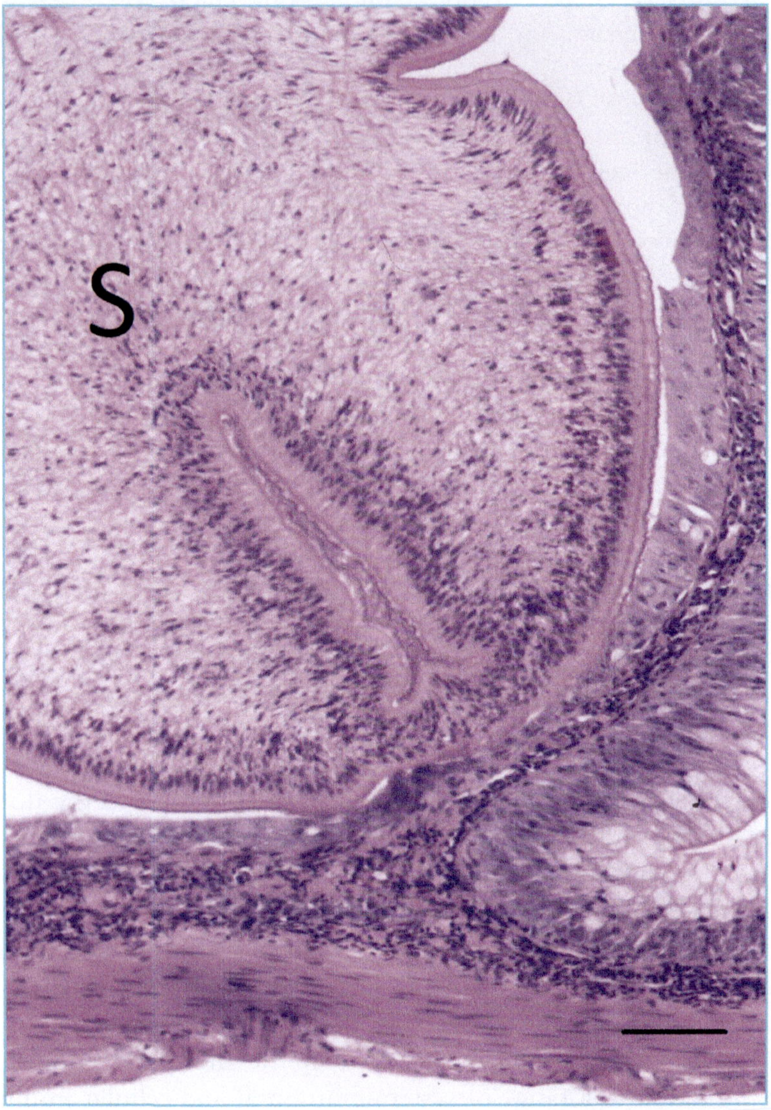

Figure 12.3 Light-microscopic, haematoxylin and eosin-stained image showing scolex (S) of *Schyzocotyle acheilognathi* attached to the gut wall of a carp. Indicative scale bar 30 μm. Courtesy of the Environment Agency, National Fisheries Laboratory, UK.

immune status, as infection can induce a range of changes within the internal organs of the fish. These include, for example, decreases in organ weight and changes in organ enzyme activity.

12.2.4 Detection

Whilst external signs of infection are an indication of ill health, the non-specific nature of the symptoms means that infection can only be confirmed by dissection of the fish. This is carried out by removal of the entire intestine and placing the organ within a suitable solution, e.g. teleost ringer. The gut is carefully opened, preferably using blunt-nose scissors or scalpel, and the gut contents are gently removed. The tapeworm can be initially characterised by the form of the scolex, but to undertake this the parasite must be removed from the gut wall. Care must be taken as the tape of the parasite can be easily separated from the embedded scolex, and extraction of the scolex is best achieved by removing the host tissue from around the attachment site. Preparation of standard whole mounts can be undertaken with, for example, fixation in formalin and treatment with a carmine-based stain. The characteristic scolex structure and internal segment body form can assist in diagnosis. The widening and extensive host and geographical range of the *Schyzocotyle* has stimulated analysis of genetic diversity and diagnostic techniques. A range of techniques have been applied, such as sequencing of ITS2 and V4-18S rRNA genes, RAPD marker-related electrophoretic patterns, and the analysis of mitochondrial genomes and nuclear rRNA operons using an Illumina sequencing platform.

12.2.5 Prevention and treatment

The sensitivity of the intermediate host and tapeworm eggs to environmental extremes means that management strategies can be utilised to reduce or eliminate infection. Ponds can be drained dry and disinfected with lime, whilst in areas of the world that experience seasonal cold temperature, exposure to winter temperatures and freezing can increase the success of treatment. Prevention is always recommended, and infection control can be achieved by slaughter of a representative sample of proposed imported fish, perhaps in compliance with local legislative requirements.

Treatment of infected fish is achievable using a range of anthelminthics, but the cost-effectiveness of such an approach needs to be ascertained, as indeed should any local restrictions of such chemotherapy. A range of treatments have been utilised. For example, in crucian carp, mebendazole at 0.042 mg/g body weight was found to be effective, and for 2-year-old grass carp some success has been obtained using praziquantel and nitroscanate treatment at 30 mg/kg and 2 mg/kg, respectively. Medicated feed has also been utilised, such as a 2% Microsal feed, but multiple treatments may be required.

FACTS SHEET

Disease: Bothriocephalosis
Pathogen: *Schyzocotyle acheilognathi* previously *Bothriocephalus acheilognathi*, *B. gowkonongensis*, *B. opsariichthydis*, *B. phoxini*
Susceptible hosts: Many carp species, including *Cyprinus carpio*, *Carassius auratus*, *Carassius carassius*, *Ctenopharyngodon idella*,
Mortality: Mortalities can be high in heavy infections, particularly in young fish
Temperature: Parasites intensities can be greater with increasing temperature
Prevention and treatment: Infection can be reduced or prevented by good husbandry practices and avoiding introduction of contaminated fish stock or water. Reduction of intermediate hosts. Treatments such as the application of anthelminthic have varying success.

12.3 Khawiosis

12.3.1 Pathogen

One of the most prominent orders of tapeworm found in fish is the order Caryophyllidea, which comprises a monozoic body form and only one set of genital organs. Although there is much controversy regarding the classification within this order, it is generally agreed that members of the genus *Khawia* are important intestinal tapeworms of cyprinids, and the type species *Khawia sinensis* was first located in common carp in the Amur river basin in China. The adult tapeworms of this genus are unsegmented and have a typically clover-shaped scolex, which merges with the rest of the parasite

Figure 12.4 Several *Khawia sinensis* attached to the gut wall of carp. Note the scolex (arrow) by which the parasite attaches loosely to the gut wall and the presence of a monozoic body form. Indicative scale bar 1 mm. Courtesy of the Environment Agency, National Fisheries Laboratory, UK.

body and thus lacks a well-defined neck region (Fig. 12.4). The life cycle comprises two hosts: a definitive fish host in which the adult tapeworm resides and an intermediate host, usually an annelid, which contains the procercoid stage within the host haemocoel. Transmission occurs when the fish consumes the infected annelid. A range of annelid species such as *Tubifex tubifex*, *Limnodrilus udekemianus*, *L. hoffmeisteri*, *Ilodrilus hammoniensis*, *I. templetoni* and *Euilyodrilus hammoniensis* have been noted as suitable hosts. The parasites have an operculated egg, which is approximately 0.04×0.03 mm in size, and some species (e.g. *K. sinensis*) have some characteristic architecture, such as a knob at one end. Eggs are liberated in the fish's faeces.

12.3.2 Epidemiology and host range

Speciation within the genus is complex and has led to much debate. It is generally accepted that one of the most pathological species is *Khawia sinensis*, which has a somewhat limited definitive host range, being primarily located in common carp, although it has occasionally been found in grass carp and goldfish. In addition, *K. japonensis* has been recorded from common carp in Japan. Both *Khawia* species have extended their geographical range, and in addition several other species of *Khawia* (e.g. *K. rossittensis*, *K. parva*, *K. iowensis*) have been noted in a range of carp species. However, it is now believed that many of these *Khawia* species (e.g. *K. iowensis*) are in fact synonymous with *K. japonensis*. The majority of the studies have been carried out on *K. sinensis*, primarily because of its important effects on carp production. Transmission within a carp population is affected by environmental temperature, which affects both the parasite development and the number of intermediate hosts. Embryonic development of the egg takes about 16 days at 24°C, whilst at cold temperatures, such as 10°C, a period of 57 days is required. Temperature also affects the adult worm and at 20°C the parasite can grow at an astonishing 5 mm a day. The adult parasite is not usually long lived, surviving approximately 40 days, although some authorities suggest that the parasite can overwinter in the fish for many months. There is much controversy on the seasonal dynamics of the disease in carp and it appears to be dependent on geographical location and farming practices. In Europe a peak has been located in breeding ponds with one or two peaks occurring in carp growing-on waterbodies. In some areas of the world high incidences occur in May with a second peak occurring in autumn.

12.3.3 Pathology

The adult parasite is usually located in the anterior portion of the intestine where a somewhat loose attachment to the gut wall occurs. Infection often causes loss of appetite and reduced growth, and infected carp have an emaciated and anaemic appearance, and exhibit sluggish movements. The gut can become blocked by parasite material and this can be so severe that there are records of the tapeworm protruding from the anus. It has been suggested that as few as three parasites can cause death in carp fry.

Whilst attachment to the gut is relatively weak, compared with other tapeworm species, the extensive clover-shaped scolex has been noted to induce lesions and inflammation, with mucosal proliferation, blood loss and an increase in mucus production. The parasite has also been noted to affect a range of systemic organs, such as the kidneys, spleen and liver where metabolism and enzyme activity is affected. Alterations have also been noted in a range of blood parameters, such as serum protein profiles, as well as leucocyte and erythrocytic content. The parasite has been associated with the generation of a range of immune cellular and humoral parameters in carp, but it is not known if these are protective.

12.3.4 Detection

The somewhat characteristic nature of the scolex and general body plan of the adult parasite aids in diagnosis, which requires the extraction of the worm from the fish's gut. This can be achieved, as with the Asian carp tapeworm, with the exception that the scolex is somewhat easier to detach from the gut wall. Whole mounts of the parasite can be prepared using standard techniques, although the recommended fixation method for fish tapeworms involves hot 4% neutral buffered formaldehyde solution, which prevents contraction of the specimen. Parasites can be stained in a range of standard stains, including Mayer's hydrochloric carmine solution. The monomeric body form, scolex morphology and the location of the vitellaria in the cortex and testes in the medulla usually make identification fairly easy. Recently there have been extensive molecular studies on *Khawia* spp., and Scholz (2011) drew up a reclassification of the genus using nuclear ribosomal (complete small subunit, ssrDNA, 1571–1655 bp, and partial large subunit, D1–D3 lsrDNA, 2100–2104 bp) and mitochondrial (complete nicotinamide dehydrogenase subunit iii, *nad*3, 345 bp, and partial cytochrome *c* oxidase subunit I, *cox*1, 672–675 bp) genes. Other studies have utilised the sequence structure of the ribosomal internal transcribed spacer 2 (ITS2).

12.3.5 Prevention and treatment

Control of khawiosis is usually targeted to the prevention of the introduction of infection and usually entails rigorous testing of fish prior to transfer and, if necessary, the application of anthelminthic treatments. Chemotherapy can also be applied to existing infections in ponds, but an assessment of the economic profitability is recommended. Several anthelmintics have been trialled with some success, for example medicated feed mixture (granules) with 1% fenbendazole, and praziquantel mixed in heat-treated amyloid vehicle (at doses of 50 mg/kg body weight). There is a range of management strategies that can be applied to reduce the effects of the parasite. These include segregation of carp fry from adults, prevention of overcrowding and controlling water quality. The dependence of an intermediate host means that control measures can be targeted at this stage of the parasite life cycle. Ponds can be drained to at least 12% moisture content and treated with lime or left to freeze.

FACTS SHEET

Disease: Khawiosis
Pathogen: *Khawia sinensis, K. japonensis, K. rossittensis, K. parva, K. iowensi*
Susceptible hosts: Many carp species, such as *Cyprinus carpio, Ctenopharyngodon idella*
Mortality: Mortalities can be high in heavy infections, particularly in young fish
Temperature: Parasite intensities can be greater with increasing temperature
Prevention and treatment: Infection can be reduced or prevented by good husbandry practices and avoiding introduction of contaminated fish stock or water. Reduction of intermediate hosts. Treatments such as the application of anthelminthic have varying success

12.4 *Atractolytocestus* Spp.

Whilst khawosis is still recognised as a problematic disease in carp culture, recent documentary evidence suggests that in Europe over the past decades *K. sinensis* prevalence may be decreasing. There is speculation on why this may be the case but over a similar period there has been an increase in another caryophyllidean tapeworm, *Atractolytocestus huronensis*. The parasite was originally described from common carp from North America, and has recently been introduced into Europe, China and South Africa. The parasite has two sister species, *A. sagittatus* and *A. tenuicollis*, which also occur in carp. *A. huronensis* is a relatively small tapeworm, 25 mm × 3–7 mm, which is unsegmented and has a bulbous scolex with a somewhat pointed anterior and a neck region (Fig. 12.5). The form of the scolex region can be used to distinguish the parasite from the other two *Atractolytocestus* species, although recently molecular characterisation has taken place using ribosomal internal transcribed spacer 2 (ITS2) and partial mitochondrial cytochrome c oxidase subunit I (cox1).

The parasite can be located throughout the carp gut, but on occasions has been found restricted to the proximal section of the intestine (Fig. 12.6). The attachment of the scolex can induce a range of pathological changes, such as atrophy and disruption of the intestinal epithelium, displacement of the mucosa, hyperplasia and haemorrhaging (Fig. 12.7).

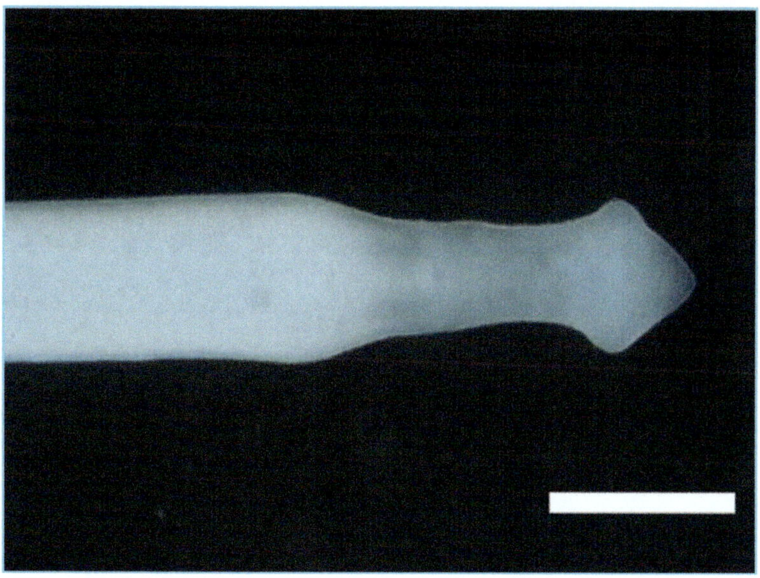

Figure 12.5 Whole preparation of *Atractolytocestus huronensis* showing characteristic scolex with pointed anterior and neck region. Indicative scale bar 100 μm. Courtesy of the Environment Agency, National Fisheries Laboratory, UK.

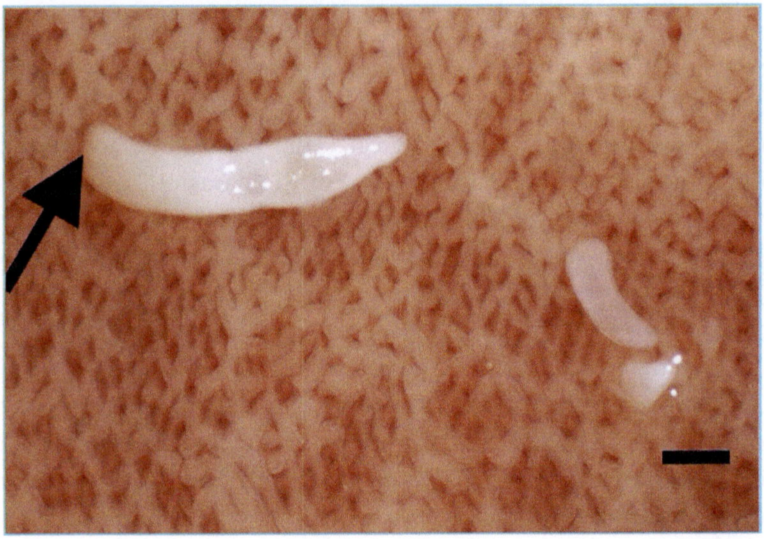

Figure 12.6 Three *Atractolytocestus huronensis* attached to the gut wall of carp. Note that the scolex is deeply embedded within the gut wall (arrow). Indicative scale bar 1 mm. Courtesy of the Environment Agency, National Fisheries Laboratory, UK.

FACTS SHEET

Disease: No common name for disease
Pathogen: *Atractolytocestus huronensis*
Susceptible hosts: Many carp species thought to be host
Mortality: Increasing in prominence and mortalities can be high in heavy infections, particularly in young fish
Temperature: Not clearly established but parasite intensities thought to be greater with increasing temperature
Prevention and treatment: Infection can be reduced or prevented by good husbandry practices and avoiding introduction of contaminated fish stock or water. Reduction of intermediate hosts

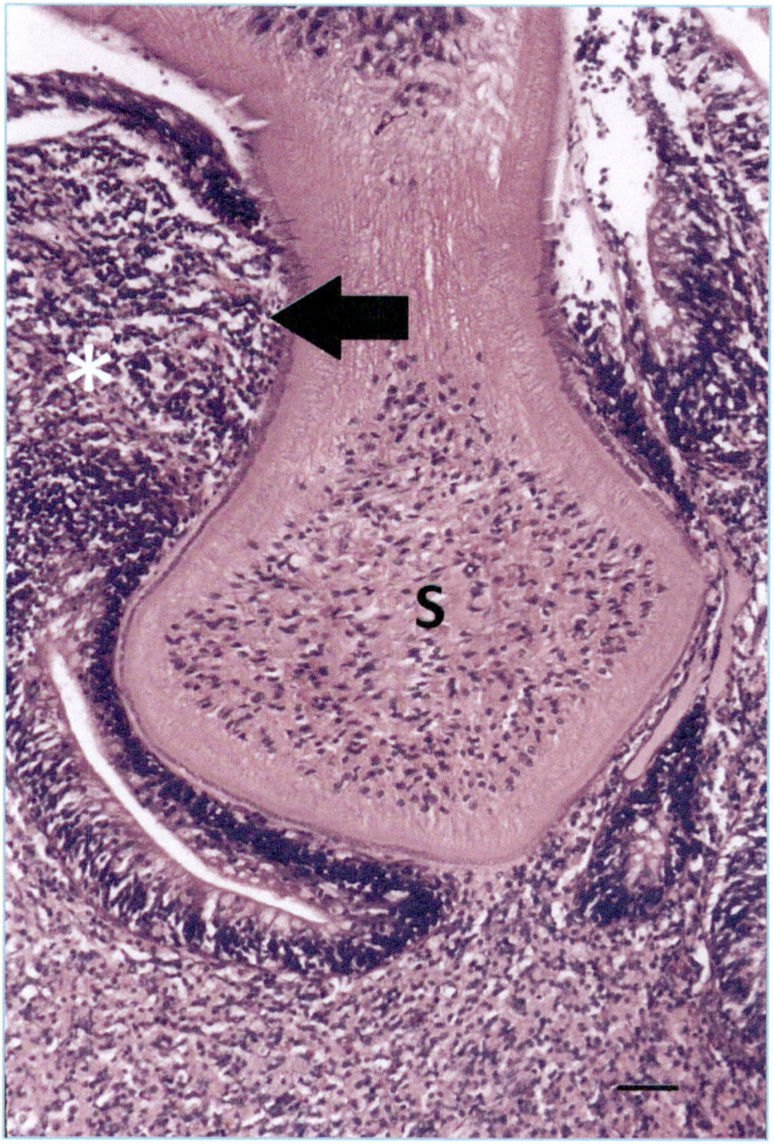

Figure 12.7 Light microscopic, haematoxylin and eosin-stained image showing scolex (S) of *Atractolytocestus huronensis* embedded in gut wall of carp. Note infiltration of mononuclear cells (*) and disruption of epithelium (arrow). Scale bar 30 μm. Courtesy of the Environment Agency, National Fisheries Laboratory, UK.

12.5 Selected References for Further Reading

General cestode biology

Borucinska, J.D. (2008) Diseases caused by Cestoda. In: Eiras, J.C., Segner, H., Wahli, T. and Kapoor, B.G. (eds) *Fish Diseases*, Vol. 2. Science Publishers, Enfield, New Hampshire.

Hoole, D., Bucke, D., Burgess, P. and Wellby, I. (2001) *Diseases of Carp and Other Cyprinid Fish*. Blackwell Science, Oxford, UK.

Schäperclaus, W. (1991) *Fish Diseases*, 5th edn, Vol. 2. Amerind Publishing Co./Oxonian Press, New Delhi.

Trujillo-Gonzalez, A., Becker, J.A. and Hutson, K.S. (2018) Parasite dispersal from the ornamental goldfish trade. *Advances in Parasitology* 100, 239–281.

Bothriocephalosis

Ahmad, F., Fazili, K.M., Sofi, O.M., Sheikh, B.A. and Sofi, T.A. (2018) Distribution and pathology caused by *Bothriocephalus acheilognathi* Yamaguti, 1934 (Cestoda: Bothriocephalidae): A review. *Revista Veterinaria – Corrientes* 29, 142–149.

Brabec, J., Kuchta, R., Scholz, T. and Littlewood, D.T.J. (2016) Paralogues of nuclear ribosomal genes conceal phylogenetic signals within the invasive Asian fish tapeworm lineage: Evidence from next generation sequencing data. *International Journal for Parasitology* 46, 555–562.

Brabec, J., Scholz, T. and Stefka, J. (2018) Development of polymorphic microsatellites for the invasive Asian fish tapeworm *Schyzocotyle acheilognathi*. *Parasitology International* 67, 341–343.

Britton, J.R., Pegg, J. and Williams, C.F. (2011) Pathological and ecological host consequences of infection by an introduced fish parasite. *PLoS ONE* 6, e26365.

Hansen, H. and Alarcon, M. (2019) First record of the Asian fish tapeworm *Schyzocotyle (Bothriocephalus) acheilognathi* (Yamaguti, 1934) in Scandinavia. *Bioinvasions Records* 8, 437–441.

Kline, S.J., Archdeacon, T.P. and Bonar, S.A. (2009) Effects of praziquantel on eggs of the Asian tapeworm *Bothriocephalus acheilognathi*. *North American Journal of Aquaculture* 71, 380–383.

Luo, H.Y., Nie, P., Zhang, Y.A., Wang, G.T. and Yao, W.J. (2002) Molecular variation of *Bothriocephalus acheliognathi* Yamaguti, 1934 (Cestoda: Pseudophyllidea) in different host species based on ITS rDNA sequences. *Systematic Parasitology* 52, 159–166.

Khawiosis

Kralova-Hromadova, I., Bazsalovicsova, E., Oros, M. and Scholz, T. (2012) Sequence structure and intragenomic variability of ribosomal ITS2 in monozoic tapeworms of the genus *Khawia* (Cestoda: Caryophyllidea): Parasites of cyprinid fish. *Parasitology Research* 111, 1621–1627.

Oros, M., Hanzelova, V. and Scholz, T. (2009) Tapeworm *Khawia sinensis*: Review of the introduction and subsequent decline of a pathogen of carp, *Cyprinus carpio*. *Veterinary Parasitology* 164, 217–222.

Scholz, T., Brabec, J., Kral'ova-Hromadova, I., Oros, M., Bazsalovicsova, E. *et al.* (2011) Revision of *Khawia* spp. (Cestoda: Caryophyllidea), parasites of cyprinid fish, including a key to their identification and molecular phylogeny. *Folia Parasitologica* 58, 197–223.

Atractolytocestus spp.

Bazsalovicsova, E., Kralova-Hromadova, I., Stefka, J., Scholz, T., Hanzelova, V. *et al.* (2011) Population study of *Atractolytocestus huronensis* (Cestoda: Caryophyllidea), an invasive parasite of common carp introduced to Europe: Mitochondrial cox1 haplotypes and intragenomic ribosomal ITS2 variants. *Parasitology Research* 109, 125–131.

Bazsalovicsova, E., Kralova-Hromadova, I., Xi, B.-W. and Stefka, J. (2018) Tour around the globe: The case of invasive tapeworm *Atractolytocestus huronensis* (Cestoda: Caryophyllidea), a parasite of common carp. *Parasitology International* 67, 366–374.

Molnar, K., Majoros, G., Csaba, G. and Szekely, C. (2003) Pathology of *Atractolytocestus huronensis* Anthony, 1958 (Cestoda, Caryophyllaeidae) in Hungarian pond-farmed common carp. *Acta Parasitologica* 48, 222–228.

Chapter 13

The Nematodes

13.1 Introduction

Nematodes, commonly known as roundworms, have a characteristic body plan and are typically elongate and cylindrical in shape, and lack body segmentation. They are sexually dimorphic, with an anterior mouth and posterior anus, and can occur in a variety of colours. Whilst some species are transparent, others can appear opaque or reddish in colour. The latter may be due to the presence of haemoglobin within the pseudocoelomic fluid. The mouth may be surrounded by a complex pattern of lips which, together with the internal body form, assist in macroscopic diagnosis. Whilst ubiquitous in nature and many species of nematodes are free-living, they are also common parasites in both freshwater and marine fish. At least 17 families have representatives located in fish, with five apparently restricted to these aquatic vertebrates. They have somewhat complex and varied life cycles, which comprise adult and larval forms. All nematodes have four larval stages, commonly referred to as L1, L2, L3 and L4. These develop through moults between each larval stage, with the L4 stage developing into the adult form. Fish can serve as definitive, intermediate and carrier hosts, i.e. paratenic hosts. Adult parasites can produce either eggs or L1 stage, depending on species. Egg survival time is governed by a range of parameters, e.g. temperature. Development through the larval stages to the L3 stage occurs, and at this stage genital primordia develop and the animal becomes the main larval parasitic stage.

Adults parasitise a range of organs in their fish definitive hosts, and have been found located in the gut, abdominal cavity, viscera, swim bladder and liver. Larval forms are particularly common in

 Carps: A Practical Approach to Disease Control and Health Management (eds David Hoole *et al.*)
DOI: 10.1079/9781836990024.0013

the viscera and musculature of fish, and tend to be more widely distributed than their adult counterparts, partly because the larval stages tend to be less host-specific than the adult stages. The potential to infect a wide range of fish hosts has resulted in the spread of several nematode species around the world and the colonisation of many waterbodies. For example, the pathogenic nematode of carp, *Philometroides cyprini*, which is native to East Asia, has spread widely in Europe over the past 60 years. A range of nematode parasites such as *Eustrongylides excisus*, *Pseudocapillaria tomentosa* and *Procamallanus* spp. have been found in carp, and in *Labeo* spp. several nematodes have been noted, such as: *Rhabdochona kharani* from the intestine, stomach and swim bladder of the Mashkal river labeo (*Labeo gedrosicus*); *Cucullanus khalili* and *C. gonii* from the intestine of rohu and kuria (*Labeo rohita* and *Labeo gonius*, respectively); and *Rhabdochana bilqeesae* found in the musculature of rohu (*Labeo rohita*). In carp, perhaps the most widely distributed and most pathogenic are those species belonging to the family Philometridae, i.e. genera *Philometra* and *Philometroides*, and the family Anisakidae, i.e. genus *Contracaecum*.

13.2 *Philometroides* Spp.

13.2.1 Pathogen

This is an extensive group of parasites, which in the past some authorities have described as the cause of the disease 'philometroidosis'. There are two important genera involved, i.e. *Philometra* and *Philometroides*, but there is much debate on the classification of species in both genera that occur in carp. For example, *Philometra lusii*, found under the scales of common carp in Lithuania, was re-described as *P. lusiana* and subsequently moved to the *Philometroides* genus. In addition, *Philometroides cyprini*, again found in carp, was thought by some authorities to be synonymous with *Philometroides lusiana*, although others consider them as distinct species. Whilst the use of molecular studies may clarify the speciation in this family, attention is drawn to the structural studies carried out by F. Moravec.

13.2.2 Epidemiology and host range

Notwithstanding the debate on speciation, a range of philometrid species have been found in carp. For example, *Philometroides cyprini* has been found in common carp (*Cyprinus carpio*) and Prussian carp (*Carassius gibelio*), along with *Philometroides lusiana*, and *P. cyprini* has also been located in crucian carp (*Carassius carassius*). Crucian carp also harbour *Philometroides sanguineus*.

Within the philometrid nematodes the genders are separate and, after copulation in the fish, there is migration of the sexes within the fish (Fig. 13.1). In *P. lusii* the male moves to the swim bladder, whilst the larger female leaves the body cavity and migrates into the scale pockets. Here there is extensive growth of the parasite, which can attain lengths of 110 mm. L1 stage larvae are then expelled into the waterbody. In the case of *P. sanguineus*, after copulation males tend to remain in the fish's body cavity, whilst the females congregate in the caudal and dorsal fins, from where L1 larvae are liberated. Philometrids have an intermediate host, which is usually a copepod. The species used can vary depending on parasite species; for example, *P. lusiana* infects *Acanthocyclops viridis*, *A. vernalis*, *Eucyclops macruroides*, *Cyclops strenuus* and *Mesocyclops leuckarti*. The last three species also serve as intermediate host for *P. sanguinea*, and for *P. carassii*, which also infects *Tropocyclops prasinus* and *Cyclops vicinus*, in addition to *C. strenuus*. The L2 stage develops in the intermediate host and transforms into the invasive L3 stage which, when the intermediate host is eaten by the carp, penetrates the gut wall and migrates to a range of organs.

Whilst parasites can be found in the fish host throughout the year, it is recognised that infection can be seasonal, dependent on geographical location, and the appearance of the adverse conditions of philometroidosis can also vary. Carp have been reported to become infected with *P. lusii* at the end of June, and females become visible usually at the end of June and beginning of July. After a period of growth on the fish that can last several months, L1 larvae are expelled at the end of May. The life cycle of *P. sanguineus* also requires a year to complete, with copulation occurring in the body cavity between May and October, and females remaining on the fins

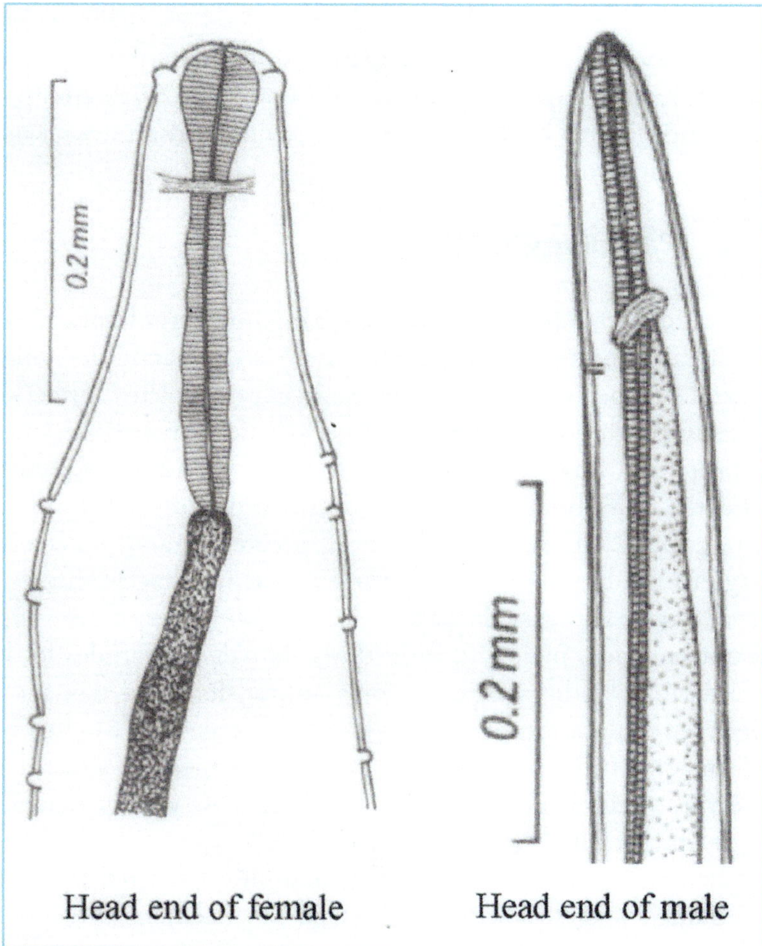

Figure 13.1 Diagrammatic representation of the morphological difference of the anterior of the female and male *Philometroides*. Scale bar 0.2 mm. Courtesy of the Environment Agency, National Fisheries Laboratory, UK.

until June, after which they leave the host and liberate L1 larvae. Gravid females of *P. carassii* have also been found in the caudal fin of crucian carp in the early spring.

Development within the environment, in either the waterbody or the copepodid intermediate host, is temperature dependent. For example, the development of the larval stages of *P. lusiana* in its intermediate and definitive hosts kept at 20–22°C varies, i.e. L1 3–4 days, L2 7–8 days, L3 14–17 days and L4 20–21 days. The survival of the L1 larvae of *P. carassii* in water is also temperature dependent and varies from 7 days at 15°C to several hours at 37°C. This apparent susceptibility to colder conditions provides a useful 'weak link' to aid in control of the parasite.

13.2.3 Pathology

Philometrid nematodes can induce a range of pathological effects on their carp hosts primarily due to their feeding activities, migrations and, in the case of the females, attachments sites. Effects are dependent on the size of the host, level of infection and the species of parasite and host. Migrations of the young and adult parasites around the fish's body can result in inflammation of the internal organs penetrated, damage to the periosteum and mechanical damage to the blood vessels resulting in haemorrhages. Effects on internal organs can vary. For example, in *P. sanguineus* infections in wild crucian carp, male and unfertilised female nematodes located on the swim bladder wall cause only minor, localised damage. In contrast, the migration of the gravid female stage into the fins provokes an acute inflammatory response. This can lead to a distortion of the fin, swelling of the dorsal and caudal tissues, degenerative changes and localised oedema (Fig. 13.2).

When encapsulation occurs, for example in the overwintering of the female stages, additional damage can occur in the connective tissue. Emergence of gravid nematodes can also cause necrosis, hyperplasia and loss of skin integrity. Nematode species that infect the skin can lead to raised scales and the induction of an inflammatory response in the scale pocket (Fig. 13.3).

Initial infection at these sites can be associated with a reduction of pigmentation, which develops further into ulceration. An immunological reaction may be associated with infection in carp as there are accumulations of neutrophils, eosinophils and lymphocytes at sites of infection, and qualitative and quantitative changes in the protein composition of blood serum of infected carp. These pathological responses appear to have negative consequences for growth, condition and energetics of the carp.

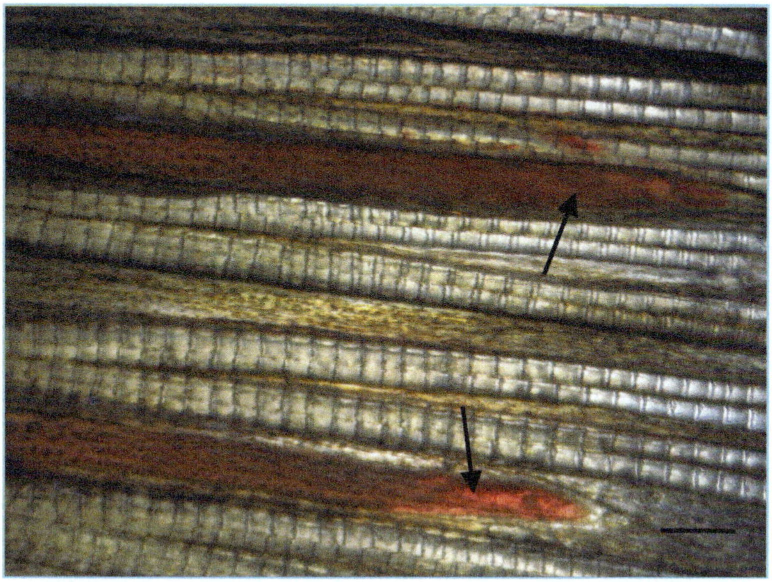

Figure 13.2 Carp fin infected with a female *Philometroides sanguineus* (arrow). Indicative scale bar 1 mm. Courtesy of Dr I. Wellby, Blueroof Environmental Consultancy, UK.

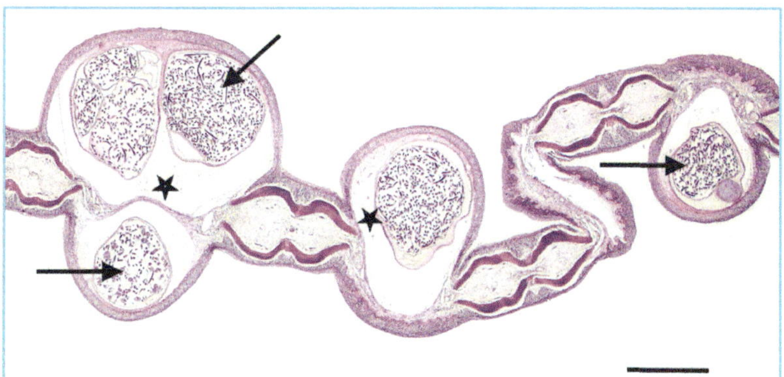

Figure 13.3 Section through the fin of a crucian carp infected with *Philometroides sanguineus*. Note presence of female nematodes (arrows) between fin rays causing a raised infection site comprising eosinophilic material, macrophages, lymphocytes and host cells (stars). Indicative scale bar 400 μm. Courtesy of the Environment Agency, National Fisheries Laboratory, UK.

13.2.4 Detection

It is important that recovery procedures for the nematodes from fish are followed, as identification of the parasite is a challenge and can lead to controversy and much debate. Ideally nematodes should be obtained intact and preferably both sexes should be acquired. Specimens need to be washed in 0.6% saline or phosphate buffered saline, and any debris removed by gentle agitation with a pipette. Fixation should ensure that the specimen is in a relaxed position, and hence hot 10% neutral buffered formalin or hot 70% alcohol should be used. For histopathological investigations, the nematode and infected tissue should be prepared in routine fixatives such as 10% neutral buffered formalin or Bouin's solution. For identification using light microscopy, the nematodes are best 'cleared' with 5% glycerine in 70% alcohol, allowing the solution to evaporate at room temperature. Whole mounts can then be prepared using glycerine. An alternative method using lactophenol is also available, which is a more rapid form of preservation but may cause damage to the parasite.

As indicated above, classification of the species within the family Philometridae can be problematic and is currently primarily based on morphological appearance. Moravec, in an extensive review of these nematodes published in 2013, used several structural features, such as cephalic form and appendages, shape and structure of the tail region, cuticular ornamentations and the shape and structure of the gubernaculum, to classify the philometrids. Recent studies have applied molecular analysis of this group of nematodes, utilising small subunit rRNA (SSU rRNA), and have led to the suggestion that *Philonema* should not be included in the philometrids. In addition, internal transcribed spacer (ITS2) of the ribosomal DNA and cytochrome oxidase I (COI) mitochondrial gene have been utilised to assist in classification, but the general consensus of opinion appears to be that the genera *Philometra* and *Philometroides* should be re-evaluated.

FACTS SHEET

Disease: Philometroidosis
Pathogen: Speciation causes debate. *Philometroides lusiana, P. cyprini, P. carassii, P. sanguinea*
Susceptible host: Many carp species, including *Cyprinus carpio, Carassius auratus, Carassius carassius, Carassius gibelio*
Mortality: Mortalities rare but can occur in heavy infections, mainly debilitating effects
Temperature: Parasites greater with increasing temperature
Prevention and treatment: Infection can be reduced or prevented by good husbandry practices and avoiding introduction of contaminated fish stock. A range of chemotherapeutic means of controlling infection are available but have varying success

13.3 *Contracaecum* Spp.

13.3.1 Pathogen

Contracaecum spp. are parasitic nematodes belonging to the family Anisakidae and are known to have a range of pathogenic impacts on wildlife such as fish, birds and marine mammals, and in some cases humans. The life cycle varies, depending on parasite species, but those that infect carp usually use this fish as an intermediate host, the definitive host being a fish-eating bird, such as a cormorant *Phalacrocorax carbo* or pelican *Pelecanus crispus*.

13.3.2 Epidemiology and host range

The complexity of the genus *Contracaecum*, and in particular the classification of the larval stages that occur in carp, has meant that in many parasite surveys, infections are referred to simply as *Contracaecum* spp. As such it would appear that this parasite genus in carp species has a worldwide distribution and has been recorded in common carp, Prussian carp, crucian carp and goldfish. When full classification has occurred several parasite species predominate: *Contracaecum microcephalum* and *Contracaecum spiculigerum* in common carp and *Contracaecum rudolphii* in common carp and

Prussian carp. Recently *Contracaecum bancrofti* has been located in the intestinal tissue of common carp in Australia.

Based on survey analysis there is evidence that *Contracaecum* spp. have seasonality within their transmission. It has been suggested that, in the wild, nematodes prefer to develop in warm waterbodies and that the definitive host (e.g. cormorant) becomes infected during their wintering season. The influence of temperature on parasite development has also been supported by laboratory studies. In the case of *Contracaecum rudolphii* eggs kept at 20–22°C, the developing larvae undergo two moults on days 4–5, and the ensheathed third-stage larvae hatch on days 5–6. These third-stage larvae, which occur inside the eggs or are free-living, are infective to copepods and fish, both of which can be considered paratenic (meta-paratenic) hosts. A range of copepods harbour the infection, for example *Cyclops strenuus*, *Acanthocyclops vernalis*, *Megacyclops viridis*, *Ectocyclops phaleratus* and *E. serrulatus*, and the carp become infected by consuming either the larvae or the infected copepod. The larvae penetrate the intestinal wall, become encapsulated and can survive for at least 15 months in carp fry.

13.3.3 Pathology

Pathology is associated with the L3 larvae, which, after penetration of the gut wall of the carp, can be found in a range of internal organs such as the gut wall, mesenteries, liver, visceral fat bodies, pancreas and connective tissue. *C. rudolphii* in gibel carp (*Carassius gibelio*) induces a granulomatous response, which comprises mononuclear cells and fibrous tissue. On some occasions necrosis of the granuloma occurs with the infiltration of melanomacrophages. In common carp, infection with the L3 stage has been associated with an infiltration of host immune cells (e.g. macrophages, granulocytes, lymphocytes) into the affected internal organs. Whilst some authorities have suggested that infection of carp with the L3 stage of *Contracaecum* spp. may not have adverse effects on the host, others have noted that, for example in gibel carp, a reduction in Fulton's condition factor occurs. It is generally considered that the parasite can have more severe effects in carp fry.

13.3.4 Detection

Diagnosis of infection with *Contracaecum* spp. is based on morphological studies and, more recently, molecular characterisation. The location of the larvae, which may be encysted in a range of internal organs of the fish host, requires extensive scrutiny of the host. The organs of the fish should be removed, placed in physiological solution such as teleost ringer PBS and examined under a binocular microscope. If the larvae are not released, then they should be gently extracted using a needle and placed under a coverslip either alive or fixed in formaldehyde. For histological preparation, the larvae with the associated tissues should be fixed by standard techniques, i.e. fixation in 10% buffered formalin for 10–20 days or in Bouin's solution for 24 hour. Specimens are then dehydrated through a graded series of ethanol and embedded in paraffin wax, whereafter sections are stained with haematoxylin and eosin.

The identification of *Contracaecum* spp. is usually based initially on the location of encysted larvae found in the dissected carp. Some authorities have noted that in fluorescent light *Contracaecum* appears yellow. Larval stages are usually more problematic to identify than their adult counterparts, but perhaps the most convenient way is to use morphological examination on whole mount preparations. A range of identification keys are available, which usually use structural features, such as position of the excretory pore, alae striations, and relative positions of body structures such as oesophagus, ventriculus and caecum. *Contracaecum* spp. have characteristically an intestinal caecum that is directed anteriorly and a posteriorly directed ventricular appendix.

More recently a range of biochemical and molecular techniques have been applied to this genus. Allozyme profiles have been used to identify three genera, i.e. *Contracaecum*, *Anisakis* and *Pseudoterranova* in the Anisakidae, and there are several investigations using molecular analysis, for example PCR and subsequent sequencing of ribosomal DNA (18S, 5.8S, 28S, ITS1 and ITS2) using the middle section of the nematode larva, which should be isolated aseptically. Such studies, with the support of morphological investigations, have highlighted the complexities of the genus *Contracaecum*. For example, in addition to *C. rudolphii*, four other congeneric species, *C. osculatum*, *C. micropapillatum*, *C. microcephalum* and *C. ovale*, occur in Europe, although it would appear that *C. rudolphii* occurs worldwide, where it is now thought to be a species complex, comprising at least two strains, named A and B.

FACTS SHEET

Disease: No widely accepted common name
Pathogen: *Contracaecum spiculigerum*, *C. microcephalum*, *Contracaecum bancrofti*, *C. rudolphii* with other congeneric species, *C. osculatum*, *C. micropapillatum*, *C. microcephalum* and *C. ovale*
Susceptible host: *Cyprinus carpio*, *Carassius gibelio*, *Carassius carassius*, *Carassius auratus*
Mortality: Mortalities rare but can occur in heavy infections, mainly debilitating effects
Temperature: Parasites can be greater with increasing temperature
Prevention and treatment: Infection can be reduced or prevented by good husbandry practices and avoiding introduction of contaminated fish stock. Restricting access of birds to ponds. A range of chemotherapeutic means of controlling infection available but have varying success

13.4 Prevention Strategies and Treatments of Philometridae and Anisakidae

The somewhat complex life cycles of nematode parasites that involve infection in carp enable the application of management strategies targeted at the different stages of the cycle. Since *Contracaecum* utilises a bird definitive host it is, in theory, possible to reduce infection in the fish by restricting access of the bird definitive host to the ponds. Whilst a range of bird protection devices can be utilised, none are completely successful. In addition, the presence of an invertebrate intermediate host, for example copepods in Philometridae and Anisakidae, allows control of the infection by eradication of the invertebrate by draining the pond and the application of lime. Such actions also kill free-living larval stages of the parasite. There is also a range of chemotherapeutic means of controlling infection, although, as with all such treatments, trials must be undertaken to ascertain any adverse effects on the fish. In addition, applications must comply with any local legislative requirements. Levamisole has been found to be highly effective against philometroidosis in carp. The response is dose dependent; for example, death of females and males of *P. lusiana* occurred at 100–115 mg/kg, but there was no effect at 5 mg/kg, whilst reduced feeding activity in fish was noted at 100 mg/kg.

13.5 Selected References for Further Reading

Cakalic, I.T., Vuic, N. and Cerba, D. (2020) First finding of roundworms (*Contracaecum* sp.) in Prussian carps (*Carassius gibelio* Bloch, 1782) in Lake Sakadas. *Hrvatske Vode* 28, 21–28.

Choudhury, A. and Cole, R.A. (2008) Phylum nematoda. In: Eiras, J.C., Segner, H., Wahli, T. and Kapoor, B.G. (eds) *Fish Diseases*, Vol. 2. Science Publishers, Enfield, New Hampshire.

Dziekonska-Rynko, J., Rokicki, J. and Gomulka, P. (2010) Development of larval *Contracaecum rudolphii* Hartwich 1964 (Ascaridida: Anisakidae) in experimentally infected goldfish (*Carassius auratus* L., 1758). *Journal of Helminthology* 84, 234–240.

Gavrillin, K.V., Bychkova, L.I., Dmitrieva, S.N. and Linnik, A.V. (2015) Laboratory studies of antiparasitic activities of levamisole against philometroidosis on carp (*Cyprinus carpio*, L. 1758). *Russian Journal of Parasitology* 3, 61–64.

Hoole, D., Bucke, D., Burgess, P. and Wellby, I. (2001) *Diseases of Carp and Other Cyprinid Fish*. Blackwell Science, Oxford, UK.

Innal, D., Stavrescu-Bedivan, M., Ozturk, M.O. and Ozmen, O. (2020) First record of *Contracaecum rudolphii* Hartwich, 1964 in *Carassius gibelio* (Bloch, 1782) from Turkey. *Aquatic Sciences and Engineering* 35, 105.

Molnar, K., Szekely, C., Baska, F., Muller, T., Zuo, S. *et al.* (2019) Differential survival of 3rd stage larvae of *Contracaecum rudolphii* type B infecting common bream (*Abramis brama*) and common carp (*Cyprinus carpio*). *Parasitology Research* 118, 2811–2817.

Moravec, F. (2009) Experimental studies on the development of *Contracaecum rudolphii* (Nematoda: Anisakidae) in copepod and fish paratenic hosts. *Folia Parasitologica* 56, 185–193.

Moravec, F. and Cervinka, S. (2005) Female morphology and systematic status of *Philometroides cyprinid* (Nematode: Philometridae), a parasite of carp. *Diseases of Aquatic Organisms* 67, 105–109.

Moravec, F. and de Buron, I. (2013) A synthesis of our current knowledge of philometrid nematodes, a group of increasingly important fish parasites. *Folia Parasitologica* 60, 81–101.

Moravec, F., Halajian, A., Tavakol, S., Nyagura, I. and Luus-Powell, W.J. (2015) *Philometroides khalili* n. sp., a new philometrid nematode (Philometridae) from the operculum of the cyprinid fish *Labeo rosae* in Zimbabwe. *Helminthologia* 53, 113–117.

Nakajima, K. and Egusa, S. (1977) Studies on the philometrosis of crucian carp Part 1 Morphological characteristics of gravid female and taxonomy. *Fish Pathology* 12, 111–114.

Nakajima, K. and Egusa, S. (1977) Studies on the philometrosis of crucian carp Part 2 Some ecological observations of gravid female. *Fish Pathology* 12, 115–120.

Nakajima, K. and Egusa, S. (1977) Studies on the philometrosis of crucian carp Part 3 Activity and resistance of the 1st stage larvae. *Fish Pathology* 12, 185–190.

Naqvi, S.M.H.M., Ghazi, R.R. and Khan, A. (2014) Light and scanning electron microscopic study of *Rhabdochona bilqeesae* sp n. (Nematoda: Rhabdochonidae) parasitizing a common edible fish *Labeo rohita* in Thatta, Sindh, Pakistan. *International Journal of Biology and Biotechnology* 11, 483–489.

Pegg, J., Williams, C.F., Cucherousset, J. and Britton, J.R. (2011) What are the consequences of infection by the introduced parasite *Philometroides sanguineus* for threatened crucian carp, *Carassius carassius*, populations in England?. *Ecology of Freshwater Fish* 20, 598–604.

Schäperclaus, W. (1991) *Fish Diseases*, 5th edn, Vol. 2. Amerind Publishing Co./Oxonian Press, New Delhi, India.

Shamsi, S., Turner, A. and Wassens, S. (2018) Description and genetic characterisation of a new contracaecum larval type (Nematoda: Anisakidae) from Australia. *Journal of Helminthology* 92, 216–222.

Sokolov, S.G. and Kazakov, B.E. (2008) On morphology and taxonomic status of some species of the genus *Philometroides* (Nematode, Philometridae). *Zoologichesky Zhurnal* 87, 1420–1424.

Szostakowska, B. and Fagerholm, H.-P. (2007) Molecular identification of two strains of third-stage larvae of *Contracaecum rudolphii* sensu lato (Nematoda: Anisakidae) from fish in Poland. *Journal of Parasitology* 93, 961–964.

Williams, C.F., Moravec, F., Turnbull, J.F. and Ferguson, H.W. (2012) Seasonal development and pathological changes associated with the parasitic nematode *Philometroides sanguineus* in wild crucian carp *Carassius carassius* (L.) in England. *Journal of Helminthology* 86, 329–338.

Zhu, X.Q., D'Amelio, S., Gasser, R.B., Yang, T.B., Paggi, L. *et al.* (2007) Practical PCR tools for the delineation of *Contracaecum rudolphii* A and *Contracaecum rudolphii* B (Ascaridoidea: Anisakidae) using genetic markers in nuclear ribosomal DNA. *Molecular and Cellular Probes* 21, 97–102.

Chapter 14

The Acanthocephala

14.1 Pathogen

Acanthocephala, or thorny-headed worms, are a relatively small group of gut parasites in a variety of fish species, and are also found in the intestine of amphibians, reptiles, birds and mammals. Adult acanthocephalans have an elongate, tube-like body shape of white or yellow-orange colouring, and fish parasites range in length from a few millimetres to about 3 cm. Acanthocephalans have an indirect development, which includes arthropods, mainly insects and crustaceans, as intermediate hosts and a vertebrate as definitive host. In some species, the life cycle also includes a paratenic host, in which infective larvae accumulate and the parasite is transmitted when the final host feeds on the infected paratenic host. Adult acanthocephalans attach to the intestinal wall of their host by means of an eversible thorny proboscis and the tube-like trunk extends into the intestinal lumen of the host (Fig. 14.1). The parasites lack a mouth, intestine and circulatory system and absorb nutrients from the intestinal content of the host across their body wall, which contains a characteristic lacunar system.

The attachment organ, the proboscis, is of globular or cylindrical shape and carries sclerotised hooks, which vary in size, number, shape and arrangement – characteristics that have some value in the identification of species. The proboscis can be totally inverted into a muscular sac, the proboscis receptacle, by proboscis inverter muscles. Eversion of the proboscis is by hydraulic pressure in the lacunar system when the proboscis receptacle contracts. The proboscis and associated structures are referred to as prosoma, while the rest of the body is called the trunk, which encloses reproductive and excretory

DOI: 10.1079/9781836990024.0014

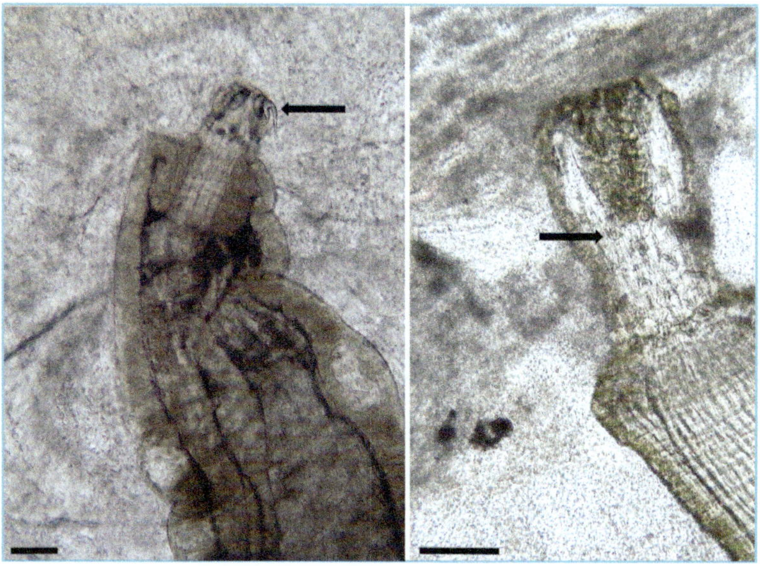

Figure 14.1 Two examples of adult acanthocephalans with a thorny proboscis. Thorns are indicated (arrow). Indicative scale bar 100 μm.

organs. The space between the viscera and the body wall is a fluid-filled pseudocoel. The fluid in the proboscis receptacle is separated from the fluid in the body cavity, which means the parasite can move its body without losing its attachment to the host gut mucosa. Acanthocephalans are dioecious and exhibit a sexual dimorphism in size, with females usually being larger than males. During copulation in the host intestine the male releases spermatozoa into the female's uterus, which fertilise the oocytes. After deposition of the sperm, the male occludes the exit of the female's vagina. This means that for some time after copulation loss of sperm and fertilisation of the female by another male is prevented.

Fertilised eggs undergo embryonic development in the female's pseudocoel and are excreted through the gonopore when a hooked larva, called the acanthor, has developed. This egg has to be eaten by an invertebrate host, usually a benthic amphipod (e.g. *Gammarus* spp.), an isopod (e.g. *Asellus aquaticus*) or an ostracod (e.g. *Cypris* spp.). In the intermediate host the acanthor hatches from the egg and develops into acanthella and subsequently into a juvenile or an encysted cystacanth. The life cycle is completed when the infected invertebrate carrying a fully developed acanthella or cystacanth larva is eaten by an appropriate vertebrate host. Sexual development occurs in the vertebrate host. Snails or aquatic insect larvae may serve as secondary intermediate hosts, which carry infective stages after having ingested an infected intermediate host, but there is no further development of the parasite in these hosts.

14.2 Epidemiology and Host Range

Cyprinid fish are infested by acanthocephalans from the class Palaeacanthocephala, including species from the genera *Acanthocephalus* and *Echinorhynchus*, as well as from the class Eoachanthocephala, including species from the genera *Neoechinorhynchus* and *Acanthogyrus*. These acanthocephalans have been recorded from crucian carp (*Carassius* species), as well as from grass carp (*Ctenopharyngodon idella*) and Indian major carps from ponds in the Holarctic, China and the Indian subcontinent. In addition, these acanthocephalan species infest various other fish species from the same freshwater habitat, including eels, salmonids, perch and sticklebacks.

All these acanthocephalans have aquatic life cycles. While species from the class Palaeacanthocephala mainly use amphipod or isopod intermediate hosts and infect hosts from all vertebrate classes, those from class Eoacanthocephalida are primarily transmitted by ostracods to fish and amphibians. In addition, the host range of the acanthocephalans appears more limited for the intermediate host than for the vertebrate host.

14.3 Pathology

In the intestine of the definitive host, the cystacanth larval stage everts its proboscis and attaches in the intestine between the mucosal folds. Many species remain at the first site of attachment, while others migrate to a particular site in the intestine where they develop to maturity and where, after fertilisation, the females remain and shed their eggs. Since the acanthocephalans embed their spiny proboscis into the intestinal mucosa, the tissue may be destroyed at the attachment site, where a chronic inflammation and a thickening of the connective tissue is induced. In some species, such as *Pomporhynchus laevis*, the proboscis may penetrate the intestinal wall and induce severe damage. It has been suggested that acanthocephalans might harm their host by the excretion of toxins, but as yet there is no evidence for the production of such substances by the parasite. In most cases acanthocephalans harm their host by food competition, as they absorb nutrients from the intestinal contents of the host. Hence, heavily infected hosts may exhibit signs of malnutrition, such as a slim dorsal line.

14.4 Detection

An infection with acanthocephalans can be detected during post-mortem examination, when the intestinal tract is opened and worms about 1–2 cm long are detected which are attached to the intestinal wall. After careful dissection from the attachment site the proboscis should be visible. Species diagnosis largely relies on the morphology of the proboscis, as well as the number, arrangement and form of the hooks. In addition, an infection with acanthocephalans can be diagnosed from eggs passed out in the faeces of an infected host, but this method is difficult and not widely used.

14.5 Prevention and Treatment

Acanthocephalan parasites can be easily distributed by movement of infected hosts. Therefore, the introduction of the parasite into an aquaculture unit or a natural waterbody should be avoided. In particular, care should be taken when stocking fish with unknown infection status into waterbodies, such as natural lakes, reservoirs or carp ponds, which have a rich invertebrate fauna and could therefore serve as intermediate hosts. This would prevent the introduction of the parasite into the invertebrate community and thereby prevent the parasite from establishing its life cycle in the production unit or waterbody. Treatment of the infection is difficult. A reduction of the infection might be achieved by reducing the population of infected intermediate hosts by liming the pond floor after draining the water. A chemical treatment of the infection has not been established.

FACTS SHEET

Disease: Acanthocephalosis
Pathogens: *Acanthocephalus* spp., *Echinorhynchus* spp., *Neoechinorhynchus* spp., *Acanthogyrus* spp., and others
Susceptible hosts: Common carp, goldfish, *Carassius* species, and other cyprinids
Mortality: Presumably low, harmful because of negative impact on condition and growth
Temperature: 2–25°C
Prevention and treatment: Prevention by stocking of aquaculture units with parasite-free fish. In infected ponds a reduction of the infection pressure could be achieved by liming the pond floor after draining the water in order to reduce the population of infected intermediate hosts. A chemical treatment is not established.

14.6 Selected References for Further Reading

Amin, O. (2013) Classification of the Acanthocephala. *Folia Parasitologica* 60, 273–305.

Bush, A.O., Fernandez, J.C., Esch, G.W. and Seed, J.R. (2001) Acanthocephala: The thorny headed worms. In: Bush, A.O., Fernandez, J.C., Esch, G.W. and Seed, J.R. (eds) *Parasitism. The Diversity and Ecology of Animal Parasites.* Cambridge University Press, Cambridge, UK, pp. 197–214.

Gibson, D., Bray, R., Hunt, D., Georgiev, B., Scholz, T. *et al.* (2014) Fauna Europaea: Helminths (Animal Parasitic). *Biodiversity Data Journal* 2, e1060.

Kennedy, C.R. (2006) *Ecology of the Acanthocephala.* Cambridge University Press, Cambridge, UK260

Nickol, B.B. (2006) Phylum Acanthocephala. In: Woo, P.T.K. (ed.) *Fish Diseases and Disorders: Protozoan and Metazoan Infections*, 2nd edn, Vol. 1. CAB International, Wallingford, UK, pp. 444–464.

Roy, C., Dey, S.R. and Bhowmik, B. (2023) Freshwater fish parasite diversity in West Bengal – a review. *Ecology Environment and Conservation* 29, 161–169.

Taraschewski, H. (2008) Acanthocephala. In: Eiras, J.C., Segner, H., Wahli, T. and Kapoor, B.G. (eds) *Fish Diseases*. Science Publishers, Enfield, New Jersey, pp. 1025–1062.

Chapter 15

The Crustacea

15.1 Introduction

Many crustaceans are free-swimming, and indeed some are commercially cultured, but a few have adopted a parasitic existence that can cause serious problems in carp production worldwide. Many carp farmers around the world recognise the ectoparasitic genus *Argulus* as the 'fish louse' and its extensive colonising ability. The large size of many parasitic crustaceans and the sometimes unusual body form means that, compared with other infections of carp, their presence is relatively easy to ascertain, being visible to the naked eye. Most parasitic crustaceans that infect carp are representatives of two classes: the Branchiura and the Copepoda. Within the latter class the families Ergasilidae and Lernaeidae are particularly important in carp production, whilst in the Branchiura the family Arguluidae, which includes the genus *Argulus*, is a major threat to the productivity of fish culture worldwide.

Whilst there is a great variation of adult body form within the parasitic crustaceans, there is some commonality within some aspects of their biology. All are surrounded by a hard exoskeleton and comprise a complex set of appendages on their head. These include antennules, antennae, mandibles and maxillae and are useful within the identification process. These appendages can be adapted, in various ways, to the parasitic life; for example, in *Argulus* spp. the maxillae have suckers and in *Ergasilus* spp. the antennae have claws. Crustacean parasites have a direct life cycle involving only one host and the genders are separate. Eggs are either released by the females into the pond environment and attach to nets, plants and cage sides within the production units, or in some genera are retained by the

DOI: 10.1079/9781836990024.0015

female on the fish. Egg development is dependent on temperature and larval crustaceans emerge. The parasite undergoes a range of successive moults to form larval, juvenile and pre-adult stages. The stage in which the parasitic mode develops varies between species. For example, in *Argulus* spp. it is the first larval stage, whilst in copepods it is usually the mated female stage. The ectoparasitic nature of crustaceans means that they are primarily associated with pathological changes on the skin, gills and fins, and wounds formed can be sites for opportunistic infections of bacteria and fungi.

15.2 *Argulus* Spp.

15.2.1 Pathogen

The genus *Argulus* (Fig. 15.1) comprises over 150 recorded species, which have an extensive distribution and host range. All the species

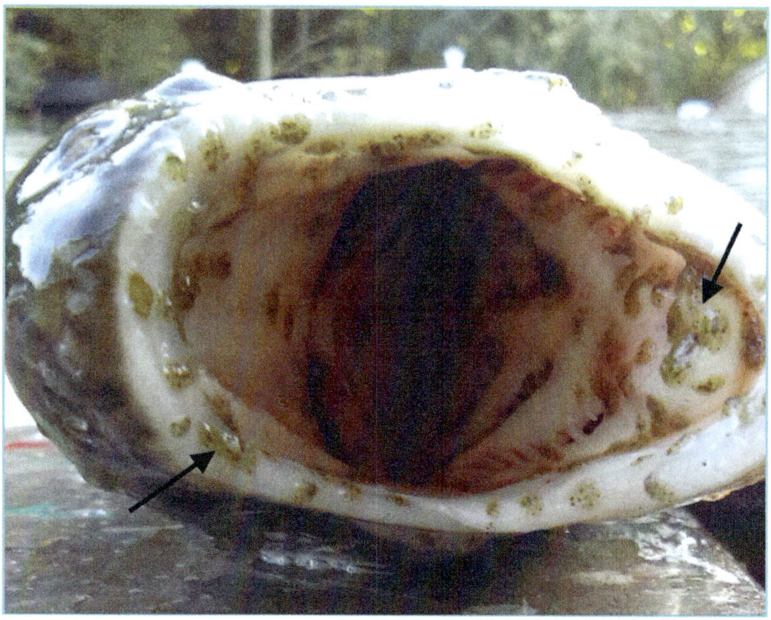

Figure 15.1 Mouth of carp showing an extensive infection of *Argulus* attached to the fish's lips (arrows). Courtesy of Dr Ian Wellby, Blueroof Consultancy, UK.

that are important in carp production have a dorsoventrally flattened body, which is covered by a large carapace. The body comprises three regions, a cephalothorax, thorax and abdomen, and has a series of swimming legs. The crustacean can leave the host and is frequently found within the waterbody. *Argulus* spp. are very proficient swimmers and leave the host for a variety of reasons, for example egg deposition, mate location, new host location and by accidental activities, which include some management practices such as grading, vaccination and application of anaesthesia.

15.2.2 Epidemiology and host range

Several *Argulus* species have been associated with carp culture worldwide. *Argulus foliaceus* has a wide range of hosts, including common carp (*Cyprinus carpio*), Indian major carp mrigal (*Cirrhinus mrigala*), crucian carp (*Carassius carassius*), grass carp (*Ctenopharyngodon idella*) and ornamentals such as goldfish (*Carassius auratus*) and koi carp. *Argulus japonicus* tends to have a more limited range, having been found on common carp/Japanese koi carp (*Cyprinus carpio*) and goldfish. Some argulids have a more limited geographical and host range, such as *Argulus siamensis*, which is found on major carp catla (*Gibelion catla*) and rohu (*Labeo rohita*), with *Argulus sindhensis* also being described from the latter major carp species. Major carp rohu, catla and mrigal have also been found to be infested with fish lice *Argulus indicus* and *A. japonicus*, and in Asia *Argulus coregoni* and *Argulus ellipticaudatus* have also been located in common carp. One species that is causing increasing concern is *Argulus mongolianus* (Fig. 15.2) which has been found in common carp, crucian carp and Japanese white crucian carp (*Carassius cuvieri*). The parasite appears to be extending its geographical range and in 2022 was recorded for the first time in the UK and Japan.

The direct life cycle of *Argulus* spp. involves the female depositing eggs on a range of suitable surfaces such as plants, gravel, stones, tank sides and aquarium glass, in numbers ranging from fewer than ten to clutches of several hundred. Development of the egg stage is temperature dependent; for example, *A. foliaceus* eggs hatch in 8 days at 26°C but hatching can take several months at temperatures

Figure 15.2 Adult female (F) and male (M) *Argulus mongolianus*. Note carapace (arrow C), two black eye spots (arrow E), proboscis (arrow P), suckers (arrow S) and swimming legs (arrow L). Indicative scale bar 2 mm. Courtesy of the Environment Agency, National Fisheries Laboratory, UK.

below 10°C. On hatching, a parasitic metanauplius larva is released and a series of developmental stages exist, which may be dependent on the species. Development is thought to be temperature dependent, which has led to suggestions that *Argulus* infections may exhibit seasonality. In temperate regions infections on carp tend to increase in late summer/early autumn and decrease in winter months, during which the parasite is thought primarily to survive in the egg stage, although in milder winters it is also suggested that adult parasites may occur on the fish or in the waterbody.

15.2.3 Pathology

The effects of infection by *Argulus* on their fish hosts can be varied and range from behavioural changes to damage caused by attachment and feeding activities of the lice.

The first behavioural changes noted in infected fish is one that is fairly common to most infections with ectoparasites. This behaviour, sometimes called 'flashing', entails the carp rubbing their flanks against submerged objects in a manner assumed to attempt to dislodge the parasite and/or ease the irritation caused by the louse. In culture systems where additional aeration is applied through air

Figure 15.3 Adult *Argulus* attached to the gill of carp. Courtesy of the Environment Agency, National Fisheries Laboratory, UK.

stones, common carp infected by *A. japonicus* have been noted to be crowded in areas of water turbulence. Other behavioural changes associated with infection include formation of denser shoals, loss of appetite and lethargy.

Argulus spp. are located on a range of external surfaces of the carp, but tend to predominate on the skin and gills (Fig. 15.3). Whilst attached to the fish, the parasite moves its thoracopods in order to maintain a water flow over the body, thus aiding respiration. In doing so the parasite can induce pressure atrophy and small bruising on and around the attachment site. It is, however, the feeding activity and attachment process that create the most damage and pathological responses. Mild infarctions such as small petechial haemorrhages and fin erosion can occur, but in severe cases loss of skin, the formation of large lesions, nuclear pycnosis in epithelial cells and an increase in mucus production have been noted. Such adversities can lead to osmoregulatory distress, anaemia and stress. Initial stages of infection can cause proliferation of fibrillar elements of the skin with oedema and tissue destruction, which extends up to the dermis. Although such pathology may be tolerated by large individuals, in small fish the host may become moribund and large mortalities can occur.

Because of the economic impact of *Argulus* infections in carp production and the accessibility to the parasite material, recently there have been extensive investigations on the immune response to *Argulus* infections in carp. Infections have been noted to be associated with an increase in leucocytes, particularly neutrophilic granulocyte numbers, at the site of attachment. In addition, up-regulation of gene transcription of the chemokine CXCa, the chemokine receptor CXCR1 and the cytokine TNF alpha have been suggested as possible indicators of early skin damage. Both CXC chemokines, as well as CXCR1 and CXCR2, have also been associated in acute *Argulus japonicus* infections. Investigations on major carp have highlighted the complexity of the *Argulus*–host interactions. In the Indian major carp *Labeo rohita*, naturally infested with different loads of the freshwater fish louse, *Argulus siamensis*, there is an increase in glucose level and total antiprotease activity of serum, whilst alpha-2 macroglobulin and complement activities are significantly lower in lice-infested fish. In contrast, in the skin of infected fish, genes controlling natural killer cell enhancing factor (NKEF-B), superoxide dismutase (MnSOD), lysozyme G and microglobulin (M-2) are significantly increased. There may also be gender differences, as the immune response of female fish may be more significantly affected by the louse than that of the male. Among Indian farmed carp, catla is thought to be relatively more resistant than rohu to *Argulus siamensis* infection, and this may be partly related to the Toll-like receptor 22 (TLR22). Studies have also indicated that there is an antibody response. *Argulus* infection in South-east Asian rohu induced a significant expression of IgD in the skin, kidney, liver, spleen and gill, whilst in catla the expression of IgM in the blood was associated with *Argulus* infection.

15.2.4 Detection

Because of the parasites' large size, the characteristic body form and presence of an exoskeleton, diagnosis can be achieved based on macroscopic analysis. The adult parasites can be removed from the fish surface using forceps, preserved in alcohol and then prepared for whole mounts using standard techniques. Recently diagnosis and species characterization have been achieved using a range of molecular techniques such as species-specific primers designed from unique sequences cloned from RAPD fragments that could identify *A. siamensis* and *A. japonicus* separately. Utilising 28S rDNA (GenBank accession number MF926591) revealed 94% sequence identity with *A. japonicus* and *A. foliaceus*.

15.2.5 Prevention and treatment

Prophylactic measures that prevent the establishment of the parasite in a pond system are always encouraged. Since the development of a new generation of parasite infection takes approximately 8 weeks, the most dangerous period, particularly in juvenile carp production, is when the fish are exposed to populations of impregnated female parasites that are ready to spawn. It is therefore essential that introduced stock are obtained from parasite-free sources and that the water supply is also parasite-free. In addition, the cosmopolitan nature of the *Argulus* species means that infections can reside in other animals, such as three-spined sticklebacks and some tadpoles, and therefore reduction in 'carrier' species may be useful as part of a control strategy. Although in the egg stage the parasite is thought to be somewhat resistant to low temperatures, significant destruction of eggs, larvae and adult parasites can be achieved in infected waterbodies by draining and treating with lime.

In salmonid farming the concept of cleaner-fish such as wrasse has been successfully applied, but such forms of biological control have not gained general acceptance in carp production units, and indeed, cleaner-fish can also be the source of other infections.

A range of chemical treatments are available for *Argulus* infections, although the application of such treatments should be undertaken in accordance with local and national legislation requirements.

Several treatments have been successfully utilised, such as a 7 day regimen of oral administration of emamectin benzoate in a controlled experimental infestation of *Argulus* when administered to goldfish at 50 μg/kg fish biomass/day and to koi at 5 μg/kg fish biomass/day. Several treatments utilise organophosphate pesticides like Nuvan and Ekalux, sometimes in combination with common salt. Administration was found to be effective against *A. japonicus* when fishponds received four treatments at weekly intervals of Nuvan (0.1 ppm) or Ekalux (0.0025 ppm) or Nuvan + common salt (0.1 ppm + 15 ppm) or Ekalux + common salt (0.0025 ppm + 10 ppm). Ekalux–common salt combination proved to be the best, as it not only eradicated the parasites with a minimum deleterious effect on the plankton, but also controlled the *Argulus* egg-laying capacity. There have also been several studies utilising extracts from naturally occurring plants; for example, papaya leaf extract contains the active alkaloid ingredient, carpain, which at six ppt reduces *Argulus* infection on common carp by 85%. Applications can also be applied directly to infected waterbodies; for example, 10 kg of powdered quicklime per acre (2.6 ppm) when applied in three treatments over 10 days, and potassium permanganate solution applied twice at 5 kg/acre (1.3 ppm) to ponds used for the culture of major carp (*Labeo rohita*, *Gibelion catla* and *Cirrhinus mrigala*) eradicated infestations with *A. indicus* and *A. japonicus*.

FACTS SHEET

Disease: Argulosis

Pathogen: *Argulus foliaceus, A. siamensis, A. japonicus, A. indicus, A. coregoni, A. ellipticaudatus*

Susceptible hosts: Many carp species, including *Cyprinus carpio, Cirrhina mrigala, Carassius carassius, Ctenopharyngodon idella, Carassius auratus, Labeo rohita, Gibelion catla*

Mortality: Mortalities primarily associated with heavy infections, particularly in young fish, mainly debilitating effects

Temperature: Parasites greater with increasing temperature

Prevention and treatment: Infection can be reduced or prevented by good husbandry practices and avoiding introduction of contaminated fish stock. A range of therapeutic agents available

15.3 Ergasilosis

15.3.1 Pathogen

This disease is caused by the family Ergasilidae, of which there are more than 100 species, the majority of which are parasitic, particularly in freshwater fish. Ergasilids are perhaps the least adapted of the parasitic crustaceans to the parasitic way of life, and many resemble their free-living counterparts, the copepods. The genders are separate and only the female is parasitic, primarily on the gills.

15.3.2 Epidemiology and host range

The ergasilids have a very large host range and can be found in the Salmonidae, Percidae, Siluridae, Cichlidae, Esocidae, Clupeidae and the Cyprinidae, including carp. This cosmopolitan host range means that infections of ergasilids can reside in other fish species in multi-species carp production ponds, thus complicating control strategies. *Ergasilus sieboldi* perhaps represents the most important ergasilid pathogen of carp, where it infects the gills of common carp and grass carp. Its relatively large size (up to 1.7 mm in length) and the presence of large white egg sacs have earned the parasite the common name 'gill maggot'. Other ergasilids are also associated with infections of carp, such as *Neoergasilus japonicus*, which occurs on the fins of crucian carp, common carp and goldfish, and *Ergasilus peregrinus* on grass carp and common carp. *Sinergasilus* species (*S. polycolpus* and *S. major*) have been found on silver carp (*Hypophthalmichthys molitrix*), farmed bighead carp (*Hypophthalmichthys nobilis*) and grass carp.

The life cycle of ergasilids such as *E. sieboldi* comprises a single host, the fish, and the majority of the life cycle is spent in free-living forms. The free-living females and males mate, after which the male dies and the female, with the spermatophore, locates a suitable host fish, where eggs are produced in two egg sacs. The development of the eggs is temperature dependent and at 16°C hatch occurs after approximately 10 days, whilst at 20°C this is reduced to 6 days. The life cycle comprises 11 free-living stages: six nauplii and five copepodid. The large amount of eggs produced (100–330 eggs per individual) and the extensive reproductive potential of the parasite result in a rapid increase in infections in carp production units. In Europe several cycles of egg production can occur; for example, when the first-generation *E. sieboldi* occurs in mid-June, a second generation can occur in September, with a possible third generation occurring later in the year. If the latter does not materialise, the females overwinter before egg production occurs in the following spring. Therefore, an exceptional increase in potential infection can occur, resulting in a rapid deterioration of fish health and economic productivity.

15.3.3 Pathology

Since it is only the female that resides on the fish, pathology is primarily associated with the attachment and feeding activities, which in the case of *E. sieboldi* is on the gill (Fig. 15.4). The second antennae are modified into an attachment organ comprising powerful hooks, by which the parasite embraces the gill filaments. This can lead to an extensive pathological response, including gill hyperplasia and obstruction of blood flow. The aggressive feeding activity on the epithelial cells induces hypertrophy and coalescing of the secondary gill lamellae. Such effects can lead to secondary infections and a reduction in respiration. The latter is particularly prominent at higher temperatures, when oxygen deficiency can lead to the fish gulping for air, exhibiting respiratory distress and the occurrence of mass mortalities.

15.3.4 Detection

Diagnosis of ergasilid infection primarily relies on the isolation of the female parasite from the host. The male and the juvenile forms are particularly difficult to locate and nearly impossible to identify to species level based on morphology. Isolated females can be fixed in alcohol and prepared by standard techniques for whole mount preparations. Identification is based on host species infected and the body form of the female parasite.

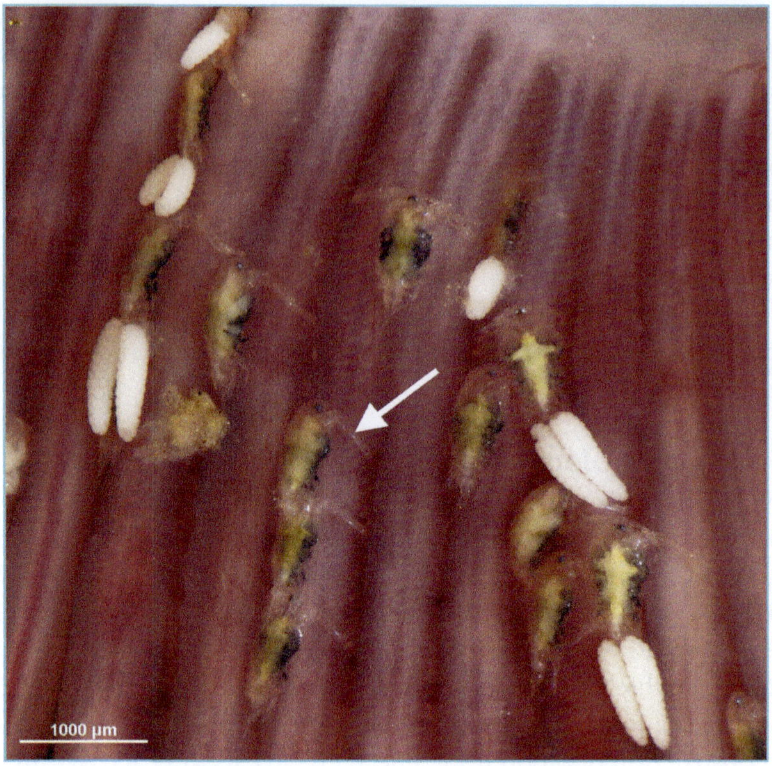

Figure 15.4 Adult female *Ergasilus sieboldi* attached to the gill of carp. Note the presence of two white egg sacs on some individuals and the modified second antennae used to attach to the gill filament (arrow). Scale bar 1000 µm. Courtesy of the Environment Agency, National Fisheries Laboratory, UK.

15.3.5 Prevention and treatment

The best strategy to control the spread of ergasilosis is the imposition of strict restrictions on the import of infected carp into the production unit. Because of the presence of a substantial free-living component in the life cycle of these parasitic copepods, an effective method of eradication of the parasite from the waterbody is by draining the pond, applying lime and preferably leaving for a period of 6 months, which includes a spring season. Other methods of reducing the impact of infection include reducing the stocking of the carp and more susceptible fish species in multispecies culture systems. The fact that the free-living stages of the ergasilids prefer open water means that some limited effects on infection can be obtained by increasing the aquatic vegetation. Since the application of organophosphate insecticides is not recommended, alternative control strategies such as biological control have been attempted. This includes the proposal to use parasitic euglenoids, for example *Paradistigma* spp. that infect the eggs of the ergasilids and kill the first nauplius stage within 2–3 days.

FACTS SHEET

Disease: Ergasilosis
Pathogen: Mainly *Ergasilus sieboldi*, but other problematic species include *Neoergasilus japonicus*, *Ergasilus peregrinus*, *Sinergasilus polycolpus*, *S. major*
Susceptible hosts: Many fish species, including *Cyprinus carpio*, *Ctenopharyngodon idella*, *Carassius carassius*, *Carassius auratus*, *Hypophthalmichthys molitrix*, *Hypophthalmichthys nobilis*
Mortality: Mortalities primarily associated with heavy infections, particularly in young fish, mainly debilitating effects
Temperature: Parasitisation greater with increasing temperature
Prevention and treatment: Infection can be reduced or prevented by good husbandry practices and avoiding introduction of contaminated fish stock. Organophosphate application not recommended. Some biological control strategies exist

15.4 Lernaeosis

15.4.1 Pathogen

This disease is caused by members of the family Lernaeidae, which comprises over 50 species within nine genera. Whilst there is much debate on the classification of this family of parasitic crustaceans, it is recognised that in carp the most important cause of lernaeosis is *Lernaea cyprinacea*, commonly called 'anchorworm', although this common name has been used for other *Lernaea* spp. It is the female of this parasite that embeds within the skin of the fish. Whilst the juvenile female is similar in structure to free-living copepods, the parasite undergoes an extensive metamorphosis after attachment to the fish. This includes obscuring of the segmentation and the growth of a complex attachment organ comprising two pairs of horn-shaped cephalic appendages, which anchor the parasite to the subdermal tissues of the carp. Closer examination of the parasite also reveals that several appendages are retained during the metamorphosis. The parasite is up to 12 mm in length and thus is easily visible with the naked eye.

15.4.2 Epidemiology and host range

Lernaea spp. have a worldwide distribution and occur on a range of freshwater fish such as salmonids, eels, catfish and several cyprinid species, including carp. The type species *L. cyprinacea* is recognised throughout the world as an economic threat and problem in carp production. *L. cyprinacea* has been found on the surface of a range of carp, including common carp, catla, rohu, mrigal, grass carp, silver carp, bighead carp, fringe-lipped peninsula carp (*Labeo fimbriatus*), and ornamentals such as Prussian carp (*Carassius gibelio*) and goldfish. Other parasitic species such as *Lernaea polymorpha* have a somewhat limited range, being recorded in Asian rohu, mrigal and catla (*Gibelion catla*); and *Lernaea oryzophila* occurring on rohu and grass carp; whilst *Lernaea ctenopharyngodonis* has apparently only been found on grass carp. The classification of *Lernaea* spp. is somewhat controversial, leading to suggestions that *L. cyprinacea* is located on crucian carp and goldfish, whilst *L. elegans* is the species that has a wider range of hosts. This has led to several reports simply noting the presence of *Lernaea* species in fish.

The life cycle of *Lernaea* spp., whilst incorporating several larval stages, is relatively simple and comprises free-living stages and primarily a single host, although occasionally some species incorporate a second fish host. There are usually three free-living nauplius stages and five copepodid stages, the latter being parasitic and residing primarily on the gills. The female, which undergoes considerable transformation, is parasitic on the surface of the fish, whilst the male usually dies after copulation. The development of the parasite is temperature dependent, although this may also rely on species, for example in *L. cyprinacea* the optimum temperature range is 20–25°C, whilst the generation time varies, for example 25 days at 20°C, 17 days at 30°C. This means that several generations of the parasite can occur within 1 year, giving rise to great reproductive potential and rapid serious impacts on carp production. Seasonal variation in infection also occurs, which appears to be dependent on parasite species and geographical location. It has also been shown that large fish are more prone to infection, although mass mortalities occur in small hosts. The location of the female on the surface of the fish is also thought to be affected by water flow. In culture ponds with slow movement of water, attachment of the female occurs over all of the body, whilst in slowly flowing waterbodies, parasites tend to congregate at the base of the fins.

15.4.3 Pathology

There are two components related to the pathological effects of lernaeosis that are linked to the two stages of parasitic infection. Since the copepodid stages reside on the gill and are motile, they can induce damage on the gill filaments by their grazing action. They feed on blood, mucin and gill tissue, and can cause tissue damage, necrosis, epithelial hyperplasia, telangiectasis and haemorrhage. In high infections, gaseous exchange is compromised; and in small cyprinids, mass mortalities can occur. It is the more conspicuous and larger female parasite, where pathological effects are more notable. The anchoring cephalic lobes, feeding mode and invasion of secondary infections can cause severe changes within the body of the infected fish. The penetration of the cephalic appendages into the subdermal muscle causes hyperaemia and local swelling (Fig. 15.5). Haemorrhagic alterations occur and a local ulcer lesion develops. Necrosis can occur throughout the attachment site from the skin to the deep muscle layers. The host reaction tends to form a capsule around the head of the parasite, possibly as a mechanism to restrict damage locally. This is not always successful and in smaller carp the parasite can penetrate into the peritoneal cavity, inducing lethal peritonitis. On some rare occasions the parasite may also affect internal organs such as the kidney, gut wall and liver. In ornamental fish the presence of external parasites and the disfiguring effects of the pathological response produced can affect their market value.

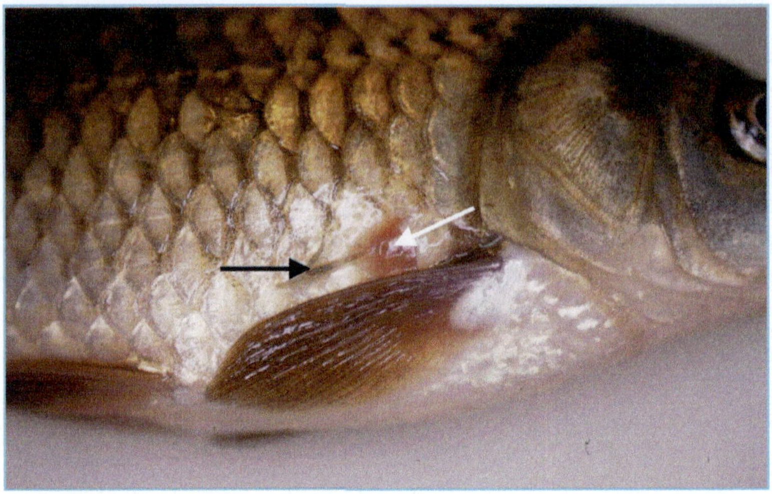

Figure 15.5 Adult female *Lernaea* (black arrow) attached to the lateral aspect of a carp. Attachment occurs via cephalic appendages, which are embedded in the musculature of the host and which invoke local swelling and haemorrhages (white arrow). Courtesy of the Environment Agency, National Fisheries Laboratory, UK.

Effects on the blood have also been noted, such as an increase in number of leucocytes, including neutrophils, eosinophils, lymphocytes, basophils, monocytes and thrombocytes, in blood of infected fish. In addition, the values of haematocrit, haemoglobin and red blood cell counts are lower in infected fish than in healthy individuals. There is also some evidence of the development of an acquired immunity, as fish that have been infected before tend to have lower levels of infection and a reduction in parasite survival on reinfection.

Infection of carp with lernaeosis has also been associated with behavioural changes, such as erratic swimming and a tendency to rub themselves against underwater objects. In addition, the pathological changes can induce emaciation and serious weight loss, affecting the economic viability of the carp production unit.

15.4.4 Detection

Whilst detection of lernaeosis is not usually a challenge, due to the large size of the female parasite on the surface of the fish, the establishment of the species involved can be problematic. The head of the parasite with the associated cephalic lobes is embedded within the fish and exhibits considerable morphological variation, depending on host species and attachment site. Such a diverse variation of structure means that identification may best be undertaken based on appendages and body form of the exposed body. The parasite can be gently removed from the skin, but care must be taken not to detach the embedded head region. Extracted parasites can be processed using standard whole mount procedures. Alternatively, histological preparations can be prepared by removing the parasite and associated skin/muscle around the attachment point, which can then be processed using standard histological preparations and appropriate sectioning. There is little information on the genome sequence of *Lernaea* species, although recently transcriptome analysis of the adult and free-living stages of *L. cyprinacea* was carried out using Illumina HiSeq 2000 sequencing. Microsatellite markers were also identified, which may aid population genetic studies in various geographical isolates of *Lernaea*.

15.4.5 Prevention strategies and treatments

As with all disease control measures, the best way of maintaining productivity of a carp production unit is the application of a rigorous management plan, which includes facilities and mechanisms for quarantining any imported fish. The generation time of *Lernaea* spp. means that the time for development of the female parasite can be fairly predictable; for example, *L. cyprinacea* takes approximately 3 weeks for the females present on the gill to mature and attach to the skin. A quarantine period of at least 21 days is therefore required. As with most external parasites, the most effective way of eradicating an infection is by draining the pond, allowing it to dry and treating with lime.

Recently there have been some interesting developments in other mechanisms of control. In Japan it has been suggested that predatory free-living cyclopoid copepods may reduce larval stages of *L. cyprinacea*, and the genome sequence of *L. cyprinacea* may, in the future, lead to the development of novel control mechanisms.

It would also appear that in some circumstances some species of carp are more resistant to *Lernaea* infection than others. For example, in monoculture conditions common carp, rohu and orangefin labeo (*Labeo calbasu*) did not appear to acquire *Lernaea* infection whereas grass carp, silver carp, catla and fringe-lipped peninsula carp were susceptible to parasitisation. Interestingly this 'resistance' was not maintained when rohu and common carp were cultured together with a susceptible species, although orangefin labeo maintained its resistance even under polyculture conditions. In addition, in polyculture conditions where carp were reared in association with rice, infection by *Lernaea* spp. was lower than in monoculture conditions. This may be due to the application of pesticides such as diazinon in rice fields

A range of chemical treatments have been proposed for the treatment of lernaeosis. However, their use should be carefully considered not only in terms of economic benefits, but also the possible impacts on the environment. In addition, the application of chemical treatments may be controlled by local legislative laws and may be toxic to the host. Common salt (NaCl) has been recommended as being effective against larval stages of *Lernaea* (0.8–1.1% for 3 days), although some fish are sensitive and trial tests should always be performed. Formalin can also kill larval stages at concentrations of 250 ppm when applied in 30–50 min baths, but again application is time consuming and can be hazardous to the farm worker and the fish. Within the ornamental sector, potassium permanganate has been used to control anchorworm in goldfish when applied at 12.5 mg/l/h in glass aquaria. Several organophosphate insecticides have been recommended, but care must be taken as these may only be effective against larval stages; hence repeat applications for at least a month may be required and there have been reports that *L. cyprinaea* may be developing resistance. Some success has been obtained, however; for example, an organophosphate (2,2 dichlorovinyl dimethyl phosphate) has been used to treat lernaeosis successfully at 0.2 ppm in rohu, catla, grass carp, silver carp and bighead carp. A commonly used organophosphate (O,O-dimethyl-2,2,2-trichloro-1-hydroxyethyl phosphonate) applied at 0.2 ppm is effective in controlling lernaeosis in ponds containing major carp (rohu, catla, mrigal), grass carp and silver carp.

FACTS SHEET

Disease: Lernaeosis

Pathogen: Mainly *Lernaea cyprinacea*, but other problematic species include *Lernaea polymorpha*, *Lernaea oryzophila*, *Lernaea ctenopharyngodonis*

Susceptible hosts: Many carp species, including *Cyprinus carpio*, *Gibelion catla*, *Hypophthalmichthys molitrix*, *Labeo rohita*, *Cirrhinus mrigala*, *Ctenopharyngodon idella*, *Carassius gibelio*, *Hypophthalmichthys nobilis*, *Labeo fimbriatus*, *Carassius auratus*

Mortality: Mortalities primarily associated with heavy infections, particularly in young fish, mainly debilitating effects

Temperature: Parasitisation greater with increasing temperature

Prevention and treatment: Infection can be reduced or prevented by good husbandry practices and avoiding introduction of contaminated fish stock. Therapeutic strategies are available

15.5 Selected References for Further Reading

General parasitic crustacean biology

Alston, S. and Lewis, J.W. (1994) The ergasilid parasites (Copepoda: Poecilostomatoida) of British freshwater fish. In: Pike, A.W. and Lewis, J.W. (eds) *Parasitic Diseases of Fish*. Samara Publishing, Cardigan, UK, pp. 171–188.

Hoole, D., Bucke, D., Burgess, P. and Wellby, I. (2001) *Diseases of Carp and other Cyprinid Fish*. Blackwell Science, Oxford, UK.

Piasecki, W. and Avenant-Oldewage, A. (2008) Disease caused by Crustacea. In: Eiras, J.C., Segner, H., Wahli, T. and Kapoor, B.G. (eds) *Fish Diseases*, Vol. 2. Science Publishers, Enfield, New Hampshire.

Schäperclaus, W. (1991) *Fish Diseases*, 5th edn. Amerind Publishing Co. / Oxonian Press, New Delhi, India.

Argulus spp.

Basu, M., Lenka, S., Paichha, M., Swain, D., Patel, B. *et al.* (2016) Immunoglobulin (Ig) D in *Lebeo rohita* is widely expressed and differentially modulated in viral, bacterial and parasitic antigenic challenges. *Veterinary Immunology and Immunopathology* 179, 77–84.

Dash, G., Parida, S.K. and Sahoo, S.N. (2009) Histopathological observations on carps infected with *Argulus* in freshwater bheries of West Bengal. *Journal of Applied Bioscience* 35, 159–162.

Forlenza, M., Walker, P.D., Vries, B.J., Wendelaar Bonga, S.E. and Wiegertjes, G.F. (2008) Transcriptional analysis of the common carp (*Cyprinus carpio* L.) immune response to the fish louse *Argulus japonicus* thiele (Crustacea: Branchiura). *Fish and Shellfish Immunology* 25, 76–83.

Hemaprasanth, K.P., Kar, B., Garnayak, S.K., Mohanty, J., Jena, J.K. *et al.* (2012) Efficacy of two avermectins, doramectin and ivermectin against *Argulus siamensis* infestations in Indian Major Carp, *Labeo rohita*. *Veterinary Parasitology* 190, 297–304.

Huising, M.O., Stolte, E., Flik, G., Savelkoul, H.F.J. and Verburg-an Kemenade, B.M.L. (2003) CXC chemokines and leukocyte chemotaxis in common carp (*Cyprinus carpio* L.). *Developmental and Comparative Immunology* 10, 875–888.

Kar, B., Mohapatra, A., Mohanty, J. and Sahoo, P.K. (2015) Transcriptional changes in three immunoglobulin types of rohu, *Labeo rohita* in response to *Argulus siamensis* infection. *Fish and Shellfish Immunology* 47, 28–33.

Kar, B., Moussa, C., Mohapatra, A., Mohanty, J., Jayasankar, P. *et al.* (2016) Variation in susceptibility pattern of fish to *Argulus siamensis*: Do immune responses of host play a role?. *Veterinary Parasitology* 221, 76–83.

Nagasawa, K., Asayama, T. and Fujimoto, Y. (2022) Redescription of *Argulus mongolianus* (Crustacea: Branchiura: Argulidae), an ectoparasite of freshwater fishes in East Asia, with its first record from Japan. *Species Diversity* 27, 167–179.

Panda, R.P., Chakrapani, V., Patra, S.K., Saha, J.N., Jayasankar, P. *et al.* (2014) First evidence of comparative responses of Toll-like receptor 22 (TLR22) to relatively resistant and susceptible Indian farmed carps to *Argulus siamensis* infection. *Developmental and Comparative Immunology* 47, 25–35.

Sahoo, P.K., Mohanty, J., Garnayak, S.K., Mohanty, B.R., Kar, B. *et al.* (2013) Genetic diversity and species identification of *Argulus* parasites collected from major aquaculture regions of India using RAPD-PCR. *Aquaculture Research* 44, 220–230.

Ergasilosis

Chang, M.X., Nie, P., Liu, G.Y., Song, Y. and Gao, Q. (2005) Identification of immune genes in grass carp, *Ctenopharyngodon idella*, in response to infection of the parasitic copepod *Sinergasilus major*. *Parasitology Research* 96, 224–229.

Nie, P. and Yao, W.J. (2000) Seasonal population dynamics of parasitic copepods, *Sinergasilus* spp. on farmed fish in China. *Aquaculture* 187, 239–245.

Lernaeosis

Hossain, Md.M.M., Ferdoushi, J. and Rupom, A.H. (2018) Biology of anchor worms (*Lernaea cyprinacea*). *Journal of Entomology and Zoology Studies* 6, 910–917.

Iqbal, Z., Shafqat, A. and Haroon, F. (2012) *Lernaea* diversity and infection in Indian and Chinese carps under semi-intensive culture conditions in Lahore, Punjab. *Journal of Animal and Plant Sciences* 22, 923–926.

Pallavi, B., Shankar, K.M., Abhiman, P.B. and Iqlas, A. (2015) Identification of putative genes involved in parasitism in the anchor worm, *Lernaea cyprinacea* by *de novo* transcriptome analysis. *Experimental Parasitology* 153, 191–197.

Pallavi, B., Shankar, K.M. and Abhiman, P.B. (2016) *In silico* prediction of potential vaccine candidates of *Lernaea cyprinacea*: An ectoparasite of Indian major carps. *Indian Journal of Biotechnology* 15, 601–603.

Tufail, H., Qureshi, N.A., Khan, N., Iqbal, K.J. and Maqbool, A. (2017) Prevalence and effects of *Lernaea cyprinacea* (anchor worm) on the growth, skin histopathology and hematology of *Catla catla*. *Indian Journal of Fisheries Science* 16, 341–351.

Chapter 16

Leeches

16.1 Pathogen

Leeches are a common ectoparasite on many fish species, and are perhaps the most well-known infection because of their conspicuous large size when fully grown. They cause a disease known as 'hirudinosis', and many belong to the family Piscicolidae (the fish leeches), genus *Hemiclepsis* (flatfish leeches). The leech body is segmented, although this segmentation may be hidden by the external rings called annuli. The number of these annuli varies depending on species; for example, *Piscicola geometra* can be up to 5 cm in length and have 14 annuli. *Hemiclepis marginata*, which is up to 3 cm long, comprises four longitudinal series of 'warts' along its dorsal surface. Perhaps the most conspicuous features of the anatomy of the majority of leeches are the anterior and posterior suckers (Fig. 16.1), which together with the mouth are associated with the induction of a pathological response in the host.

16.2 Epidemiology and Host Range

Many leech species have a very diverse range of hosts, including many carp species. For example, *Piscicola geometra* occurs on common carp (*Cyprinus carpio*), Prussian carp (*Carassius gibelio*), pike (*Esox lucius*) and even trout, whilst *Hemiclepis marginata* has been located on common carp, other cyprinids and salamanders (*Anura*). *Limnotrachelobdella sinensis*, occurring in the Far East, has been found on crucian carp (goldfish, *Carassius auratus*; and Japanese white crucian carp, *Carassius cuvieri*) and common carp.

DOI: 10.1079/9781836990024.0016

Figure 16.1 Whole preparation of *Piscicola geometra* showing conspicuous anterior (A) and posterior (P) suckers. Scale bar 1 mm. Courtesy of the Environment Agency, National Fisheries Laboratory, UK.

It has been suggested that the location of the leech species within a waterbody may affect the fish species or age infected. For example, *Piscicola* occurs more often on adult carp residing at the bottom of the waterbody, whilst *H. marginata*, which lives in shallow to moderately deep water, appears more frequently on young carp fingerlings.

The laying of eggs and the development of the leech within the waterbody is dependent on temperature, being more prominent in spring/summer months, when the temperature is high. Leeches leave their host fish primarily when the parasites are engorged with blood.

16.3 Pathology

The presence of several leeches on a fish can, over a period of time, cause adverse effects, mainly associated with the injuries occurring on the fish by the attachment of the suckers and the feeding process. It has been estimated that 100 leeches on a juvenile fish can suck 230 mm^3 of blood from the host and lead to a reduction of blood protein content from 5 to 3.5%. The location of the infection can also affect the fish; for example, large infections around the fish's mouth may impede food intake (Fig. 16.2). Leeches can be more pathogenic when attached to the gills. In such instances, degeneration and hyperplasia of the epithelial cells of the gill filament occur, leading to fusion of the lamellae.

Leeches, due to their haematophagous feeding, are also known to be a transmitter of various pathogens such as *Cryptobia*, *Trypanoplasma borreli*, *Trypanosoma carassii*, or SVCV, which results in a range of pathologies associated with these infections.

16.4 Detection

The large size of adult leeches makes them easily spotted on the fish's surface and morphological appearance is still used to make an initial diagnosis to genus level. For example, the body of *Piscicolidae* spp. is circular in cross-section, whilst in Glossiphoniidae, containing

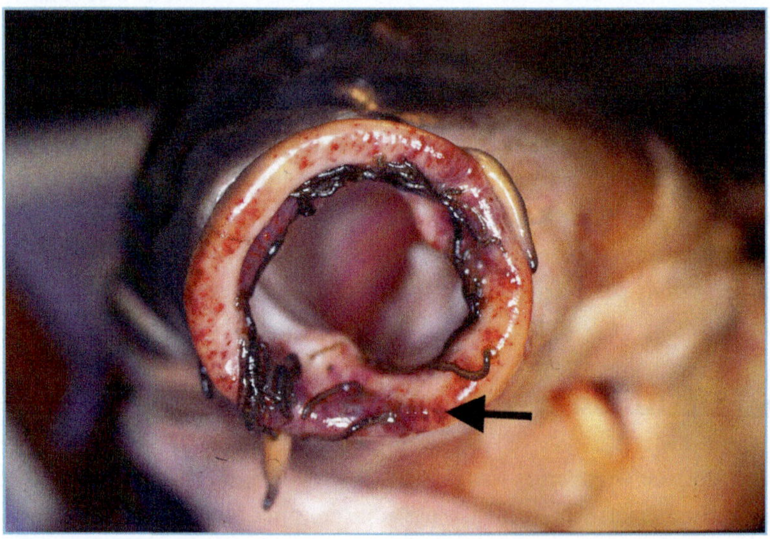

Figure 16.2 Leeches surrounding the mouth of common carp. Note haemorrhages (arrow) present on the 'lip' of the carp. Courtesy of the Environment Agency, National Fisheries Laboratory, UK.

Hemiclepis spp., the body is flat in cross-section. Recently molecular studies have been used to characterize leeches. Such techniques have utilised mitochondrial DNA, or a combination of 18S rDNA, cytochrome *c* oxidase subunit I (CO-I) and nicotinamide adenine dinucleotide dehydrogenase subunit I (ND-I), in association with morphological data. The phylogenetic relationships of the fish leech subfamily Pontobdellinae have also been studied utilising mitochondrial (12S rDNA, COI, tRNA-Leu, ND1) and nuclear (28S rDNA) DNA sequences under maximum likelihood, Bayesian inference and parsimony.

16.5 Prevention Strategies and Treatments

Prevention of an infection of leeches can be difficult, but the leech burden within the waterbody can be reduced by sanitary measures that reduce the introduction of eggs/cocoons via equipment or water. In particular, *Piscicola* can hide in underwater vegetation awaiting a fish to attach to, therefore the removal of vegetation may reduce the leech burden on the host. Leeches are often carried from one water to another by waterfowl, such as ducks or swans. The dependence of most leeches on the fish for food and the susceptibility to dry conditions means that management practices can be implemented to reduce leech burden. *Piscicola* can remain in water for approximately 3 months without the fish host, whilst *Hemiclepsis* can survive for nearly 10 months. It is therefore important that, when ponds are drained, small water pools should not remain. To facilitate this, these small water areas can be treated with lime, for example 1g of quicklime to 2 litres of water. Leeches can be removed from the fish either by using forceps, although this can damage the fish, or by the application of external treatment such as lime, potassium permanganate, concentrated chlorinated water or salt up to 10%. Carp farmers use a short bath for a few seconds in water containing quicklime in order to kill leeches attached to fish. Trial applications are strongly recommended before any large treatment process.

FACTS SHEET

Disease: Hirudinosis
Pathogen: Piscicolidae – the fish leeches and *Hemiclepsis* flatfish leeches
Susceptible hosts: Range of cyprinids such as *Cyprinus carpio*, *Carassius gibelio*, *Carassius auratus*
Mortality: Severe haemorrhaging on areas such as gills may cause death
Temperature: Disease tends to be more prevalent at higher temperatures
Prevention and treatment: Good husbandry practices, including reduction of debris in ponds. Leeches can be removed physically. Application of a range of substances such as lime, potassium permanganate, concentrated chlorinated water or salt

16.6 Selected References for Further Reading

General biology/pathology

Hoole, D., Bucke, D., Burgess, P. and Wellby, I. (2001) *Diseases of Carp and Other Cyprinid Fish*. Blackwell Science, Oxford, UK.

Kearn, G.C. (2004) *Leeches, Lice and Lampreys: A Natural History of Skin, and Gill Parasites*. Springer, Dordrecht, The Netherlands.

Schäperclaus, W. (1991) *Fish Diseases Volume 2*, 5th edn. Amerind Publishing Co./Oxonian Press, New Delhi, India.

Molecular studies/speciation

Utevsky, S.Y. and Trontelj, P. (2004) Phylogenetic relationships of fish leeches (Hirudinea, Piscicolidae) based on mitochondrial DNA sequences and morphological data. *Zoologica Scripta* 33, 375–385.

Williams, J.I. and Burreson, E.M. (2006) Phylogeny of the fish leeches (Oligochaeta, Hiruinida, Piscicolidae) based on nuclear and mitochondrial genes and morphology. *Zoologica Scripta* 35, 627–639.

Chapter 17

Co-Infections

17.1 Introduction

Fish can be hosts of a very diverse and abundant number of biological entities, which mostly are symbiotic and are thus not harmful for the host. However, the border between symbionts and pathogens can be fluid and may change rapidly in the case of stress, immunosuppression or breach of mucosal tissues. One triggering factor can be an infection with an obligatory pathogen, which enables subsequent infection with further facultative pathogenic organisms. Hence, the same fish specimen can concurrently be the carrier of multiple infections with divergent obligatory and facultative pathogens. The collection of all pathogenic microorganisms inhabiting a single host is called the pathobiome. The pathobiome can consist of obligatory pathogens, such as virulent viruses and pathogenic bacteria or parasites, as well as facultative pathogens, such as bacterial members of the natural flora of skin, gills or the intestine. During the infection, the primary pathogen can trigger changes in other elements of the symbiotic bacterial flora, making them dangerous for the host. This may be an element of virulence that the pathogen did not take into account during its evolution and which could lead to death of the host. For example, the pathology of certain pathogens involves the induction of severe stress, immunosuppression, or the capacity to dismantle the barrier function of mucosal tissues, such as skin, gill and intestine, which could be very dangerous for the host in the presence of facultative pathogens. Immunosuppression, stress or disruption of mucosal layers could create an ecological niche for additional facultative pathogens to flourish. For example, the presence of cyprinid herpesvirus 3 (CyHV-3) infection is very

 Carps: A Practical Approach to Disease Control and Health Management (eds David Hoole *et al.*)
DOI: 10.1079/9781836990024.0017

often associated with the occurrence of secondary co-infections with *Aeromonas* spp. or *Flavobacteria* spp., which increase the speed and severity of mortality. As a consequence, an infection with several ectoparasites, such as *Ichthyobodo necator*, *Ichthyophthirius multifiliis*, *Dactylogyrus* spp. and *Gyrodactylus* spp., may be associated with CyHV-3 infection. The abundance of *Flavobacterium branchiophilum* may also be associated with infection with carp oedema virus (CEV) resulting in increased pathological changes as well as mortality rates.

On the other hand, certain pathogens can induce immune responses that can protect the host from infection with other pathogens. To curtail viruses, an antiviral response can be induced, which can protect the fish from the infection with additional pathogens. Certain microbes can also produce substances, such as antimicrobial peptides and proteases, which can limit the growth of other pathogens. This makes interactions between potential pathogens dynamic and sometimes difficult to interpret.

In the case of a co-infection, a cumulative impact of the infection by virus, parasites and bacteria and the aquatic environment, which is hostile for most of the fish, creates a complex challenge for the host to overcome. Multiple very complex interactions occur between all of the elements of the pathobiome, causing a multifactorial complex disease. Such multifactorial diseases require treatment strategies beyond classical protocols designed to contain a single pathogen.

17.2 Complex Diseases of Mucosal Tissues

The mucosa, which is an interface between the surrounding environment and the body of the fish, is a place of a complex network of interactions. The mucosal tissues have their associated bacterial flora, which can be a source of facultative pathogens. Therefore, so-called 'complex diseases' of a multifactorial aetiology are most often present in gill, skin and intestine. Such a multifactorial disease is very often associated with the presence of multiple secondary bacterial and/or parasitic infections (Fig. 17.1). This is often related to a

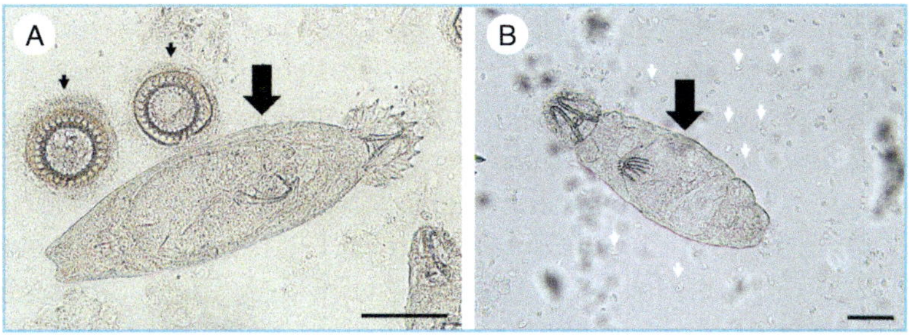

Figure 17.1 Light microscopy of the skin mucus wet mount collected during onset of complex skin diseases in common carp. (A) *Gyrodactylus* (black arrow) and *Trichodina* (black arrowheads) co-infection. (B) *Gyrodactylus* (black arrow) surrounded by multiple *Ichthyobodo necator* (white arrowheads) co-infection. Indicative scale bar 100 µm. (A) courtesy of the Environment Agency, National Fisheries Laboratory, UK.

disruption of mucosal layers in skin, gills and intestine, causing an increased challenge for the infected fish.

17.3 Complex Gill Diseases

One example of a multifactorial disease on gills can be found in certain cases of koi sleepy disease in which the primary viral pathogen CEV and multiple secondary pathogens are flourishing at the same time. CEV can be associated with secondary co-infections with *Flavobacterium* spp. that grow rapidly on virus-infected gills. This increases the speed and severity of the pathological changes, and raises the probability and the rate of mortality. In field cases, several ectoparasites, including *Ichthyobodo necator*, *Ichthyophthirius multifiliis* and *Dactylogyrus* spp., are also located on the gills of CEV-infected fish, showing that both bacterial and parasitic infections can be promoted by this poxvirus infection.

17.4 Complex Skin Diseases

Multiple infectious agents can occur on the skin of carp. For example, CyHV-3 infection is very often associated with the occurrence of secondary coinfections with *Aeromonas* spp. or *Flavobacteria* spp. contributing to the severity of pathological changes in skin. Several ectoparasites, such as *Ichthyobodo necator*, *Ichthyophthirius multifiliis*, *Dactylogyrus* spp. and *Gyrodactylus* spp., can increase the intensity of disease presentation. Interestingly, vaccination for CyHV-3 protects the fish from disruption of mucosal layers and from bacterial co-infections, even in an environment with high infection pressure by secondary pathogens. This suggests that CyHV-3 is the primary pathogen in this example of a complex skin disease, and is required for the disruption of skin mucosal barrier, allowing the secondary bacteria to grow.

17.5 Complex Intestine Diseases

Disruption of barrier integrity and permeability of the gut mucosa can lead to increased infection pressure from the bacteria residing in the intestine. For example, certain parasites can contribute to the development of bacterial enteritis. The bacteria can penetrate the gut wall, increasing inflammation and possibly spreading to additional organs, causing a systemic disease. The infection of juvenile common carp or goldfish with the gut-dwelling coccidian parasite, *Goussia carpelli*, causes a severe enteritis with destruction of the intestinal mucosa. Affected specimens suffer from an electrolyte loss and secondary bacterial infections. In addition, the infestation of carp with tapeworms, such as *Khawia* spp., alters the composition of the intestinal microbial community towards an increased ratio of pathogenic bacteria such as *Plesiomonas* spp.

17.6 Detecting and Treating the Co-Infections

The co-infections leading to the occurrence of multifactorial diseases caused by complex pathobiomes bring unique challenges to diagnosis and treatment of diseases. The detection of the presence of one pathogen should not stop the diagnostician from looking further. Most pathogens induce clinical signs only when a certain threshold of abundance is crossed. Therefore, more emphasis should be put on a quantitative evaluation of pathogen abundance, which can help to establish which pathogen is a culprit of the visible pathology, and this should determine the treatment strategies. In certain co-infections some pathogens are difficult to treat directly, for example infections with viruses; however, the treatment of the co-infecting pathogens, such as bacteria, can limit fish losses. Some pre-treatments of co-infecting pathogens are required, for example getting rid of an *Ichthyobodo necator* infestation before increasing the water temperature to help the fish to clear the viral infection. This will prevent *Ichthyobodo necator* from flourishing in increased temperature and possibly killing the fish. In some cases the treatment of such secondary infections is needed to increase the survival chances of the infected host by lowering the pathology related to the secondary pathogen burden.

17.7 Selected References for Further Reading

Adamek, M., Teitge, F., Jung-Schroers, V., Heling, M. and Gela, D. (2018) Flavobacteria as secondary pathogens in carp suffering from koi sleepy disease. *Journal of Fish Diseases* 41, 1631–1642.

Adamek, M., Teitge, F. and Steinhagen, D. (2019) Quantitative diagnostics of gill diseases in common carp: Not as simple as it seems. *Diseases of Aquatic Organisms* 134, 197–207.

Gorgoglione, B., Bailey, C., Fast, M., Bass, D. and Saraiva, M. (2020) Co-infections and multiple stressors in fish. *Bulletin of the European Association of Fish Pathology* 40, 4–19.

Okon, N.M., Okocha, R.C., Taiwo, A.B. and Michael, F.B. (2023) Dynamics of co-infection in fish: A review of pathogen–host interaction and clinical outcome. *Fish and Shellfish Immunology Reports* 4, 100096.

Ren, W., Pan, X., Dai, C., Shu, T. and Li, L. (2021) Investigation of cyprinid herpesvirus 2 and bacterial coinfection in carassius gibel. *Aquaculture* 537, 736521.

Chapter 18

Zoonosis

18.1 Introduction

Human contact, either with fish or with the water that they inhabit, can present several health hazards. The World Health Organization defines such hazards as zoonotic diseases, which are infectious diseases that have jumped from a non-human animal into humans, particularly when humans are an accidental host. These zoonotic pathogens may be viral, bacterial or parasitic. When the original host is a fish, these diseases usually enter the human either by contamination, as in the case of many bacterial infections, or by the consumption of raw fish, particularly associated with parasitic infections. Several authorities also define diseases as 'infections in common' where animals and humans acquire infection from common environmental sources. The increase in aquaculture and associated interactions between humans and fish has resulted in fish-borne zoonotic diseases becoming of increasing importance throughout the world, with estimates of over 20 million people being infected. Human infections arise in a range of societies, based on local customs and practices, but can occur in fish-production and fish-processing activities.

18.2 Bacterial Zoonosis

18.2.1 *Mycobacterium*

Mycobacterium spp. are probably the most well-known bacteria that can cause harm to humans. This is manifested as a granulomatous

 Carps: A Practical Approach to Disease Control and Health Management (eds David Hoole *et al.*)
DOI: 10.1079/9781836990024.0018

inflammation of the skin and is commonly known as 'tropical fish tank granuloma', 'fish-fancier's finger', 'fish-keeper's granuloma' or 'fisherman's finger', highlighting the main infection site and fish practices it is associated with (Figs. 18.1 and 18.2). The infection is caused by acid-fast bacteria, primarily *Mycobacterium fortuitum* and *M. marinum*. In some rare cases more intensive involvement of deeper tissues can occur, resulting in a range of pathological effects, such as arthritis, osteomyelitis, tenosynovitis and bursitis. Some of these symptoms can take months to appear.

Recent studies have highlighted some of the genetic differences between *Mycobacterium* isolates from fish and humans, particularly concerning *M. marinum*. Mycobacterial interspersed repetitive unit (MIRU) typing highlights the separation of human and fish isolates, although genetic linkages between mycobacteria from fish and human sources have been demonstrated using amplified fragment length polymorphism (AFLP) and pulsed-field gel electrophoresis (PFGE). In the latter, genomic DNA was digested using restriction enzymes, SpeI and AseI. However, caution is recommended when

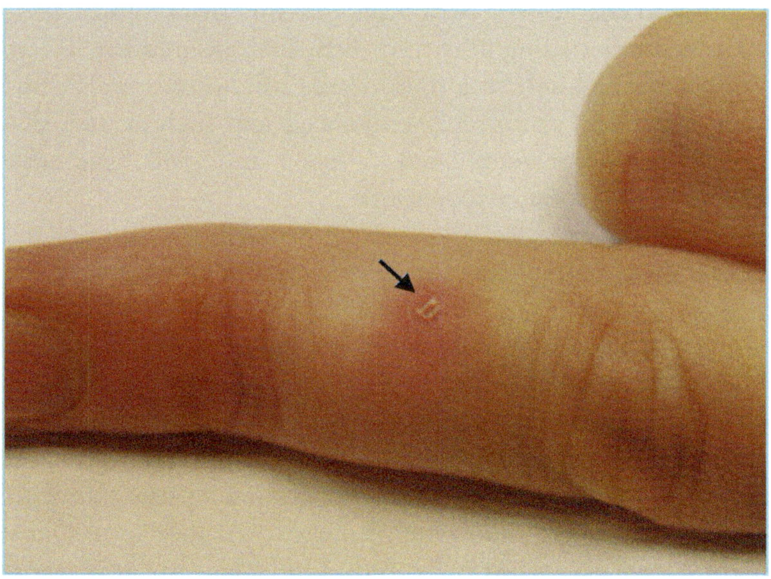

Figure 18.1 Hand showing infection of *Mycobacterium* sp. and typical inflammatory response (arrow).

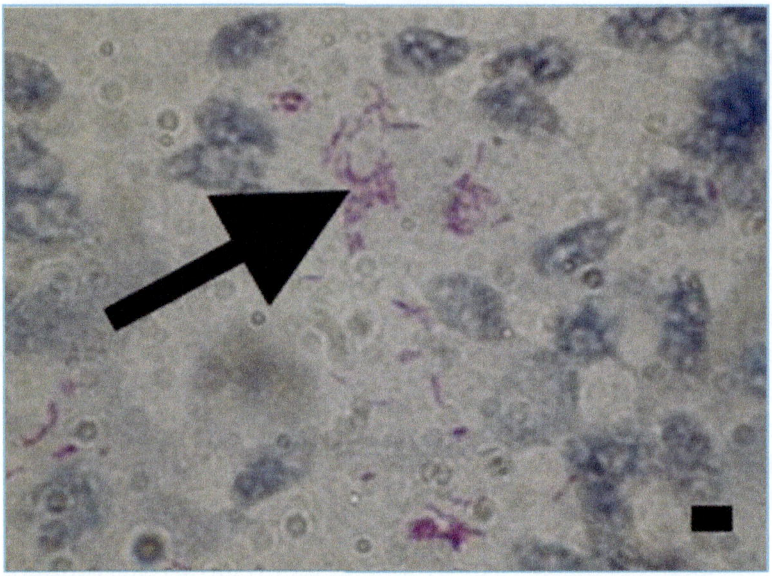

Figure 18.2 Whole stained wet preparation from a *Mycobacterium* sp. infection showing bacterial colonies (arrow). Indicative bar 1 μm.

attempting to characterise *Mycobacterium* spp. from human and fish sources, as studies using aquarium fish, such as goldfish (*Carassius auratus*), have highlighted a high degree of conservation in housekeeping genes and commonly sequenced loci such as 16S rRNA, although analysis of genes such as *hsp*65 and *rpo*B have enabled differentiation of some mycobacteria.

18.2.2 Accidental bacterial infections of humans

Many bacterial species which, whilst not being technically zoonotic, can be considered as pathogenic to humans are acquired by close contact between humans and fish or their water environment. Such bacterial species are associated with gastrointestinal disorders such as *Campylobacter jejuni*, which causes gastroenteritis, *Edwardsiella tarda* and *Escherichia coli*, which lead to diarrhoea, or *Salmonella* spp. associated with food poisoning. Other bacterial species, such as *Leptospira interrogans*, are associated with more widespread body effects, whilst others, such as *Erysipelothrix rhusiopathiae*, have been linked to skin rashes. Such infections are usually associated with ponds that contain dead or dying fish, or in food outlets where good hygiene is not always practised.

18.2.3 Treatment and prevention of bacterial zoonosis and infections

It is perhaps not surprising that the control and treatment of bacterial zoonoses has been targeted at human infections, where a range of antibiotics can be utilised, for example rifampin and minocycline/ethambutol. However, the threat of inducing antibiotic resistance and the concept that prevention is better than cure means that the implementation of strict working practices should be enforced. This applies to all aspects of carp production, including the aquaculture and ornamental sectors, as well as fish processing, veterinary practice and utilisation of fish in experimental procedures.

Good working practices include the following.

- Abstain from eating, smoking or drinking whilst dealing with fish.
- Wear eye or respiratory protection where appropriate.
- Avoid placing broken skin in contact with fish or water in ponds, tanks and other aquatic environments. Always wash and cover abrasions with clean waterproof dressings.
- Wear dedicated gloves and protective clothing when handling aquarium/pond water, fish, fish tissues, body fluids and waste, and wash hands after contact. Launder protective clothing as appropriate.
- If wounds show signs of infection such as swelling, redness, pain and draining fluids, with or without a fever, seek medical treatment.
- Always keep fish areas clean and disinfect equipment after using it on fish or in fish areas. Use cleaning techniques that do not aerosolize dirty water or other materials.

It is important that operatives should familiarise themselves with the fish and the environment they are working in and the potential zoonotic diseases that can be acquired.

The disinfection of materials contaminated with mycobacteria can be difficult. The thick cell walls of the bacteria and the formation of bacterial biofilms can make atypical mycobacteria resistant to disinfectants normally used in fish practices, such as quaternary ammonia compounds and sodium hypochlorite. Disinfectants that have been shown to be effective include high concentrations of alcohol, strong sodium chlorite solutions and phenols. In addition, care must be taken in the disposal of aquarium water.

FACTS SHEET

Disease: Tropical fish tank granuloma, fish-fancier's finger, fish-keeper's granuloma, fisherman's finger
Pathogen: *Mycobacterium fortuitum, M. marinum*
Susceptible host: Main pathogenic effects in human host. Mainly associated with tropical fish keeping
Mortality: Does not usually cause mortalities in humans
Temperature: Infection in humans can be greater with increasing temperature
Prevention and treatment: Infection of humans can be reduced or prevented by good husbandry practices and avoiding contact with contaminated fish stock. A range of antibiotic treatments for humans available.

18.3 Parasitic Zoonosis

Parasitic zoonoses occur throughout the world, but predominate in societies where raw fish is consumed. Whilst a range of parasitic groups, such as trematodes and nematodes, occur in many freshwater and marine species of fish, it is primarily the trematodes that are associated with several pathogenic diseases in humans obtained by the consumption of cyprinids, including carp. Fish-borne zoonotic trematodes (FZTs, as they are sometimes referred to) are well known to cause liver and intestinal issues in humans. It has been estimated that approximately 680 million people in the world are at risk of infection and that more than 20 million humans are already infected with the liver flukes (*Clonorchis sinensis* and *Opisthorchis* spp.). Members of the Heterophyidae family also cause significant zoonotic intestinal diseases of humans, the most important being *Heterophyes heterophyes* and *Metagonimus yokogawai*. Human infections of FZTs are frequently reported in the Middle East and Asia, and predominate in the Philippines, Indonesia, Thailand, Vietnam, China, Korea and Japan, but may occasionally occur in Central Asia and Europe.

18.3.1 Life cycles

Clonorchis sinensis, which causes the disease clonorchiasis, is perhaps the most well-known and understood FZT and has been recorded in over 100 fish species, including 71 species of cyprinids. Adult worms are usually mature in the bile ducts of a range of mammals, including rats, cats, pigs and dogs. The flukes are approximately 8–15 mm in length and 4 mm in width, and are characterised by the presence of their reproductive system, which comprises two highly branched testes and a prominent seminal vesicle. Eggs are usually liberated from faeces of the definitive host, and when they enter water the enclosed miracidium hatches from the egg. This stage then infects the snail intermediate host, which includes *Parafossarulus manchouricus*, *Bulimus fushsiana*, *Thiara granifera*, *Melanoides tuberculate*, *Bithynia longicornis*, *Semisulcospita libertina* and *Assiminea lutea*, in which the sporocyst and redial stages develop. Cercariae are liberated from the snail and penetrate the fish, the second intermediate host, in which the infective metacercaria develops in a range of organs, including the muscle tissue. Opisthorchiasis is caused by *Opisthorchis* spp., which have a similar life cycle to the more predominant *C. sinensis. Opisthorchis* species infect over 35 fish species, including cyprinids. Two parasitic species *O. viverrini* and *O. felineus* cause the major health issues in humans and also occur in the biliary ducts of a range of mammals, such as civets, cats, dogs, foxes and pigs. The two parasite species differ in that *O. felineus* has less lobate testes situated a greater distance from the ovary and a different vitellaria arrangement than *O. viverrini*. The two species differ from *C. sinensis* in the structure of the testes in the adult and in the arrangement of the excretory system in the cercarial and metacercarial stages. Snail intermediate hosts in South-east Asia include *Bithynia gonoiomphalus*, *B. funiculate* and *B. laevis*. Fish-borne trematode zoonotic infections can also result in infections of the human intestinal tract. Such flukes belonging to the families Echinostomatidae, Heterophyidae and the Nanophyetidae predominate, although it is primarily the Heterophyidae that are found in carp species. Of the approximately 35 species of Heterophyidae flukes that infect humans, *Heterophyes heterophyes*, *Haplorchis pumilio* and *Metagonimus yokogawai* predominate in carp. FZTs, therefore, infect a range of carp, including goldfish, common carp (*Cyprinus carpio*), koi carp, grass carp (*Ctenopharyngodon idella*), rohu (*Labeo rohita*), silver carp (*Hypophthalmichthys molitrix*), mud carp (*Cirrhinus molitorella*) and barbel chub (*Squaliobarbus curriculus*).

18.3.2 Epidemiology and pathology

The medical importance of AZT infections in humans has led to a range of pathological investigations being carried out. Liver flukes are perhaps the most serious, as their presence can cause chronic irritation and subsequent enlargement of the liver and bile ducts with accompanying jaundice, pancreatitis, cholangitis, anorexia and occasionally cholangiocarcinoma. In contrast, intestinal flukes can be asymptomatic, although intestinal discomfort can occur associated with diarrhoea, nausea and vomiting. Very rarely infections can be more problematic if parasite eggs become entrapped in vital organs such as spinal cord, heart and brain. Such health issues mean that care should be taken during any examination of fish thought to be infected with FZTs.

Trematode zoonoses are particularly prevalent in integrated agriculture/aquaculture systems practised in Asia. Whilst such systems are sustainable and economically viable, the use of recycled wastes (particularly manure) can result in metacercarial cysts occurring in fish and increase in prevalence in localised communities. Experimental studies utilising *Cyprinus carpio* have revealed that small fish, i.e. under 1 g in weight, are highly prone to cercarial attack by FZTs. Although immune factors cannot be ruled out, the thin skin of these fish may be an important factor relating to increased susceptibility. However, metacercariae of heterophyidae may persist in *C. carpio* for a long time and may still be present at harvesting.

18.3.3 Isolation of metacercariae from fish

Metacercariae of FZTs can occur in most tissues within carp, but it is those that predominate in muscle that cause the major issues regarding transmission to the definitive host, particularly humans. Isolation of the metacercarial stage from tissue entails a digestive process using either the entire fish or isolated infected tissue. A range of techniques have been utilised, but most are based on protocols set down by the WHO in 1995. Fish, either whole or tissue samples, are disrupted in artificial stomach acid, which comprises 0.06 M HCl and 1% pepsin in distilled water, at a ratio of 100g fish tissue per litre of digestant. Incubation takes 2–3 hour at 37°C, after which the digested material is filtered through a 1 × 1 mm mesh sieve. Isolated metacercariae are washed several times with 0.85% saline solution. Some authorities only use the 1% pepsin solution to digest the fish tissue and, in some instances, when the parasites are not easily discerned, the metacercaria can be mechanically excysted, either by physical pressure (i.e. pressing on a coverslip) or by exposing them to a digestion fluid comprising 0.4% sodium hydrogen carbonate, 1% trypsin and 0.85% NaCl. In some instances, a compression technique has proved useful in the location of the parasite in tissue. In this technique, slivers of tissue are pressed between two glass plates and observed under a dissecting microscope.

Identification of the metacercariae is usually based on morphological appearance and authorities have utilised a range of structures. For example, *Opisthorchis viverrini* has oval cysts enclosing active C-shaped metacercariae, which possess a large excretory bladder. In contrast, *Haplorchis pumilio* and *H. taichui* are enclosed in a thin-walled, oval cyst enclosing a C- or S-shaped metacercaria and a small excretory bladder. Several studies have utilised molecular biological techniques, including PCR and sequencing of ribosomal DNA (the ITS2 region) in cercariae and metacercariae to identify species of FZTs from both first and second intermediate hosts.

18.3.4 Control

Whilst the control of the flukes and the involvement of snail intermediate hosts lends itself to the utilisation of control procedures as indicated in Chapter 11 (The Flukes), the involvement of humans as a possible definitive host and the associations with integrated agricultural/aquaculture practices, particularly in Asia, does open up the application of additional control strategies. Control of the disposal of faecal material, from humans by the use of latrines sited away from water sources, or from animals such cats, dogs and pigs by stopping the use of faeces as fertiliser, should be encouraged. Several methods of treating fish meat used for human consumption have been adopted. The US Food and Drug Administration recommends freezing of fish meat at –20°C for 7 days or smoking at 65°C, whilst the European Union recommends temperature treatment, i.e. –20°C for 24 h or smoking at >60°C. Fish should be consumed preferably after cooking, but if they are to be consumed uncooked then freezing at –20°C is strongly recommended prior to consumption. Other food treatments such as salting and marinating are not considered reliable methods for the destruction of the metacercaria.

An alteration in fish management systems has also been suggested to control infection by keeping fry in cement tanks or composite tanks with filtered water for longer than a week, although in many rural settings this is not feasible.

FACTS SHEET

Disease: Clonorchiasis and Opisthorchiasis
Pathogen: *Clonorchis sinensis, Heterophyes heterophyes, Haplorchis pumilio* and *Metagonimus yokogawai*
Susceptible host: Main pathogenic effects in human host. Occurs in several carp species, including *Carassius auratus, Cyprinus carpio, Ctenopharyngodon idella, Labeo rohita, Hypophthalmichthys molitrix, Cirrhinus molitorella, Squaliobarbus curriculus*
Mortality: Does not usually cause mortalities in humans if treatment given
Temperature: Infection in fish can be greater with increasing temperature
Prevention and treatment: Infection of humans can be reduced or prevented by good husbandry practices and avoiding consumption of undercooked or raw contaminated fish. A range of treatments for humans are available. Consult medical practitioner

18.4 Selected references for further reading

Boerlage, A.S. (2013) Transmission and control of fish-borne zoonotic trematodes in aquaculture. Doctor of Philosophy, Wageningen University, Netherlands.

Chi, T.T.K., Murrell, K.D., Madsen, H., Khue, N.V. and Dalsgaard, A. (2009) Fishbourne zoonotic trematodes in raw fish dishes served in restaurants in Nam Dinh province and Hanoi, Vietnam. *Journal of Food Protection* 72, 2394–2399.

Clausen, J.H., Madsen, H., Murrell, K.D., Van, P.T., Thu, H.N.T. *et al.* (2012) Prevention and control of fish-borne zoonotic trematodes in fish nurseries, Vietnam. *Emerging Infectious Diseases* 18, 1438–1445.

Clausen, J.H., Madsen, H., Murrell, K.D., Thi, V.P., Manh, H.N. *et al.* (2012) Relationship between snail population density and infection status of snails and fish with zoonotic trematodes in Vietnamese carp nurseries. *Neglected Tropical Diseases* 6, e1945.

Dung, V.T., Waikagul, J., Thanh, B.N., Vo, D.T., Nguyen, D.N. *et al.* (2014) Endemicity of *Opisthorchis viverrini* liver flukes, Vietnam, 2011–2012. *Emerging Infectious Diseases* 20, 152–153.

Gauthier, D.T. (2015) Bacterial zoonoses of fishes: A review and appraisal of evidence for linkages between fish and human infections. *The Veterinary Journal* 203, 27–35.

Harriff, M.J., Wu, M., Kent, M.L. and Bermudez, L.E. (2008) Species of environmental mycobacteria differ in their abilities to grow in human, mouse, and carp macrophages and with regard to the presence of mycobacterial virulence genes, as observed by DNA microarray hybridization. *Applied and Environmental Microbiology* 74, 275–285.

Hung, N.M., Dung, D.T., Anh, N.T.L., Van, P.T., Thanh, B.N. *et al.* (2015) Current status of fish-borne zoonotic trematode infections in Gia Vien district, Ninh Binh province, Vietnam. *Parasites and Vectors* 8, 21.

Iqbal, Z., Ansar, F. and Huma, Z-e. (2018) Risk of importing zoonotic diseases through infected ornamental fish. *Punjab University Journal of Zoology* 33, 211–215.

Ljubojevic, D., Novakov, N., Djordjevic, V., Radosavljevic, V., Pelic, M. *et al.* (2015) Potential parasitic hazards for humans in fish meat. *Procedia Food Science* 5, 172–175.

Phan, V.T., ErsbØll, A.K., Bui, T.Q., Nguyen, H.T., Murrell, D. *et al.* (2010) Fish-borne zoonotic trematodes in cultured and wild-caught freshwater fish from the Red River Delta, Vietnam. *Vector-Borne and Zoonotic Diseases* 10, 861–866.

Phan, V.T., ErsbØll, A.K., Nguyen, T.T., Nguyen, K.V., Nguyen, H.T. *et al.* (2010) Freshwater aquaculture nurseries and infection of fish with zoonotic trematodes, Vietnam. *Emerging Infectious Diseases* 16, 1905–1909.

Schäperclaus, W. (1991) *Fish Diseases Volume 1*, 5th edn. Amerind Publishing Co./Oxonian Press, New Delhi, India.

Scholz, T., Ditrich, O. and Giboda, M. (1991) Differential diagnosis of opisthorchiid and heterophyid metacercariae (Trematoda) infecting flesh of cyprinid fish from Nan Ngum Dam Lake in Laos. *Southeast Asian Journal of Tropical Medicine and Public Health* 22, 171–173.

Yossepowitch, O., Gotesman, T., Assous, M., Marva, E., Zimlichman, R. *et al.* (2004) Opisthorchiasis from imported raw fish. *Emerging Infectious Diseases* 12, 2122–2126.

Chapter 19

Non-Infectious Diseases

19.1 Introduction

Environmental conditions, nutritional challenges and genetic factors have a strong impact on fish health that may lead to disease or may predispose fish to infectious agents. Carp, compared with other fish, can tolerate a wide range of water parameters, including temperature, oxygen content, pH and nitrogen compounds. Water parameters outside optimal ranges can predispose carp to infection or may cause disease. Many of these environmental effects on fish can induce symptoms that are similar to those caused by infection. It is therefore important to include considerations of environment, diet and genetics within analysis of fish health and mortalities.

19.2 Water Parameters

19.2.1 *Temperature*

Several carp species, including bighead carp (*Hypophthalmichthys nobilis*), *Carassius* species, common carp (*Cyprinus carpio*) and grass carp (*Ctenophoryngodon idella*), are eurythermal and tolerate water temperatures between 0.5°C and 30–35°C, whilst in comparison the Indian major carps catla (*Gibelion catla*), mrigal (*Cirrhinus mrigala*) and rohu (*Labeo rohita*) have a lower temperature limit of about 14°C. For all species, temperature fluctuations are more problematic than absolute temperature ranges. Sudden temperature shifts greater than 1–3°C may result in restlessness and a refusal to feed, and may trigger the outbreak of diseases. In order to allow an adaptation of

 Carps: A Practical Approach to Disease Control and Health Management (eds David Hoole *et al.*)
DOI: 10.1079/9781836990024.0019

carp physiology to a new temperature, temperature shifts should only be 1–2°C per day. Even in eurythermal species, such as crucian carp and common carp, temperature shifts of about 10°C above or below the adaptation temperature may cause mortality. A shift to higher temperature will increase the oxygen demand and can only be tolerated within the above-mentioned range in well-aerated water. When the tolerated range is exceeded, death may occur, often associated with convulsive movements, darkening of the skin and opened gill covers. If the temperature is too low, a disturbance of the osmotic equilibrium, oedematisation and detachment of large areas of the skin, damage to the intestine and haemolysis may occur. In warm-adapted carp, a sudden drop in temperature can lead to torpor and death due to paralysis of the respiratory centre. After a slow drop in temperature, eurythermal species, such as bighead carp, grass carp, crucian carp (*Carassius carassius*) and common carp, can hibernate in ponds with an ice cover at low temperatures and good oxygen conditions without any damage. However, such carp should not be disturbed by noise or vibrations.

For reproduction, egg and larval development, as well as for optimal growth, all carp species require elevated water temperatures of at least 18–20°C, although Indian carp grow best at 25–32°C. Developing eggs and larvae tend to be more vulnerable than adult fish to a sudden drop in temperature, which may result in developmental disorders such as spinal deformities.

19.2.2 Dissolved oxygen

In both pond aquaculture and recirculation aquaculture systems, dissolved oxygen levels are critical for development and survival of fish. The oxygen requirements of pond-dwelling cyprinids are a minimum of above 3 mg/L, which is low compared with coldwater-adapted species, like salmonids (above 5 mg/L), although lower lethal levels (e.g. below 1 mg/L) occur for many carp. Some species, like *Carassius* (crucian carp, goldfish (*C. auratus*)), may even survive exposure to oxygen at 0–0.3 mg/L for a few hours. The optimal range for rearing cyprinids is generally considered to be above 5 mg/L and, in particular, for the incubation of eggs, fry and alevins, the oxygen concentration should be close to saturation. In most cyprinids on-growing specimens survive a prolonged exposure to oxygen levels between 1 and 5 mg/L but then they are more sensitive to diseases, and have a lower growth rate and a poorer food conversion rate. Fish exposed to such chronic oxygen shortage may be anorexic. Although oxygen levels in water can be replenished by diffusion from air and from photosynthesis by phytoplankton, it should be noted that oxygen consumption can increase through the respiration of various aquatic organisms, including plankton, fish and benthic organisms, as well as by diffusion from water to air. There are important balances that need to be considered when determining oxygen availability to carp. For example, in ponds, oxygen levels are highest during late afternoon in upper water layers, where photosynthesis is most active, and lowest during the night, especially at dawn, after oxygen is consumed by respiration by pond organisms. Oxygen shortage may also occur during warm temperatures, when there is an algal bloom during the night, or when an algal bloom collapses and decomposes. Because the oxygen consumption of fish per weight units increases with the quantity of food ingested, farmers should adjust the feeding level according to the expected weather condition, in particular when it is not possible to improve the oxygen level in the fish culture units. Generally, fish exposed to low oxygen levels are lethargic, concentrate at the water surface or in shallow water and breathe rapidly. In ponds with too low oxygen content, usually the largest fish die first, because these have the highest oxygen demand. Aeration of pond water is the major action to compensate for oxygen shortage.

19.2.3 Alkalinity

During photosynthetic activity, plants incorporate CO_2 or HCO_3^-, which results in a raising of pH during daylight and the reverse effect during the night. In particular, during algal blooms, water pH may fluctuate daily in the range of 7.0–11.0. This effect is more prominent in ponds with a low level of carbonates and bicarbonates (low alkalinity), i.e. with a low buffering capacity. High buffering capacity prevents prominent pH fluctuations caused by strong photosynthesis. At alkaline pH fish are stressed, showing milky skin or mucus and frayed fins, and the gill epithelium may be hypertrophic. When water alkalinity is very low, pH can exceed 11 during a phase with intense photosynthesis, which may be lethal for fish. As a preventive action, especially when alkalinity is below 30 mg/L, it can be raised by the addition of slaked lime ($Ca(OH)_2$) or quicklime (CaO). This should be done in autumn or at the beginning of winter before the fertilisation of ponds with phosphorus is carried out.

19.2.4 Gas supersaturation

When underground or spring water is used for a hatchery, this water may be oxygen deficient and therefore may need oxygenation. It may also contain other dissolved gases, like nitrogen and carbon dioxide, which are toxic above 100% saturation or above certain levels (e.g. CO_2 above 10 mg/L). Such a gas supersaturation often occurs when air is drawn into pipes or cold groundwater is piped into fish culture units containing warmer water with a lower gas saturation level. The effect is that fish may develop gas bubbles in the skin, fins (Fig. 19.1), gills or eyes during severe gas supersaturation. In the case of slight gas supersaturation, chronic mortalities may occur whose cause is not readily apparent. To eliminate gas supersaturation, a degassing of incoming water by passing it over a cascade or a degassing/oxygenation column may be needed, before it enters hatchery units.

19.2.5 Nitrogen compounds

A range of nitrogen compounds, such as ammonia, nitrite and nitrate, are excreted by fish into the water and represent end products of protein metabolism or originate from organic discharges. Problems with toxic levels of ammonia and nitrite can occur especially in hatchery water, but also during caging, retail storing or transportation. While nitrate is less toxic and is tolerated in higher concentrations, nitrite is more dangerous. Although toxic at a range of 0.1–0.2 mg/L, greater toxicity occurs in water with low alkalinity.

Figure 19.1 Whole fin of goldfish showing enclosed gas bubbles (arrow). Courtesy of the Environment Agency, National Fisheries Laboratory, UK.

Ammonia is very toxic in its un-ionised state (NH_3) and the proportion of this form increases with increasing temperature or pH. Because methods of analysis give only the total amount of nitrogen in ammonia form, the proportion of un-ionised ammonia has to be calculated from this in relation to water temperature and pH. Ammonia and nitrite are more toxic to alevins and juveniles compared with larger specimens. Affected fish tend to lie on their side on the bottom of the container, but swim up to the surface to feed. Un-ionised ammonia irritates gills (Fig. 19.2), affects the mucus cover of the skin, mouth and intestine and may cause nervous symptoms, such as erratic swimming behaviour. Levels of un-ionised ammonia as low as 0.02 mg/L may cause chronic problems, and levels of approximately 0.2 mg/L can cause rapid death. Nitrite interferes with oxygen uptake by the formation of methaemoglobin. Affected fish show heavy gill movements and may have mud-coloured brown gills. The emergency measure against high ammonia or nitrite levels is an exchange of water. Since ammonia is more toxic at higher pH, lowering the water pH might also help. Chloride ions, which compete with nitrite for transportation across the gill epithelium,

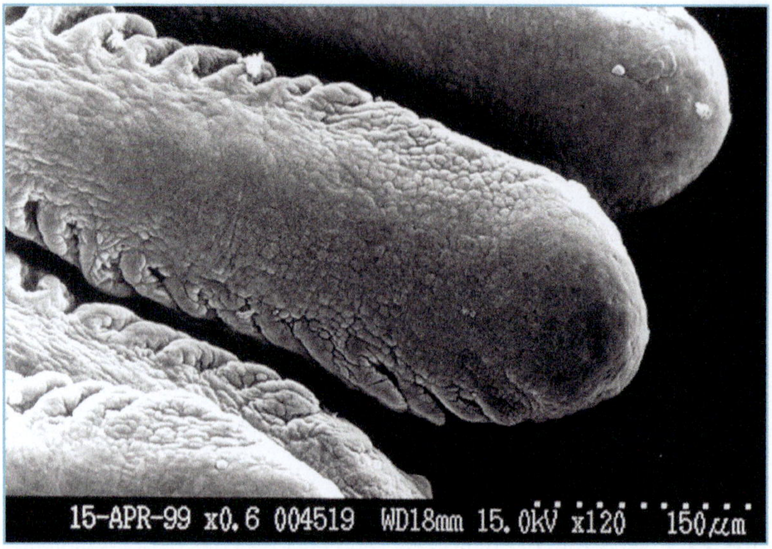

Figure 19.2 Electron micrograph image showing fusion of gill lamellae induced by ammonia. Scale bar 150 μm.

can also help, hence the addition of table salt to the water of the aquaculture unit at a concentration of up to 0.3% can be a remedy against nitrite toxicity.

19.3 Water Contaminants

19.3.1 Bacterial contamination

Heterotrophic bacteria, which occur in organic discharges, are components of the heterotrophic food web in the pond. Fish are usually well protected from infection by these bacteria with an intact skin and mucus cover; however, disease can result when fish skin is damaged during husbandry measures, such as improper sorting methods or transport. When manure is used for fertilisation of pond water, it has to be ensured that zoonotic bacteria causing for example salmonellosis or cholera are not introduced into aquaculture ponds and spread when the fish are harvested.

19.3.2 Pollution

Various toxic compounds are released during human activities, such as industrial processes, agriculture, forestry, traffic and waste disposal, and many of these compounds eventually reach the aquatic environment. These substances include, for example, various organic compounds used as pesticides and herbicides, and by-products of industrial and urban activities, such as various heavy-metal ions (cadmium, chromium, copper, mercury, silver or zinc). The effect of these substance on aquatic organisms, including fish, depends on their bioavailability, which is determined by factors such as chemical form of the compound, whether it is dissolved in water, or the temperature and ionic composition of the water. Carbonate ions, as well as organic compounds in water, form complexes with heavy-metal ions, and therefore reduce their uptake and toxicity, whilst lower pH leads to an increased free metal ion concentration and thus increases their toxicity. In addition, increased respiration activity of fish at higher water temperature leads to an increased uptake of heavy-metal ions. The uptake of heavy metals in fish occurs across the body surfaces in contact with the environment, namely skin, gills and the intestine. The mechanisms involved include passive diffusion or carrier-mediated transport of dissolved ions over the gills, as well as the ingestion of metals associated with organic material and its endocytosis in the intestine. In the fish, heavy metals can interact with various metabolic pathways, resulting in effects on respiration, haematology, plasma composition, reproduction and behaviour. Depending on the concentration and duration of exposure, these effects may be local, affecting only skin, gills or intestine, or may be systemic, involving a range of different tissues such as liver, kidney, muscle and brain. Heavy-metal-induced changes to the gill respiratory epithelium, including swelling and necrosis of epithelial cells, increase the ventilation rate and coughing activity of fish. Whilst exposure to heavy metals results in increased mucus formation on the body, this mucus may entrap the metal, thereby reducing the amount of metal that comes into contact with and is transported by the epithelium. Overall, only traces of heavy metals can be tolerated, particularly in hatchery water, and chronic exposure of fish may result in chronic losses, developmental deformities and reduced reproduction success. In hatcheries, the use of galvanised and copper pipes must be avoided.

A large variety of synthetic organic compounds are produced and used commercially and a large number of these compounds have the potential to contaminate the aquatic environment and induce disorders in fish. When a chemical is dissolved in water in a sufficient concentration, it can irritate skin and gills of the fish. It can also accumulate in internal organs, such as the liver and kidney, and induce toxic responses. Various xenobiotic organic compounds from different chemical classes have been reported to induce toxicopathic lesions in the liver. Because of the large variety of compounds, standards for toxic responses of carp under exposure cannot be given. However, eggs and larvae have a much higher susceptibility to toxic effect of organic xenobiotics than adult fish. In particular, some herbicides and pesticides are highly toxic to fish, and the risk of contamination of pond water and fish must be considered in ponds in the vicinity of agricultural land where these substances are applied, where accidental spillages may occur, or from where an influx may happen after heavy rainfall.

19.4 Feed-Related Disorders

The aquaculture of cyprinids mainly relies on the stocking of fertilised ponds and the fish prey on plankton or benthic invertebrates from the pond environment. This largely ensures a diet with an adequate supply of essential amino acids, fatty acids and vitamins. In addition to the quality of food, the quantity of food consumption is an important factor and reduction in food intake, which can originate from a range of factors, can result in emaciation (Fig. 19.3). In many ponds additional feed is offered in order to achieve additional harvests. In such situations, an unbalanced feed in respect to protein-to-energy ratio and amino-acid composition can result in a fatty degeneration of the liver and an excessive deposition of fat in the peritoneal cavity. The development of such lipoid alterations in the liver is also observed after inclusion of rancid fats

Figure 19.3 Emaciated koi carp. Note relatively large head in comparison with thin body.

in fish feed. In a temperate climate, such lipoid/adipose carp often suffer from chronic mortality when a high amount of energy has to be mobilised in order to cope with stressors or after hibernation. Another disorder linked to improper storage of feed may result from a contamination of feeds with aflatoxins, produced for example by several *Aspergillus* strains. These aflatoxins induce hepatocellular adenomas and carcinomas.

19.5 Neoplasms

An epidermal hyperplasia is caused by an infection of common carp with CyHV-1. This disease is known as 'carp pox', and the virus genome has been detected not only in active lesions, but also in gills, cranial nerve ganglia and spinal nerves of specimens after the lesions have regressed.

Gonadal neoplasms occur in a relatively high prevalence in mature ornamental koi females kept in artificial ponds without giving them the possibility for spawning.

Examples of tumours in carp are given in Fig. 19.4.

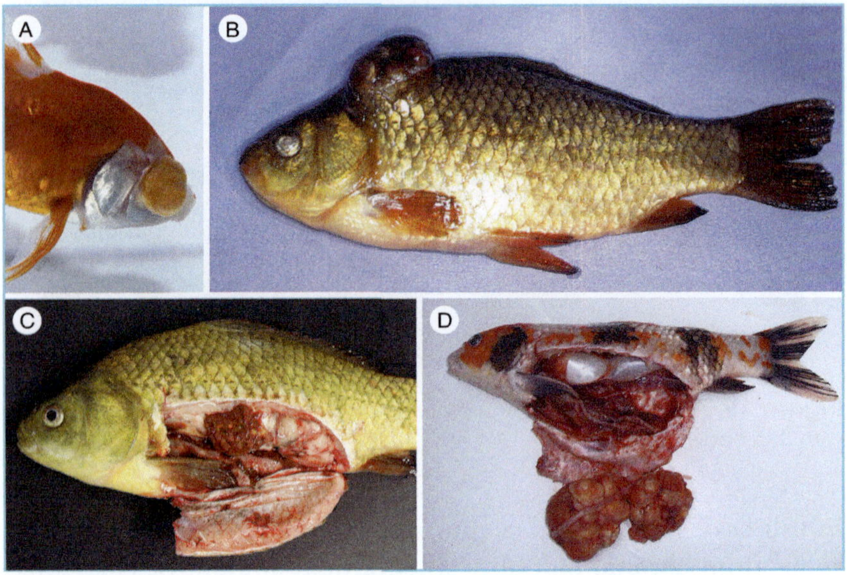

Figure 19.4 Examples of tumours occurring in carp. (A) Ocular tumour on goldfish. (B) Crucian carp neoplasia. (C) Dissected koi carp showing abdominal tumour. (D) Polycystic kidney in carp. Courtesy of the Environment Agency, National Fisheries Laboratory, UK.

19.6 Selected References for Further Reading

Billard, R. (1999) *The Carp. Biology and Culture*. Aquaculture and Fisheries series, Springer Praxis, London.

Boyd, C. and Tucker, C. (2014) *Handbook for Aquaculture Water Quality*. Craftmaster Printers, Auburn, Alabama.

Davis, D., Li, M., Gatlin, D.M.I. and O'Keefe, T. (2009) Advances in aquaculture nutrition: Catfish, tilapia and carp nutrition. In: Burnell, G. and Allan, G.L. (eds) *New Technologies in Aquaculture. Improving Production Efficiency, Quality and Environmental Management*. Woodhead Publishing, Oxford, UK, pp. 440–458.

Honda, M. and Suzuki, N. (2020) Toxicities of polycyclic aromatic hydrocarbons for aquatic animals. *International Journal of Environmental Research and Public Health* 17, 1363.

Jerdzierska, B., Witeska, M. and Häggblom, M.M. (2006) The metal uptake and accumulation in fish living in polluted waters. In: Twardowska, I., Stefanik, S. and Allen, H.E. (eds) *Soil and Water Pollution Monitoring, Protection and Remediation*. Springer, Berlin, Germany, pp. 107–114.

Kroupova, H., Valentova, O., Svobodova, Z., Sauer, P. and Machova, J. (2018) Effects of nitrite on freshwater organisms: A review. *Reviews in Aquaculture* 10, 525–542.

Leatherland, J.F. and Woo, P.T.K. (1998) *Fish Diseases and Disorders, Vol. 2: Non-infectious Disorders*. CAB International, Wallingford, UK.

Olivia-Teles, A. (2012) Nutrition and health of aquaculture fish. *Journal of Fish Diseases* 35, 83–108.

Roberts, R.J. (2023) Nutritional pathology. In: Hardy, R.W. and Kaushik, S.J. (eds) *Fish Nutrition*, 4th edn. Academic Press, London, pp. 823–856.

Chapter 20

An Ornamental Perspective

20.1 Introduction

Within the ornamental sector, carp also play a significant financial role and are held in high esteem personally. Koi carp, the coloured variety of the common carp (*Cyprinus carpio*), are highly prized and are very popular pet fish that are kept in ponds and aquaria worldwide. Also, goldfish (*Carassius auratus*), which are bred in many different varieties, are perhaps the most popular pet worldwide. Both koi and goldfish are highly valued as social and trainable pets and their owners are concerned about their health and well-being, seeking veterinary advice not only in the case of a fatal disease, but also when the well-being of their fish is affected. Owners may present individual fish specimens where they believe that there is an emotional bond between them and their fish and this particular specimen should be diagnosed and cured. In order to remove stressors related to the examination, fish can be anaesthetised, for example by using a chemical with anaesthetising effect, which is mostly administered by immersion. In this case, fish are immersed in a container of water in which the anaesthetic is dissolved (Fig. 20.1A). Tanks used for immersion anaesthesia and for subsequent recovery of the fish should be prepared ahead of time, enabling a smooth operation of the anaesthesia and recovery processes. The amount of anaesthetic agent needed has to be carefully calculated (Fig. 20.1B) and the anaesthetic solution should be properly buffered. The water utilised in the anaesthetic and recovery tanks should be of good quality, well aerated and at a temperature that has been adjusted to the temperature of the water that the fish have been kept in prior to the sedation process. Fish should be placed into the anaesthetic

 Carps: A Practical Approach to Disease Control and Health Management (eds David Hoole *et al.*)
DOI: 10.1079/9781836990024.0020

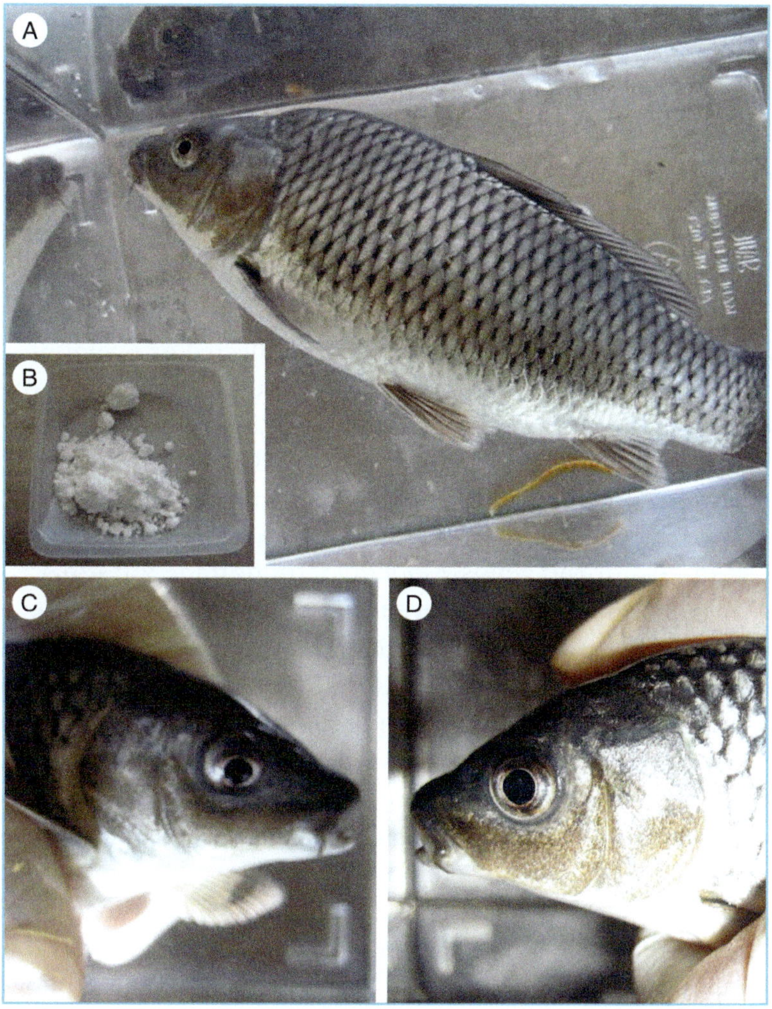

Figure 20.1 Sedation procedure. (A) Carp in the sedation tank losing equilibrium. (B) Prepared anaesthetic. (C) Fish showing eye roll reflex. (D) Fish showing loss of eye roll reflex.

tank, and the anaesthetic added gradually. The fish should be monitored continually for any changes in behaviour. Fish in light narcosis exhibit signs such as a cessation of voluntary swimming, a reduction in or absence of reaction to tactile or visual stimuli, a decrease in the respiratory rate, a loss in equilibrium and a decrease in muscle tone. The stage of deep narcosis is reached when the fish has totally lost its equilibrium and shows no effort to right itself but there is still some response to strong tactile stimuli. At this stage the fish is suitable for external sampling for disease diagnosis. A full stage of anaesthesia is reached when the muscle tonus is lost, and a total loss of reactivity and a loss of the eye roll reflex are observed (Fig. 20.1C, D). In addition, in the stage of surgical anaesthesia, respiratory movements are very low or lost and the heart rate is slow.

When the required depth of anaesthesia has been reached the fish is moved out of the water, placed on the examination table on a wet cover and the required samples are collected or operations are performed. Fish can survive for a period of a few minutes out of the water and they recover in water very well. For longer operations, constant recirculation systems must be used, which pump oxygenated water with the anaesthetic over the gills in order to provide oxygen to the fish (Fig. 20.2). Additional care should be taken that the skin remains wet. A recovery tank with well-aerated water from the original tank or from the transport container should always be at hand, so that the fish can be placed in this container when the sampling or operations are finished or in case of any incident during the manipulation. Various chemicals can be used for anaesthetising fish, for example isoeugenol, clove oil, 2-phenoxyethanol, tricaine, benzocaine, lidocaine and ethomitate.

All previously described diseases and disorders caused by environmental factors or pathogens can, of course, affect pet fish. This is especially pertinent for diseases caused by suboptimal keeping conditions that are very common in koi carp and goldfish. The sizes of the ponds or aquaria are often not sufficient for the number of fish being kept and water changes are not regularly performed in some private ponds, leading to poor water quality. The discrepancy between the demands of the fish and the owners' expectations can be an additional challenge. The fish keepers tend to prefer clear water and ponds without substrate on the bottom or plants, primarily because in such ponds the colours of the koi are best seen. Part of the normal behaviour in koi carp, like in common carp, is dabbling in the substrate on the bottom, which is not feasible in ponds with an environmentally sterile bottom. In many cases koi carp are overfed, or fed with feed that has no ideal composition regarding fatty acids

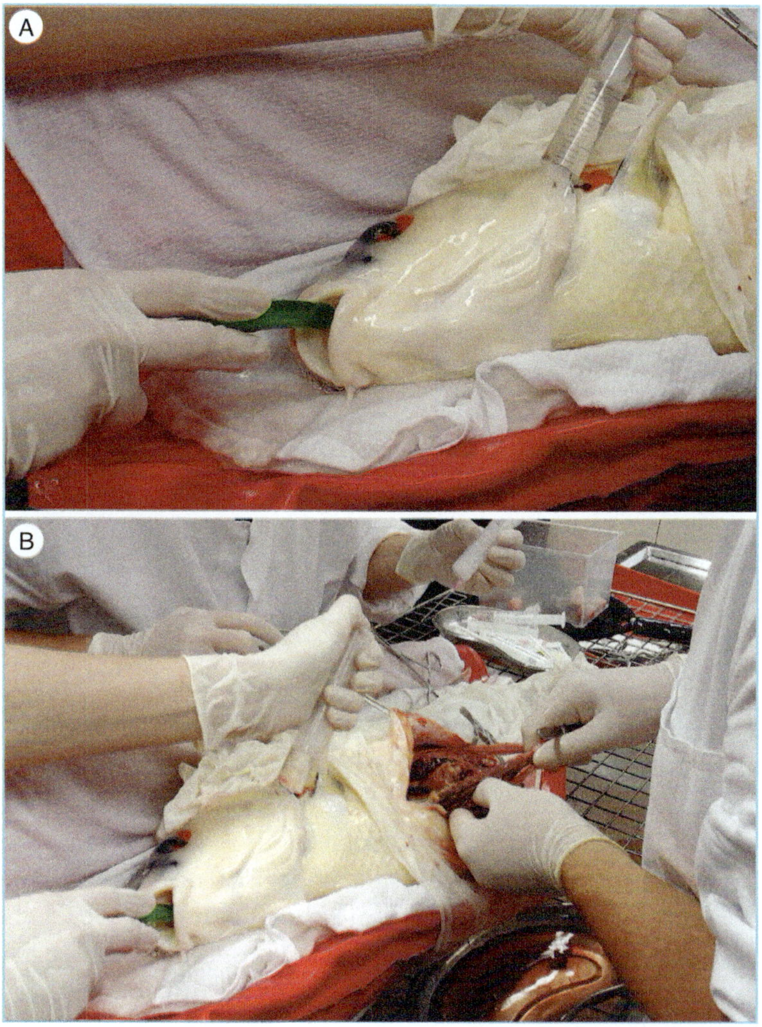

Figure 20.2 Operation on the abdomen of a koi carp. (A) Flushing of gills with water containing anaesthetic. (B) Overview of the operation scene.

and proteins. This results in fat deposits in the abdominal cavity, between the internal organs and in the liver (Fig. 20.3), leading to health problems. If koi carp are kept in outside ponds during the winter, they must be optimally fed before the cold period. If the water temperatures are rising in spring and the koi are not

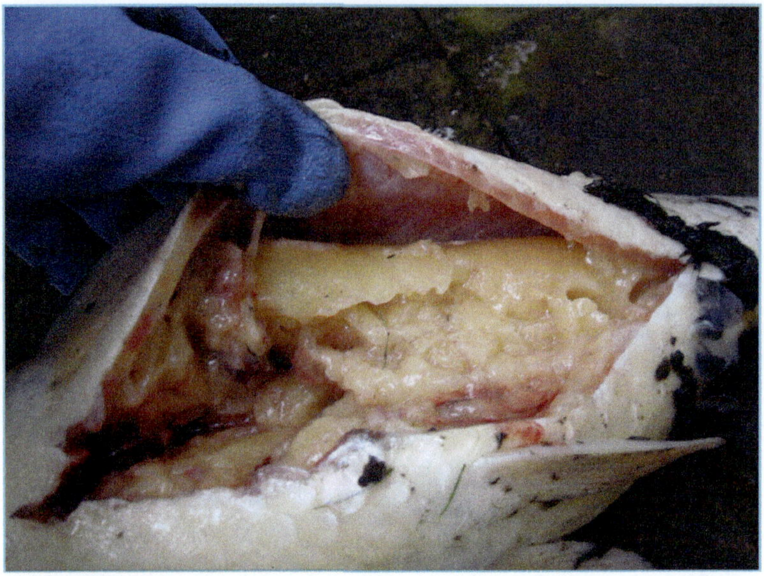

Figure 20.3 Koi showing extensive fat deposits in the abdominal cavity between the internal organs.

well-conditioned, an energy-deficiency syndrome can lead to the death of the fish in the worst-case scenario. Very often parasitic infections and secondary bacterial infections can be diagnosed in koi carp and goldfish. If the environmental conditions are not improved, treatment of these pathogens is mostly not successful in the longer term. Viral infections, especially infections with CEV or CyHV-3 in koi carp and infections with CyHV-2 in goldfish, are a major threat for ornamental fish. These pathogens are distributed through the worldwide trade with ornamental fish movements and can lead to high morbidities and mortalities in naïve fish populations. Therefore, in ornamental fish disease, surveillance programmes are recommended and retailers of pet fish should be aware of the symptoms of diseases.

In addition to the already-mentioned diseases, koi carp and goldfish can suffer from additional disorders that are usually not treated in fish for food production. Tumours are often diagnosed in koi carp and in goldfish (Figs. 20.4 and 20.5). Carp pox shows flat papillomas of the skin in carp (Fig. 20.6) that are caused by

Figure 20.4 (A) Koi carp showing location of a tumour on its tail fin. (B) Higher magnification of tumour shown in (A), showing nodular appearance.

Figure 20.5 Goldfish with a skin tumour on its lateral body.

Figure 20.6 Koi carp with a flat carp pox papilloma located on the skin of the head.

an infection with CyHV-1, a herpesvirus. The presence of the skin tumours is temperature related and at water temperatures higher than 20°C the clinical signs tend to disappear. However, in the next cold season the carp pox can proliferate again, as the virus persists in fish cells. Therefore, the surgical removal of carp pox does not result in the disease being cured and is only recommendable if one of the 'poxes' is affecting the mobility of the opercula or is disturbing the uptake of food. Carp pox can be transmitted from one infected koi to another via contamination of the pond water and therefore care should be taken when new fish are purchased and introduced into the system. In some cases, large squamous-cell carcinomas can develop out of the flat papillomas. Pigmented cell tumours of the skin regularly occur in goldfish and, depending on the appearance of the cell types affected, they are called melanomas (black), iridophoromas (silver), erythrophoromas (red) or xanthophoromas (yellow).

Koi are particularly prone to the development of tumours in the ovaries. These tumours are often very large and take up almost the

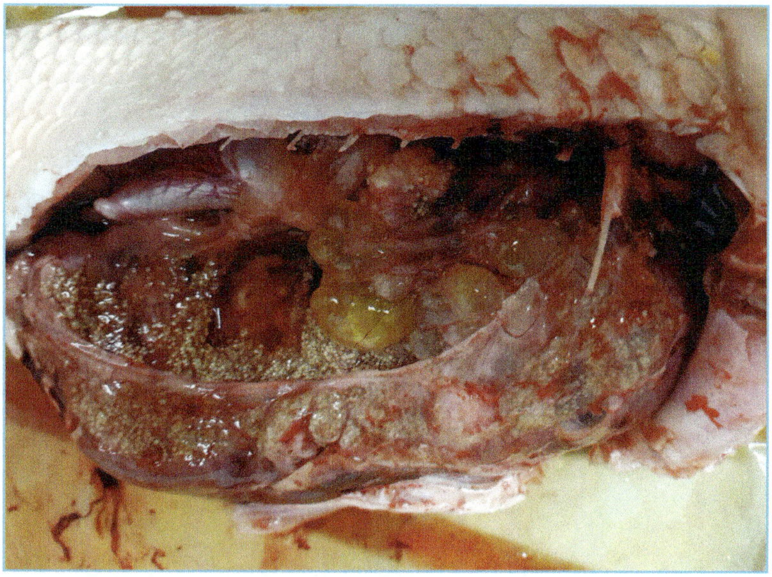

Figure 20.7 Koi carp with extensive egg deposits and a developing cystic ovarian tumour in the abdominal cavity.

whole abdominal cavity (Fig. 20.7). One reason for the occurrence and development of ovarian tumours is the retention of the eggs. Koi carp need structures, like plants, in their ponds to spawn. If these structures are missing or other adverse factors are preventing the spawning, female koi carp do not release their mature eggs into the water during spawning. The term 'egg-bound' is used for this condition. Another reason for the development of gonadal tumours might be the difficulty in carp hybrids to develop a functional gonad, which can be a form of dysmorphogenesis rather than a real neoplasia. Gonadal tumours can be diagnosed as fibromas, granulosa-cell tumours, gonadoblastomas and teratomas, depending on the tissues they are developing from. A surgical removal of the tumours is possible if they are diagnosed in an early stage of development. Unfortunately, in many cases the tumours are already very large and connected to other internal tissues when fish are presented to a veterinarian and thus the prognosis is not good. Surgical removal is often not possible in these cases, and as long as the general condition of the fish is good,

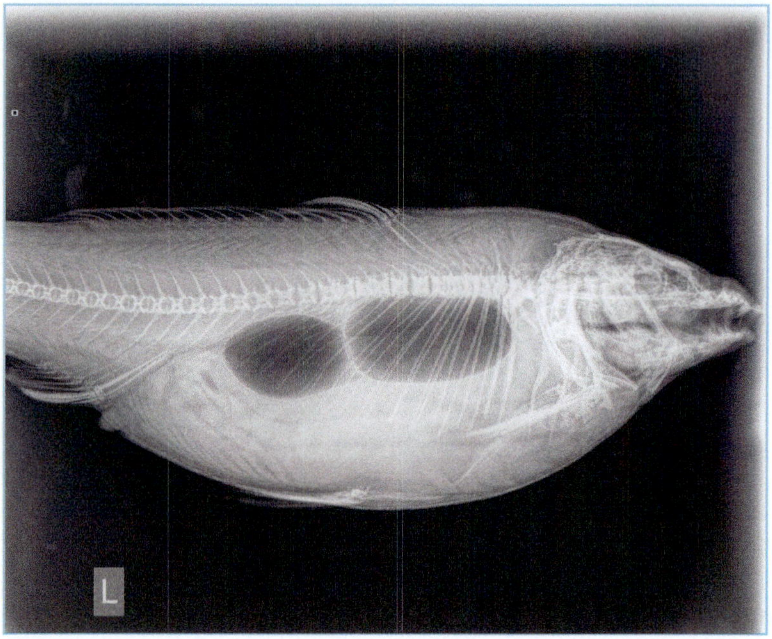

Figure 20.8 X-ray image of a koi carp with over-inflated and dislocated posterior chamber of the swim bladder.

intervention is usually not necessary. If the general condition of the fish is poor, euthanasia of the animal should be considered.

Different swim bladder disorders are regularly seen, especially in different cultivated varieties of goldfish. In fancy goldfish, for example, the body shape of the fish is shorter than in the original form, leading to a shortening of the muscle filaments and often a reduced caudal chamber of the swim bladder. Over-inflation of the swim bladder may lead to a dislocation of the caudal chamber (Fig. 20.8). The cause of this abnormality is not always clear, but in some cases a gas supersaturation has been detected. All swim bladder disorders may lead to problems with buoyancy. The degree of the swim bladder dislocation or alteration can be assessed by radiography. In cases of swim bladder over-inflation, treatment is possible by aspiration of air from the organ using a small needle and a syringe. Deformations of the swim bladder cannot be treated, but the general condition of the fish should be improved as much as possible.

Cystic changes of the kidneys also regularly occur in goldfish and koi. Different numbers of liquid-filled cysts of varying sizes are present in the abdominal cavity. By dislocation of other internal organs, the clinical signs can be worsened. Infections with myxosporea or genetic disorders are thought to be the cause of these cysts, and depending on the cause, a treatment such as removal of the fluid from the cysts by punctuation can be attempted. Ascites, also known as 'dropsy', is another disease often found in koi and goldfish. Any disease that leads to a change in osmoregulation, like any gill, kidney or heart disease, can lead to an abnormal accumulation of fluid within the abdominal cavity. Bacterial infections might also be the cause of an increase in ascites. Dropsy can occur with or without oedema of the skin and the scale pockets, and with or without exophthalmos. Depending on the cause, a treatment such as the application of antibiotics is possible in some cases.

20.2 Selected References for Further Reading

Martin, T., Valentim, A.M., Pereira, N. and Antunes, L.M. (2016) Anaesthesia and analgesia in laboratory adult zebrafish: A question of refinement. *Laboratory Animals* 59, 476–488.

Noga, E.J. (2010) *Fish Disease. Diagnosis and Treatment*, 2nd edn. Wiley-Blackwell, Ames, Iowa.

Roberts, R.J. (2023) Nutritional pathology. In: Hardy, R.W. and Kaushik, S.J. (eds) *Fish Nutrition*, 4th edn. Academic Press, London, pp. 823–856.

Stoskopf, M.K. (1993) *Fish Medicine*. Saunders, Philadelphia, Pennsylvania.

Von Krogh, K., Higgins, J., Saavedra Torres, Y. and Mocho, J.-P. (2021) Screening of anaesthetics in adult zebrafish (*Danio rerio*) for the induction of euthanasia by overdose. *Biology* 10, 1133.

Wildgoose, W.H. (2001) *BSAVA Manual of Ornamental Fish*, 2nd edn. British Small Animal Veterinary Association, Quedgeley, Gloucester, UK.

Index

Note: Page numbers in **bold** type refer to **figures**; page numbers in *italic* type refer to *tables.*

Printed and bound by CPI Group (UK) Ltd, Croydon, CR0 4YY

13/09/2026

14961083-0001